The Economics of Waste

The
Economics
of Waste

Richard C. Porter

Resources for the Future | Washington, DC

Printed in the United States of America

An RFF Press book
Published by Resources for the Future
1616 P Street, NW, Washington, DC 20036–1400
www.rff.org

Library of Congress Cataloging-in-Publication Data
Porter, Richard C.
    The economics of waste / by Richard C. Porter.
      p. c.m.
    Includes bibliographical references and index.
    ISBN 1-891853-42-2 (lib. bdg.) — ISBN 1-891853-43-0 (pbk.)
    1. Refuse and refuse disposal—Economic aspects—United States. 2. Factory and trade waste—Economic aspects—United States. 3. Recycling—Economic aspects—United States. 4. Pollution—United States—Costs. I. Title.

HD4483 .P67 2002
338.4′76284′0973—dc21   2001058872

f e d c b a

The paper in this book meets the guidelines for permanence and durability of the Committee on Production Guidelines for Book Longevity of the Council on Library Resources.

The text of this book was designed and typeset by Betsy Kulamer in Utopia and Ocean Sans. It was copyedited by Alfred F. Imhoff. The cover was designed by Rosenbohm Graphic Design.

ISBN 1–891853–42–2 (cloth)   ISBN 1–891853–43–0 (paper)

About
Resources for the Future
and *RFF Press*

Resources for the Future (RFF) improves environmental and natural resource policymaking worldwide through independent social science research of the highest caliber.

Founded in 1952, RFF pioneered the application of economics as a tool to develop more effective policy about the use and conservation of natural resources. Its scholars continue to employ social science methods to analyze critical issues concerning pollution control, energy policy, land and water use, hazardous waste, climate change, biodiversity, and the environmental challenges of developing countries.

RFF Press supports the mission of RFF by publishing book-length works that present a broad range of approaches to the study of natural resources and the environment. Its authors and editors include RFF staff, researchers from the larger academic and policy communities, and journalists. Audiences for RFF publications include all of the participants in the policymaking process—scholars, the media, advocacy groups, nongovernmental organizations, professionals in business and government, and the general public.

Resources for the Future

Contents

Part 3. Special Waste Categories

Part 4. Final Thoughts

Preface

Economics is just too complicated. It makes our heads ache. So when anything economic goes awry, we respond in a limited and personal way by searching our suitcoat pockets to see if there are any wadded up fives inside.

—P. J. O'Rourke, *Eat the Rich*, 1998

Students often ask me why I am interested in waste, and I never know quite what or how to answer. I do remember, as a kid, I collected bottle caps by scavenging around public places, and I also remember great adventures scouring for treasure at the town dump, whenever the rats, smell, fire, and smoke permitted. I also recycled (though I didn't realize it at the time) by collecting scrap metal for the "war effort" (World War II) and by converting the family's empty soda bottles (at two cents deposit on each) into my weekly allowance. I also backpacked regularly, and nothing forces itself so indelibly into one's consciousness as litter in the backcountry.

The environmental movement of the late 1960s and early 1970s attracted me, and I began to teach a course on environmental economics in the mid-1970s. I quickly found that it is hard to teach air and water pollution because the problems and the solutions involve so much science, which neither I nor my students wanted to devote much time to. The answer, I also found out quickly, was to center the course around the environmental problems that require the least science to understand and discuss—vehicles and waste. I published a much expanded version of my class notes on vehicles in 1999—*Economics at the Wheel: The Costs of Cars and Drivers* (Academic Press). Now, this book represents a much expanded version of my class notes on waste, presented for students as well as for a more general audience.

I have had two reasons for continuing to enjoy teaching environmental economics to undergraduate students for so long. First, students like the subject—they find environmental problems important and interesting. And second, environmental policies so often fail to properly attack environmental problems that studying these policy failures is an excellent way to teach students how to think economically with real-world examples. Nothing is duller to (undergraduate) students than widgets. In this book, there are some assume-therefore-theorems, but I have tried to disguise them, and there are *always* (I hope) real-world reasons for the theoretical exercises. I am proud of the fact that I am "recycling" waste into economic knowledge and analytical ability.

Three caveats to the reader. First, because this book is as much a vehicle for teaching economics as a book about waste per se, it relentlessly stresses the economic analysis and barely touches on the cultural, social, political, and technical aspects of waste. Second, because the economics of waste is a fairly new field of research, there are still many gaps in our knowledge, and I have often had to resort to an on-the-one-hand-but-on-the-other kind of discussion. And third, because waste problems and solutions depend heavily on the geographic and demographic attributes of a place, it is rarely possible to give generalizations and conclusions about waste policy that clearly apply everywhere.

To offset the generalities of the text, many chapters have appendixes, in which a specific Ann Arbor or Michigan environmental problem is more closely analyzed. I realize that most readers care much less than I do about Ann Arbor and Michigan. The purpose of these appendixes is to outline a procedure so that readers can undertake their own such analyses in other cities or states. If you do this, please send it to me, so I can learn—if you send it to me as an email attachment in Microsoft Word, I promise to add my comments and return it, so you can learn, too.

As I mentioned, this book arose from class notes. The classes have never been so large that I have been unable to ask questions and get answers. I have tried to preserve that style here. I hope you find the occasional flippancy refreshing, but if it angers you that I do not take every ounce of waste more seriously, I apologize— somewhat. I also want to remind you now, as I will regularly in the book, that *not all waste should be taken very seriously.*

The book includes boxes with side material. If you are in a hurry, you can skip them without loss of continuity, but I suggest it would be a loss if you do. Most are intended to be provocative or amusing; each one not only offers a little information about the waste periphery but also challenges readers to think through the implications of some theory, circumstance, problem, or policy.

I further encourage you to track down at least a few of the referenced works, where you disagree with what I am saying or want to know more about the subject. Some of these references are mathematical or statistical, but most people can still browse and learn from them. The full citations are all in the references section at the end of the book.

Overview of the Book

This book has four parts. In the first part, we will look at ordinary solid waste, both business and household, and the various ways of collecting and disposing of it. To start simply on a complex problem, we will assume in Part 1 that there is *no* possibility of recycling. All solid waste is simply disposed of, in a landfill, an incinerator, or illegally (e.g., litter), or through exporting (to another state or another country). Many of the actors in the waste process face prices that are below marginal private cost, not to mention below marginal social cost. This means an inducement to generate too much waste or to dispose of it in inappropriate ways.

In the second part, we add the possibility of recycling. We will look at recycling in various ways. Why do private markets fail to recycle a socially desirable amount of material? What are the social benefits and costs of recycling? What kind of prices and nonprice policies would achieve the optimal amount of recycling? Too often, people think the optimal goal for recycling is 0% or 100%. Economics can guide us toward a socially appropriate system of recycling.

In the third part, we turn to special waste categories, those that involve greater hazards to health and the environment. These wastes fall under two broad categories, hazardous waste and radioactive waste, and we will examine the economics of the handling and disposal of each. Because the social cost of inappropriate handling of such wastes is high, and hence the social benefit of appropriate handling is high, the optimal regulation of such wastes usually requires greater care and cost. There is also another category of hazardous waste: past waste disposals that pose potential future risks to health and/or the environment—what are called Superfund sites. It is too late to control their creation, but economics can help determine where, how, how much, and when to undertake their remediation. Not all contaminated sites should be immediately or completely remediated.

In the final, brief, fourth part, I will highlight the lessons we learned from our journey through the waste-land. There, I will suggest not so much detailed policies for the future but rather changes in the ways that we think about these policies. I hope that by then, it is not so much that I suggest policy approaches as that the policy approaches have suggested themselves.

Note about Prices and Discount Rates

In an effort to make money figures readily comparable throughout the book, most dollar figures have been converted to 1997 prices using the U.S. gross domestic product (GDP) deflator. An example may make clear exactly what I have done. Suppose $1.00 million was spent in 1986. This figure is converted to 1997 dollars by multiplying by the 1997 GDP deflator (112.4 with 1992 = 100), and then dividing by the 1986 deflator (80.6, with the same base year). The $1 million becomes $1.39 million—and may be rounded and reported as $1.4 million. When I used a source that did not indicate what year's prices were being used, I assumed these to be prices of the year before publication.

When present values are calculated, a real discount rate of 4% is used (unless stated otherwise). Again, an example: An estimated expenditure of $10 million in 2009 is multiplied by $(1.04)^{-12}$, where 4% is the discount rate and the power of -12 reflects the number of years from 1997 to 2009. The $10 million in 2009 has a present value in 1997 of $6.246 million—and may be rounded and reported as $6.25 million or $6.2 million.

Acknowledgements

I want to thank past students of my environmental economics courses for asking the right questions and giving the right answers, reassuring me that the lectures were on the right track, and for asking the wrong questions and giving the wrong answers, signaling to me where the material needed to be made clearer. More than that, I have drawn on many of their term papers (cited in the reference section). I especially want to thank many students in the Fall 2000 course for poring painstakingly through an early draft of this book. Through discussion, email, and notes, they have corrected errors, noted inconsistencies, and flagged awkward or incomprehensible writing.

Too many people have helped me learn about waste to list all their names. But some have been particularly helpful, and I want to thank them individually: Frank

Ackerman, Ray Ayer, Victor Bell, Michael Blumenthal, Paul Brown, Jan Canterbury, Robert Coronado, Truett DeGeare, Brian Ezyk, Henry Ferland, Kurt Friehauf , Mike Garfield, George Garland, Gene Glysson, Linda Gobler, Robin Jenkins, Graham Johnson, Richard Kashmanian, Ian Kearney, Steve Kesler, Tom Kinnaman, Katherine Klein, David Konkle, Stanley Lancey, Angie Leith, Jef Mallett, Paula Malone, Jake Miklojcik, Chaz Miller, Dick Morgenstern, Sam Morris, Don Nicholson, Mark Nowicki, Richard Owings, Pat Plunkert, Michael Podolsky, Jerry Powell, Michele Raymond, Adam Read, Don Reisman, Steve Salant, Scott Salisbury, Chris Shaefer, Wes Sherman, Hilary Sigman, Brett Snyder, John Trotti, Dave Tyler, Bob Wallace, Bryan Weinert, Warren Whatley, and several anonymous referee/readers. Chapter 16 benefited greatly from a grant from the Michigan Memorial Phoenix Project, and the material on the developing world benefited from grants from the U.S. Agency for International Development and the Fulbright Scholar Program. I am also indebted to Paul Portney for encouraging me in the writing of this book and for making the facilities of Resources for the Future available to me when I worked in Washington.

Despite all this help, errors will persist. When you find them, I will appreciate it if you call them to my attention. You bought the book; maybe others will buy it, too; who knows, maybe there'll be a second edition; and you will have helped to reduce its errors. Also welcome are comments, quibbles, quarrels, ideas, data, anecdotes, and so on. I'm always game for a good ecochat by email. I am sometimes away from email for a month or two, so my reply may be delayed, but you will definitely get a response eventually.

DICK PORTER
rporter@umich.edu
The University of Michigan

The
Economics
of Waste

Economics and Waste: An Introduction

Where complex issues are involved, we must rely on analysis to help. Intuition and goodwill alone will not suffice. It is not really important that the analysis will be accepted by all the participants in the bargaining process. We can hardly expect that information systems will be so complete, necessary assumptions so obviously true, or constraints so universally accepted, that a good analysis can be equated with a generally accepted one. But analysis can help focus debate upon matters about which there are real differences of value, where political judgments are necessary. It can suggest superior alternatives, eliminating, or at least minimizing, the number of inferior solutions. Thus, by sharpening the debate, systematic analysis can enormously improve it.

—Charles Schultze, *The Politics and Economics of Public Spending,* 1968

The United States of America has become one of the richest countries in the world, which means that it produces huge amounts of goods and services for its citizenry. Alas, along with these "goods" come "bads"—and one of the most pervasive of these is waste, things that we don't want but that can be dangerous or expensive to get rid of. Between 1960 and 1996, while the U.S. living standard (i.e., real gross domestic product per capita) was doubling, the volume of municipal solid waste (MSW) was rising from 2.7 to 4.3 pounds a day per person. As early as 1970, one author ranked waste in importance with air pollution and water pollution by calling it the "third pollution" (Small 1971).

As these numbers suggest, waste generation has positive income elasticity—that is, generation grows as income grows. The cities of Western Europe, as well as those of North America, generate more solid waste than they did a century ago, and they generate more solid waste than cities in poor countries do today (World Bank 1998, Data Table 9-3). The good news, however, is that the income elasticity of waste generation is less than one—a doubling of living standards does not double the concomitant waste. Part of the explanation is the relative shift of rich countries away from goods, and their attendant packages, toward services, which are less waste intensive. But part of the explanation is that even poor countries generate a lot of waste, albeit of a different kind (Table 1-1). The relative decline over time in the amount of putrescent matter—less euphemistically, stuff that rots—reflects the increasing industrial treatment of foodstuffs and the increasing use of sanitary wrapping and refrigeration; and the decline in ash reflects the movement away from coal and wood for heating and cooking.

Not only do we produce more solid waste as we get richer, but also we produce more of it in our growing cities, where it presents special problems. Waste, as with many other forms of pollution, is not much of a problem if it is produced in small quantities and then spread out thinly. But as countries grow richer, agricultural productivity rises and farmers are freed for nonagricultural occupations. These new occupations are usually urban occupations, which means that cities grow and waste becomes a more serious esthetic and health problem. There is evidence of extensive concern with waste as early as 2500 B.C., in the Indus Valley city of Mohenjo Daro (in what is now Pakistan). Drinking water flowed into houses in troughs; liquid wastes went out in separate troughs; and solid waste was dumped into a pile outside the house or into street-corner bins and was then taken out of the city by regular municipal employees (Niemczewski 1977). The first dumps began to appear in Greece as early as 500 B.C., and edicts forbidding littering were promulgated two centuries later.

The modern era of waste concerns began in the latter part of the nineteenth century, as slums grew and the connection of waste to public health problems became established (Melosi 1981; Alexander 1993). Water treatment, sewage systems, regular solid waste collection, compactor trucks, trash dumps, and waste incinerators all emerged in the early part of the twentieth century in Western Europe and North America.

In the latter part of the twentieth century, new concerns emerged. We thought we were producing too much waste. Indeed, most international estimates show the United States generating half again as much municipal solid waste per capita as most other industrial countries (OTA 1989b; Weddle 1989; Alexander 1993). We also began to recognize dangers in our incinerators and landfills. These concerns led to new waste policies—and economic analysis of these policies is what this book is all about.

What Is Waste?

Waste is stuff we don't want—and hence we are willing to pay to get rid of it. We are going to ask these kinds of questions: How much waste do we generate? How do we get rid of it? How much does it cost to get rid of? We are going to distinguish between private and social cost. (The private cost of waste disposal is what the waste generator pays. The social cost is what it really costs society to dispose of that waste. The two concepts differ because of external costs and hidden subsidies. If these concepts are not familiar, read Appendix A on page 11.) But before attacking these bigger and more interesting questions, we must attend to a few definitional matters.

Waste goes by many names—trash, rubbish, garbage, refuse—and for some purposes it is important to make fine distinctions. Here, however, we shall use just the word "waste" to refer to anything that is no longer privately valued by its owner for

Category of material	Urban areas, India	All of United States
Putrescent matter	0.68	1.66
Paper	0.02	1.64
Plastic, rubber, leather	0.01	0.41
Glass	0.00	0.29
Metals	0.00	0.36
Ceramics, dust, ash, stones	0.17	n.e.
All other	0.03	n.e.
Total	0.91	4.36

Table 1-1

Relative Composition of Waste in Urban India and the United States (pounds per day per capita)

Note: n.e. means not estimated (though surely near zero).

Sources: Diaz et al. 1993; U.S. Bureau of the Census, various years; 1997 data.

use or sale, and the word "trash" for household waste. From an economic viewpoint, what is interesting is not the exact source or composition of the waste, but rather the cost of handling it and its effect on health and the environment if incorrectly handled.

Once the law becomes involved, of course, exact definitions become important. In the Resource Conservation and Recovery Act (RCRA) of 1976, solid waste is defined as "any garbage, refuse, sludge from a waste treatment plant, water supply treatment plant, or air pollution control facility and other discarded material, including solid, liquid, semisolid, or contained gaseous material resulting from industrial, commercial, mining, and agricultural operations, and from community activities" (http://www4.law.cornell.edu/uscode/unframed/42/6903.html).

To an economist, two other solid waste distinctions are more interesting. The first is the distinction between municipal solid waste and other solid waste. MSW is the trash of households and small businesses that is picked up by municipalities— or by private waste haulers where municipalities either do not provide the service or contract it out. This service is usually provided cheaply and often at no *marginal* private cost (MPC) at all to the waste generator. Typically, any business that is satisfied with a weekly pickup and generates only modest waste can qualify for this MSW collection. Those businesses that produce putrescent matter or large quantities of waste generate so-called other solid waste and must dispose of it themselves or hire a private hauler to remove it. We will deal with this other solid waste in the next chapter and thereafter focus on MSW (in Parts 1 and 2).

The second distinction is between hazardous and nonhazardous solid waste. Under RCRA, all solid waste is defined as either hazardous or nonhazardous, and there are special sets of rules regarding the handling and disposal of each type of solid waste. In fact, of course, almost all solid waste is potentially hazardous, because improperly handling and disposing of it can pose an environmental or health danger. These hazards can range from the slight esthetic disutility of newspaper litter to the formidable carcinogenic capacity of dioxins. While recognizing that all waste lies on a continuum of hazard, we will nevertheless accept the RCRA polarization for organizational purposes. (Hazardous waste will be left to Part 3.)

Tight definitions are necessary, not only for legal purposes, but also for quantitative ones. It would be impossible to estimate the total amount of MSW if we did not have a precise definition of it. But the measurement problem transcends definition.

In the fall of 1997, the City of Chicago announced that it was meeting the State of Illinois goal of 25% recycling. Half of this "recycled" material is yard waste that contains so much glass and other debris that it cannot be composted. So it was sent to a nearby (closed) landfill despite a state law prohibiting the disposal of yard waste in landfills. But city officials maintain that the yard waste is not being put *in* the landfill, but rather *on* the landfill—as landfill cover. And landfill cover qualifies as recycling (Greenwire 1997a).

On 2 July 1996, the mayor of New York City declared the city's legal goal of 25% recycling to be "absurd ... irresponsible ... and impossible" (Toy 1996a). On 3 July 1996, the mayor declared that the city was meeting this goal. What happened between these two otherwise uneventful days? The mayor decided to count as recycled the 30,000 abandoned automobiles that the city towed and sold to junkyards each year and also the debris from construction sites that the city ground up into gravel to construct the internal roads at its Fresh Kills landfill. These two redefinitions lifted the recycling proportion from 14% to the required 25% (Toy 1996b).

Almost nobody closely measures the weight or volume of MSW as it is collected and disposed of, so roundabout estimates are needed. The U.S. Environmental Protection Agency (U.S. EPA) estimates the total weight of our MSW by estimating materials flows through the economy. What goes into products is assumed to later come out as MSW. (It is of course not quite that easy; input data must be adjusted for exports and imports, the average lifetimes of different products need to be estimated, and wastes for which production data are not available—particularly food and yard waste—need to be estimated by sampling.)

Private estimates of MSW use surveys, asking relevant state agencies how much MSW has been generated. The U.S. EPA estimate is always much smaller—for example, the 1998 EPA figure was 220 million tons, whereas that of the *BioCycle* survey was 340 million tons, half again larger (Franklin Associates 2000; Glenn 1999). The difference may come from wastes missed by the EPA, but it is probably mostly due to overstatements by the states, which mistakenly include some industrial, mining, agricultural, or commercial wastes in their MSW reports.

Solid waste is usually measured by its weight—as in the previous paragraph—but its volume also matters, and the two are not related one to one. What the typical U.S. household puts out as trash has a density of about 10 cubic yards per ton, but the collection truck then compacts that to under 4 cubic yards per ton. (Trash is much denser in poor countries, where it consists of less paper and more food waste and ash.) If the trash goes through a transfer station, it may be further compacted to less than 2 cubic yards per ton. (Why trash might be transferred from one vehicle to another will be discussed in Chapter 3.) Some landfills and incinerators charge by the ton, some by the cubic yard. Here, we will use the two measures interchangeably, but keep in mind that conversion between them is not straightforward.

Why Think Economically about Waste?

Before launching into the serious business of thinking economically about waste, we should stop for a few minutes and ask *why* we should think economically at all. It is certainly not everybody's cup of tea. Many would prefer to think about protecting our environment, saving our planet, or ensuring our future. To the purest environmentalists, thinking about economics seems a little mundane, if not misguided and misguiding. At the other extreme, to the truest believers in market efficiency, economics seems unnecessary since waste markets work well and there is little, if any, need for public policies to correct failings. Such market believers tend also to think that if you were (somehow) to discern a market failure in waste, you would risk greater government failure if you tried to correct it.

I feel very strongly both ways. Economics is not everything, and most markets do work pretty well. But I still think that economics can help to give direction to our waste policies—by showing where, in what ways, and how badly waste markets fail, and by estimating the costs and benefits of various actual and proposed policies to correct these failures.

To think economically, one must start with the assumption that people are basically rational, knowing what they have, what they want, and how to go about making themselves better off through their economic choices. If you don't believe that is roughly true, prepare to stop reading at the end of this paragraph. But before you stop, think about what alternative assumption you want to make about people's economic behavior. The assumption of knowledge and rationality may not be very good, but as Winston Churchill said of democracy, it is the best we have been able to come up with. Furthermore, if people are making irrational choices on the basis of imperfect information, then how are we going to set policies aimed to make people better off? The very thought invites replacing consumer sovereignty with benevolent dictatorship. If you don't like the economic paradigm, then you are skeptical of the values of democracy.

With these assumptions about rationality and information, we can conclude that anything that makes one consumer feel better off, and nobody else worse off, is a good thing. Making people better off is what thinking economically is all about. When we have reached a point at which it is impossible to make anyone better off without simultaneously making someone else worse off, we have reached a situation of "economic efficiency." That's the goal of thinking economically about policies and projects.

Economic efficiency is the goal of economic thinking, but often, especially with waste issues, we have neither the knowledge nor the will to choose the policies that will get us there. So thinking economically in a practical sense usually means looking, not for *the best way* to do things, but for small changes from where we are

Trouble in Happyville

You are mayor of Happyville, a town of 1,000 people. The drinking water in Happyville is naturally contaminated with a substance that scientists unanimously agree is harmless, but which residents are convinced causes the cancers that periodically strike the town. They are so concerned that they insist you install a special water purification system that costs $1 million a year. Should you install it? There are three possible answers (Portney 1992).

First, you could declare the inhabitants to be ignorant and/or irrational, refuse to install the purifier, and get retired from the mayor's job at the next election. Second, you could go ahead with the system on the grounds that if the people want it and are willing to see their local taxes go up by $1,000 each (to cover the $1 million cost), they should have it. Or third, you could scream to your congressional representatives about the contaminant and try to get the EPA to install the purifier—at an annual cost of about $0.004 per U.S. citizen.

The rationality assumption means that you should choose the second action, but most mayors would choose the third. By the way, just in case you think this is a not a problem in the real world, try this: The U.S. public ranks cleaning up Superfund sites (which we will get to in Chapter 15) as the country's number 1 priority, but the EPA's experts rank Superfund as 16th in importance out of a list of 31 environmental problems faced by the United States (Clymer 1989).

that will make people better off on average. Consider a policy change that makes Alice $9 better off while making Bob and Carla each $3 worse off. Looking at the average change in welfare means that we will count this policy change as a winner, because $9 minus $3 minus $3 is greater than zero. (Another way of looking at this is to note that Alice could give $4 to each of Bob and Carla, with Alice having $1 left over. Then, each of the three people would be $1 better off than before.) Thus, any policy that is *potentially* a step toward economic efficiency, we will *usually* embrace.

Note the word "usually" in the preceding sentence. Thinking economically means concern not only for the size of the pie, but also for its distribution among different people. Even when a policy change makes people better off on average, there will be losers from the change as well as gainers. In the example just given in the text, two of the three people are made *worse off* by the suggested change—a democratic society would vote against this change! Clearly, our economic thinking must also worry about who gains and who loses, about whether the gainers are already rich or the losers already poor, and about ways for the gainers to compensate the losers for their losses.

Seeking economic efficiency, or more modestly, seeking steps toward economic efficiency, is one activity we shall undertake throughout this book. But there is a second activity, even more modest but no less important in practice. We will also continually ask whether we are doing whatever we are doing at least cost. This is called *cost-effectiveness* analysis. Cost-effectiveness is clearly a necessary, though not sufficient, ingredient of economic efficiency. If some policy is not being implemented cost-effectively, then we could do it more cheaply and have resources left over with which to make somebody better off. But remember, doing something as cheaply as possible is no guarantee that we are doing the right thing, just that we are doing it at the least possible cost.

Thinking economically thus comes down to asking two questions about markets or policies. First, can we make changes in what we are doing so as to make people better off on average? Second, can we do what we are doing with fewer resources? There are lots of ways that markets can fail or policies can fail when they are addressed by these two questions. But, as we will see over and over again, there are two kinds of failure that are endemic to waste problems.

Hidden subsidies are the first kind of failure. Consider, for example, another government policy that gives something to Alice—that is, the government produces it, sets a price of zero on it, and allocates it to Alice. Suppose it costs $5 worth of resources to produce it, but Alice would be willing to pay no more than $2 for it. This is an inefficient policy, because we would on average be better off not producing this product, just giving $2 to Alice, and having $3 worth of resources left over to pass out among Alice, Bob, Carla, and everybody else. Everyone could be made better off. This policy failure happens whenever somebody sells something for less than it costs to produce it. Usually, of course, this "somebody" is the government, because private businesses rarely find it profitable to sell things below cost (for more on hidden subsidies, see Appendix A on page 11).

Externalities are the second kind of failure. An external cost arises whenever anyone does anything that directly makes someone else worse off without compensating that person for it. Because we are going to use this idea a lot in this book, let's be more precise about it here. The *marginal private cost* of something is what it costs the *producer* to produce it. Its *marginal social cost* is what it costs *society* to produce it. The difference between the two is the *marginal external cost*—costs to society that are not costs to the producer.

Consider the following (extended) example. Alice hires Bob for $5 to remove her trash each week. There is no external cost (unless Bob creates a lot of noise, odor, or congestion while doing the collection). Alice is better off, because she would not hire Bob if his service were not worth $5 to her. And Bob is better off, because he would not take the job if the $5 were not worth more than his time spent in leisure or at his best alternative trash-collection opportunity. The market works! This simple transaction makes two people each better off, and nobody else is affected—a clear step toward economic efficiency.

But suppose that instead of hiring Bob, Alice just throws each day's garbage over the back fence into Carla's yard. Alice is better off because she doesn't have to store the garbage in the kitchen or garage until Thursday and then lug it to the curbside. Carla, however, is made *directly* worse off by this action—the damage to her welfare is an external cost of Alice's action. If Alice is $1 better off every day using Carla's backyard as a disposal site, and Carla is $3 worse off every day because she must look at the trash and clean it up, then Alice and Carla could *both* be made better off if Alice ceased this practice; $3 minus $1 yields a positive number that Alice and Carla could share so that each was better off. Once an external cost appears, there is no guarantee of economic efficiency.

There are many ways of handling external costs. If you don't remember them, read Appendix A to this chapter. Even if you do remember them, take a glance through that appendix. Directly restricting the activity that generates external cost is the EPA's favorite approach. Another approach, which we

How about Government-Generated Waste?

Waste is waste, regardless of who generates it. Waste should be handled, disposed, recycled, or whatever, up to the point at which the marginal benefit of the treatment activity no longer exceeds the marginal cost of that treatment activity. It should not matter who generates the waste.

Nevertheless, when the government generates the waste—when it is both the cop and the criminal—very few of the federal regulations and other policies we will be discussing apply (Armstrong 1999). Federal agencies are completely exempt from state environmental laws. Federal agencies have been required since 1992 to comply with federal environmental laws, but there are few consequences for those that don't.

For example, in 1988, the EPA gave owners of the 3 million underground storage tanks 10 years to either remove them or upgrade them, so that toxic leaks were less likely to occur and more likely to be spotted when they do occur. Most private owners completed the upgrade by 1998, but many city and state agencies have so far ignored it. Why aren't they fined? As one person observed, "Are you going to go around to the state police, or are you going to check private citizens? … How smart would you look to be the one to shut down an ambulance's tanks?" (Zielbauer 2000).

Even if all government evasion of EPA regulations were to completely end today, the legacy of past actions by federal agencies is huge. Cleaning up the environmental problems on U.S. military establishments, according to the Department of Defense, will cost more than $30 billion. And cleaning up other federal agencies' past contamination, according to the Environmental Protection Agency, will cost nearly $300 billion. Because policies so partially apply to government-generated waste problems, we will not be dealing much with such waste, but we should not forget the inefficiency that this problem causes.

will talk about in this book, is taxation. Levying a tax equal to the marginal external cost—called a Pigovian tax after A. C. Pigou, who first wrote about it—has the effect of "internalizing the externality" (Pigou 1920). The tax makes the perpetrator of the external cost think more carefully about continuing the practice. In the garbage-dumping example of the preceding paragraph, a Pigovian tax (or fine) on Alice of $3 (i.e., the damage) per illegal garbage dump would induce her to stop this activity, because gaining $1 in utility is not worth losing $3 in tax payments.

Basically, then, thinking economically means trying to get prices right. The right price of anything is the marginal social cost (MSC) of producing it. People will buy something only if their willingness to pay (WTP) is greater than the price (P). Therefore, setting P = MSC guarantees that WTP ≥ MSC. Nobody will end up consuming a

Table 1-2	Relationship between P and MPC	Relationship between MPC and MSC	
		MPC < MSC (external cost)	MPC = MSC
Taxonomy of Market Failure from Subsidy or External Cost	P < MPC (subsidy)	WTP ≥ P P < MPC < MSC WTP ⪌ MSC ???	WTP ≥ P P < MPC = MSC WTP ⪌ MSC ???
	P = MPC	WTP ≥ P = MPC MPC < MSC WTP ⪌ MSC ???	WTP ≥ P = MPC = MSC WTP ≥ MSC !!!

Note: MPC = marginal private cost; MSC = marginal social cost; P = price; WTP = willingness to pay.

product unless they are willing to pay at least as much as the social cost of producing it. A hidden subsidy means underpricing; that is, the price does not cover the marginal private cost (P < MPC). An external cost means that the marginal social cost is greater than the marginal private cost; that is, MSC > MPC. If either P < MPC or MPC < MSC, there is no guarantee that WTP ≥ MSC. It may help to lay out all these possible market failures schematically. This is done in Table 1-2.

Finally, we will have to think about environmental problems that *kill* people. Because we cannot prevent every possible environmentally caused death in the United States, we will also have to ask how much environmental safeguarding should we do. The answer to this always depends upon how highly we value the lives of Americans. By "value" of life, I mean simply the maximum amount we collectively are willing to spend to save one expected life. By "save one expected life," I mean, for example, reducing the probability of dying of cancer by 0.000001 for each of the 1 million residents of a city. (A probability of 0.000001 means 1 chance in 1 million; multiplying 0.000001 by 1 million people means 1 *expected* life will be saved.) If you think every American life is priceless and that we should never stint where anyone's environmental health is involved, then please read Appendix B on page 15 of this chapter. (And even if you fully understand the necessity of putting a ceiling on how much we should spend per expected life saved, it won't hurt to glance over the appendix to see how life-saving environmental policies are analyzed.)

Is Economics Everything?

Economics is not everything—hardly. I shall be surprised if, as you read along, you don't find yourself often mumbling "that may be the economics, but there is a lot more to it than that." My point is not that economics is everything, but that economics must *also* be considered when thinking about handling waste. Before proceeding further, we should also stop for a minute and clarify where economics is at its best and where it is at its weakest in helping us answer waste questions.

Economics is at its best when dealing with a small, immediate problem where all the relevant costs and benefits are accurately and readily quantifiable. Consider Figure 1-1. On the horizontal axis is some environmental activity—call it pollution abatement. Its possible range is from 0% (no abatement) to 100% (total abatement). Without economics, there is a temptation to call for either 0% or 100% abatement.

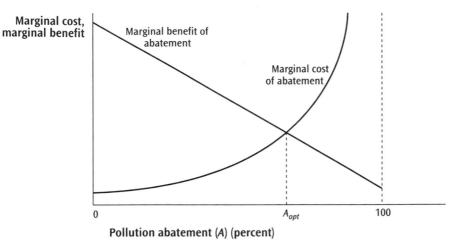

Figure 1-1

Optimal Pollution
Abatement

Without economics, we have no criterion on which to base a judgment to stop somewhere between nothing and everything. With economics, we can estimate both the marginal benefit of abatement and the marginal cost of abatement. The marginal benefit is usually downward sloping—as the environment gets ever cleaner, there is an ever smaller benefit in taking the next step of cleaning up. And the marginal cost is usually upward sloping—as the environment gets ever cleaner, there is an ever higher cost in taking the next step of cleaning up. Where the marginal benefit of abatement falls to equality with the marginal cost of abatement, we have reached the optimal abatement (A_{opt})—or put another way, the optimal pollution (100% minus A_{opt}). The optimal abatement is usually somewhere between 0% and 100%. (Note the word "usually"; if the marginal cost exceeds the marginal benefit at 0%, then 0% is optimal, but if the marginal benefit exceeds the marginal cost at 100%, then 100% is optimal.) Economic analysis can help to tell us when to stop.

When is economic analysis not so helpful? First, economics is not at its best when intangible outcomes matter a lot. We want to quantify all the benefits and costs of a policy into dollar values, so that we can compare them. When important benefits or costs are difficult or impossible to assign dollar values, the resulting comparison becomes less useful as a policy guide. Indeed, it may become completely useless. This problem is what Ezra Mishan called "horse-and-rabbit stew":

> … the outcome of all too many benefit–cost studies follows that of the classic recipe for making horse-and-rabbit stew … one horse to one rabbit. No matter how carefully the scientific rabbit is chosen, the flavour of the resulting stew is sure to be swamped by the horseflesh. (Mishan 1988, 154)

Economists have proved imaginative in finding ways to quantify seemingly intangible benefits and costs. For example, how much people are willing to pay to see Old Faithful can be estimated by observing how much they will spend to get there. How much people are willing to pay for a safe working environment can be estimated by determining how much lower the wage is in safer factories. Such estimates, however, are rough and arguable. If the analysis turns on them, things other than economics are probably more important to the decision.

A second area where economics provides weak guidance is for events that are distant in time. Recall that we assume that people know what they want and are rational and informed in their pursuit of their own utility. That is a pretty good assumption if we are dealing with a person at the supermarket choosing between asparagus and broccoli for today's dinner. But for choices years from now—especially when we know our tastes and our technology will change in unknown ways, and when the choice is about a highly unlikely event which carries catastrophic implications—the calculation of marginal costs and benefits becomes ever more difficult and ever less appropriate.

The third area where economics helps little is where our assumption that people know what they want and how to go about getting it breaks down. Again, the assumptions are fine when people are choosing between broccoli and asparagus for dinner. They are much less dependable when people are choosing between the potential air pollution of an incinerator and the potential groundwater pollution of a landfill. Where there is a high degree of scientific uncertainty or complexity, economics cannot overcome that uncertainty.

You will notice throughout this book that the policy advice is less confidently offered when intangibles, distant outcomes, and scientific uncertainty and complexity are involved. As an example of this abdication, the siting of a U.S. high-level nuclear waste repository is not going to be done through benefit–cost analysis and market incentives, and (almost) no economist would insist that it should be.

Finally, economists like to assume that tastes are given. This means you will not find much in this book urging that people be "educated" to litter less or recycle more. I am not against education on moral values—it may be cheaper to teach people not to litter than to achieve the same result by spending resources on anti-litter-law enforcement. But I do not think that it is a good idea to rely totally on educated altruism, for two reasons. The first and obvious reason is that it may not work. Some people will be selfish and stubbornly refuse to do the "right" thing. The second and less obvious reason is that it may work too well. Once one is taught that recycling is a "good" thing, there is no moral stopping point until everything is recycled. The nice thing about prices and taxes is that they use the motive of selfishness to get people to undertake socially correct actions. They don't beg or teach people to do the right thing, and they don't require or force people to do the right thing. They make people *want* to do the right thing.

When it comes to waste handling—and other activities that are potentially damaging to the environment—economists are almost alone in seeking market-oriented incentives for correct action. To almost everyone else, pollution is a moral issue. People should do the environmentally correct thing because it is the right thing to do. I agree. I don't dump my garbage in my neighbors' yards, even though it would be a quick and easy way to get rid of it. And the reason I don't do that is not (only) from fear of getting caught and penalized. I like my neighbors, and it would be a bad thing to do to them. Ethics works for most people, at least for small groups. But, as George Orwell reminded us, "on the whole, human beings want to be good, but not too good, and not quite all the time" (Orwell 1941). Prices and taxes can be a powerful reinforcement, especially in groups too large to rely on cooperation, respect, trust, and peer pressure. As Alfred Marshall said, "Progress mainly depends on the extent to which the strongest, and not merely the highest focus of human nature can be utilized for the increase of the social good" (quoted in Robertson 1956, 148).

Usually, economic incentives—that is, prices and taxes—can be devised to reinforce intrinsic moral motivation. But they can also "crowd out" moral motivation. Sometimes, once one has paid the price of pollution, one feels morally free to pollute to one's heart's content (Weck-Hannemann and Frey 1995). This book is about economics, but we will periodically consider the noneconomic incentives involved in waste-handling activities, especially when they may conflict with economic incentives or may be altered by the use of economic incentives.

Where Next?

The approach of this book is to start simple, and then add complexity one layer at a time. In Part 1 (Chapters 2 through 7), on solid waste creation, collection, and disposal, we will assume that there is *no such thing as recycling*. We will worry about waste—how businesses create it, how households handle it, whether it should go to a landfill or an incinerator (the only two choices for now), the possibility of illegally disposing of it ("litter"), and the possibility of exporting it to other states or countries. Not until Part 2 (Chapters 8 through 13) will we look at the possibility of recycling (in much detail).

I have adopted this approach because (I think) it makes it easier to follow the economic reasoning. Moreover, it has a chronological virtue. In the 1960s and 1970s, Americans began to concern themselves with waste disposal problems—and many other pollution problems—but there were, for all practical purposes, no recycling programs to aid in solving them. Part 1 addresses the situation of these decades. In the 1980s and 1990s, recycling programs blossomed all over the United States, and solid waste policymakers began very much to consider recycling—as well as all the issues of the previous decades. Part 2 addresses the changes to waste handling brought on by the advent of the possibility of recycling. In short, for the next six chapters, don't keep saying "but what about recycling"—although you are encouraged to note the page whenever that thought occurs, and later to check that the question has indeed been covered (in Part 2).

APPENDIX A

External Costs and Hidden Subsidies

One person's trash basket is another's living space.
—National Academy of Sciences, *Waste Management and Control*, 1966

With perfect competition and equilibrium, price equals marginal cost (MC) in the short run, and price equals the minimum possible long-run average cost in the long run. Thus, all those potential consumers whose willingness to pay for a good exceeds the lowest possible opportunity cost of producing the good get to consume it. As a corollary, no resources are wasted producing goods for people whose WTP is less than the opportunity cost of production. As you well know if you took the usual introductory microeconomics course, these are the great virtues of the perfectly competitive equilibrium, that it channels scarce resources into their most highly valued uses. If all producers were in perfectly competitive industries, there would be no way to make anybody feel better off without making somebody else feel worse off. This is what is meant by economic efficiency.

But all the above is premised on some assumptions. Because we are going to be thinking economically throughout this book, we had better think for a moment about these assumptions. In addition to the basic assumptions that people know what they want, know how to get it, and go about their consumption choices rationally, there is another big assumption behind the desirability of the perfectly competitive equilibrium. We assume that all valuable resources bear prices that reflect their scarcity. If people, either producers or consumers, are able to use a scarce output or input without having to pay for it, all the optimality attributes of the competitive equilibrium break down.

External Costs

Consider a consumer who consumes just two things: food and a neighboring river to dispose of his or her garbage. The consumer pays a price for the food; but in an unregulated environment, the consumer pays nothing for the use of the river. If the countryside abounded in rivers, so that people were not inconvenienced by the presence of the garbage, there would be no problem. (You might be thinking that the fish in the river might be inconvenienced. But remember that economics is extremely anthropocentric—the value of everything depends upon how much *currently living human beings* are willing to pay for it.) Rivers would be a "free good" if they were so abundant that river pollution caused no inconvenience to any human being. And the proper price of free goods is indeed zero.

If, however, rivers are scarce, downstream producers and consumers will be inconvenienced or even sickened or killed by our hypothetical consumer's method of garbage disposal. Commercial and recreational fishing will be damaged, swimming will be less pleasant and healthful, drinking water will need to be treated, and so on. In short, the use of the river for waste disposal burdens society, even though it does not cost anything to the person who does the burdening. There are social costs even though there are no private costs. This excess of social cost over private cost is the *external cost*.

The private cost is the scarcity signal that this consumer receives and responds to; if the private cost of the river is less than the social cost, the consumer will overuse the river. In general, where there are external costs attached to the production or consumption of something, too much of it will be produced or consumed. In this example, one can say either that too much river garbage is being produced or that too much river quality is being consumed.

Correcting External Costs

There are several ways of correcting this misuse of resources. First, downstream residents could band together and negotiate with this consumer, offering a payment in return for reduced waste dumping. This solution was suggested and analyzed by Ronald Coase and is called *Coasian negotiation* (Coase 1960). It has the advantage of yielding economically efficient outcomes because people will continue negotiating as long as there are ways to make one of them better off without making the other worse off. And, if you don't like the idea of the blameless downstream residents having to pay the antisocial upstream resident to reduce his or her pollution, you can make river pollution illegal unless the polluter has permission from all the downstream residents; then the upstream polluter will have to negotiate with the downstream residents and will end up either stopping the pollution or paying those downstream residents for the right to continue polluting. But the Coasian solution has a big disadvantage: The downstream residents must be sufficiently few and organized that they can band together and negotiate. For most waste-related externalities, the number of people affected is too great for a negotiation to take place at reasonable cost.

Second, if the downstream victims of pollution have a clear right to clean water, they could take their case to the courts, suing any upstream polluter for violating that right. In principle, this also solves the pollution problem, for the courts should assess the monetary value of this damage and require the polluter to recompense the victims. If the polluter knows that any external costs must be compensated, these external costs become private costs, and the polluter will stop unless the value of the activity exceeds its full social costs. In reality, the solution is not so clear. Courts can be slow, court outcomes are uncertain, damage assessment is not straightforward, and where there are many polluters, it may be difficult to figure out just who is responsible for what. For most waste externalities we will encounter, the courts are at best a very partial solution.

Third, the government could pass a law banning or restricting the generation of external costs. This is the usual approach of the U.S. EPA, once an external cost is recognized as a serious misallocation problem; it is called command and control. It overcomes many of the problems of negotiations and legal suits among large numbers of affecting or affected parties. But there is no reason for thinking that the politically, legally, bureaucratically, and adversarially determined command-and-control allocation is efficient. To see the possible inefficiency, think of two factories, ABC Corporation and XYZ Incorporated, alongside each other on a river. Each disposes of the same waste in the same quantities into the river. ABC has a cheap alternative waste disposal process, but XYZ only has a very costly alternative. For the sake of "fairness," command-and-control regulations would probably order each to reduce its pollution by the same amount, say by half. But the same result, a 50% reduction in total pollution, could be achieved more cheaply by somehow arranging to have ABC cease polluting altogether while XYZ continues to pollute as before. It would seem that a switch to this latter policy would make ABC worse off, but XYZ could pay all the extra costs to ABC plus a little something, and then both would be better off than under the equal 50% cut policy. This example, though terribly simplified, indicates why we usually find that command-and-control approaches are not cost-effective.

Fourth, the government could estimate the external cost being done per unit of river waste disposal and tax each river waste disposer that amount per unit of his or her disposal. This solution, first suggested by Pigou (1920), has the great virtue of "internalizing the externality." Consumers and producers now face a new private cost, the pollution tax, and they reduce their pollution in response to it. Though the producers and consumers may not recognize it, if the marginal tax exactly equals the marginal external damage, they are effectively responding to the external cost of the pollution as if it were a private cost. There are three disadvantages: the Pigovian tax may levy what is politically seen as an unfair or inequitable tax burden on some people; the Pigovian tax must be levied precisely on the variable that is causing the external cost, not on some roughly correlated or administratively convenient other variable; and the government must be confident that it has correctly measured the damage done by the externality, for an incorrect Pigovian tax will lead to an inefficient level of pollution. Despite these possible problems, we will often be looking at the potential for Pigovian taxation as a means of handling waste problems.

(An additional point on this taxation: Instead of taxing the creation of pollution, the government could subsidize the abatement of pollution—such abatement after all is an external benefit, the exact opposite of an external cost. This is almost the same thing as taxing the pollution, because rational, optimizing firms and people must now recognize that each unit of pollution means the loss of a unit of subsidy, and so the subsidy is just like the tax in that it creates an opportunity cost of polluting. But there is one important asymmetry: The tax makes firms and people worse off and causes exit, whereas the subsidy makes firms and people better off and causes entry. Abatement subsidies could end up increasing pollution if the newly entered firms or newly immigrated people add more pollution than the old established firms and residents subtract; Porter 1974.)

Fifth, the government could issue to all current polluters a number of "rights" or "permits" to pollute—presumably, fewer total rights than the previously excessive levels of pollution. Firms and people that could reduce their pollution cheaply would presumably sell their rights to firms and people that could reduce pollution only at great cost. This is called a system of marketable permits. It is cost-effective in the sense that the government achieves its desired reduction in pollution at least cost—agents will trade permits as long as cost reduction is possible. Possibly costly Coasian negotiations are not needed; possibly inefficient command-and-control allocations of the permits are corrected by the subsequent trades; and the possibly objectionable income redistribution of Pigovian taxation can be avoided. There are a few places in our waste discussions where marketable permits can be considered.

Sixth, if the pollution is industrial, the industry could be nationalized and run in a socially profitable, though possibly privately unprofitable, manner. For ideological reasons, the United States does not usually choose this path. Waste, however, often provides an exception to the U.S. aversion to "socialism." Most U.S. cities operate their own waste collection and disposal services, the states are being required to handle all low-level radioactive waste disposal, and the federal government is planning to supervise the burying of U.S. high-level nuclear waste. Unfortunately, there is no guarantee that such government-operated waste services will be run with a close eye to maximizing *social* profit.

Hidden Subsidies

Even where there are no external costs (MPC = MSC), subsidies may create a price that is below the cost of the activity (P < MPC) and hence induce people to demand more of that activity. This violates the "benefit principle" of public finance, which obligates those who benefit from a public activity to pay for that activity. The benefit principle is not only a means for equitably distributing the cost of a public activity but also a way of ensuring that people's willingness to pay for every unit consumed is at least as great as the cost of providing that unit (WTP ≥ P ≥ MPC). (More accurately, there is a hidden subsidy whenever P < APC, the average private cost. The MPC may be small once a facility is built, especially if it is being used below capacity, and efficient use may dictate a low price. But you shouldn't have built the facility unless the average user's willingness to pay for it exceeded the average cost of both its construction and its operation.)

Exceptions to the benefit principle may be warranted, basically on two grounds. First, pure public goods should not be priced. Once they are produced, the MSC of their "consumption" is zero, and so the price should be zero. Clean air, for example, is costly in smokestack scrubbers, catalytic converters, and fuel treatment, but breathing it imposes no further cost on society. Subsidies are appropriate for pure public goods. Second, goals other than efficiency may be sufficiently important to urge subsidies—free health clinics, free voting booths, and free primary and secondary education are examples that quickly come to mind. But the exceptions should not be accidental. We will see many hidden subsidies in the handling of waste that do not involve public goods and that serve no higher purpose. Indeed, we will see the opposite: Subsidizing waste disposal encourages waste creation.

APPENDIX B

The Economic Value of an Expected Life Saved

We pay a large price for our current haphazard pattern of social investments in life-saving.

—Tengs and Graham (1996, 180)

Most people have an instinctive distaste for thinking about "the value of life"—especially when we are talking about the termination of some particular person's life. When we talk about expected deaths caused by mishandled waste, however, there are two important differences between those lives and the lives of, say, convicted murderers on death row. One, the lives in question are not yet identified; and two, the policy changes we are considering will move the probability of death by small amounts, not from one all the way to zero. These two differences make small changes in the rate of fatalities due to different degrees of waste handling much easier to discuss than capital punishment for a convicted murderer.

Still, it is hard for most people to say that we should undertake a policy if it costs less than *x* dollars per expected life saved and not undertake the policy if it costs more. It may be hard, but it is *essential* that we make ourselves willing to do this, for three reasons:

1. When we make decisions that marginally affect our expected longevity, we do not act like we value our lives infinitely—we play contact sports, we scuba dive, we hang glide, and so on—we even cross streets and drink tap water. If there are finite private pleasures that are worth some risk to your life, then this means that you are implicitly placing a finite value on your life. If we value small changes in longevity finitely in our private behavior, it would be inconsistent of us to value such small changes infinitely in our public decisions.
2. If all life were socially of infinite value, then any public policy that saves a life is worth doing, regardless of its cost. For example, banning the use of cars or

requiring all households to drink distilled water would be justified! The corollary of this is that all policies that do not save lives will be pushed to the bottom of the public agenda—and probably just crowded out of consideration, given budgetary realities. Do we want to see the end of all free primary education in order to find the cure for a disease that strikes one person in a billion? If we are to make choices between life-saving policies and non-life-saving policies, we must put a finite value on the lives that the life-saving policies save.

3. When we undertake public policies to save lives, we pay for them through taxes, or through forgone public expenditures elsewhere. This makes people worse off, and hence many eat less, work harder, drive older cars, worry more, and spend less on health care or smoke detectors. All of these changes affect their expected longevity. Estimates, albeit very rough, suggest that every time the government taxes people about $5–15 million, it causes people to reduce their private spending on health so much that it ends up costing one expected life—exactly how much depends upon how the taxes are levied (Wildavsky 1979; Keeney 1990, 1997; Viscusi 1996).[1]

Probabilities and Expected Lives

A brief review of what is meant by "expected life": Suppose a cutback in the annual budget of the fire department of a city of 1,000,000 people raises the annual probability of each person suffering a fire-related death from 0.00003 to 0.00004, then the cutback causes (0.00004 − 0.00003) (1,000,000) = 10 "expected deaths" per year. Alternatively, an equivalent increase in the fire budget that reduces the annual probability of death from 0.00003 to 0.00002 could be said to "save" 10 expected lives a year.

When dealing with waste problems, however, the probabilities are not usually stated on an annual basis, but rather as a "lifetime risk" of, say, cancer from exposure to some waste-related risk. We shall often want to convert lifetime risk to annual risk. This is easily done.

Consider a lifetime risk of cancer of 0.01. If the probability of dying of cancer during one's lifetime because of some hazard is 0.01, the probability of not so dying is 0.99. Let p be the probability of dying *each year* of cancer because of this hazard. Then, during an 80-year expected lifetime, the probability of not so dying is $(1 - p)^{80}$. Thus, $(1 - p)^{80} = 0.99$, and p is 0.00013. If you want to make the conversion quickly and roughly, just move the decimal point two places to the left. The annual risk is roughly 1/100th of the lifetime risk. For example, the probability of your being killed in a highway accident when you are *not* the driver is about $1*10^{-4}$ (i.e., 0.0001) each year but about $1*10^{-2}$ (i.e., 0.01) in your lifetime—that's 1 chance in 100 (Porter 1999).

Given that we need estimates of the value of life, how do we get them? Actuaries might suggest that the amount of life insurance tells us something about how a person values his or her own life, but use of such a number would mean that public policies are worthless if they save the lives of unmarried people, because most people without partners or dependents neither need nor carry life insurance. Lawyers might suggest that the present value of all future earnings tells us the value of one's life, but use of such a number would mean that public policies are worthless if they save the lives of the retired or the disabled, because these people anticipate no further earnings.

Economists reject both these approaches to the value of life and ask, instead, how much are people willing to pay for their *own* lives. Not with a gun in your ear and a mugger growling "your money or your life." That would bring us back to the identified life and a zero-versus-one probability. Rather, economists are interested in knowing how much people are willing to pay for a *small reduction* in the probability of dying (Schelling 1968).

Suppose, for example, that every person has one chance in a million of dying each year of the Arctic Flu. Suppose further that an inoculation is available that would reduce that probability to zero. Suppose finally that the average person is willing to pay $5 for such an inoculation. Then if a million people were inoculated, one expected life would be saved, and their total willingness to pay would be $5 million. In a WTP sense, $5 million is what these million people would be willing to pay to save one expected life.

This suggests how we could evaluate a policy of Arctic Flu inoculations. Let Δp be the reduction in each person's probability of death (as a result of the inoculation), let N be the number of people to be inoculated, and let V_L be the monetary value each person implicitly puts on his or her own life. Each person's V_L will of course be different, but for public policy purposes we must pick some average, socially acceptable value. Then the social benefit of this inoculation policy is $(\Delta p)(N)(V_L)$. In the inoculation example of the preceding paragraph, Δp was 0.000001 (i.e., one in a million), N was 1 million people, and V_L was $5 million, and the product of those three numbers was $5 million. In a full policy analysis of this inoculation program, we would go on to look at what the program costs (C) and ask whether the cost is greater or smaller than this expected-life-saving benefit of $5 million. The *net* benefit of the policy is positive or negative as

$$(\Delta p)(N)(V_L) \gtrless C \qquad (1\text{-}1)$$

If the net benefit is positive, the policy passes its benefit–cost test. If it is negative, it does not.

But how big is V_L? Evidence is all around us. Whenever people demand money in order to undergo additional risk of death or spend money to reduce their risk of death, they are offering information about their WTP for expected longevity. Studies have looked at a wide variety of such choices: How much higher are wage rates in more risky occupations? Who buys smoke detectors? Who speeds? Who uses seat belts? How much would people pay for cancer-detecting check-ups? How much more do houses cost in

Present Value and Annualizing

Economists compare values from different years by discounting. A dollar is not a dollar if it appears at a different date. Why? A dollar, or more precisely, a dollar's worth of resources, can be invested and if invested wisely, will return more than a dollar's worth of resources in the future. Suppose $1.00 worth of resources can be invested so as to yield $1.04 in a year; then $1.00 today is "worth" $1.04 in a year. Put another way, $1.00 in a year has a present value of only $$(1.04)^{-1}$ today, where $(1.04)^{-1}$ is about $0.96. By compound interest, the present value of $1.00 in T years is only $$(1.04)^{-T}$ today. Many projects cost C today and then yield a stream of benefits of B during the next T years. How do economists decide if B is big enough relative to C to make the project worthwhile? One way is to compare their present values. The present value of the cost, since it is incurred today, is simply C. The present value of the benefit stream (PVB) is

$$PVB = B(1 + i)^{-1} + B(1 + i)^{-2} + B(1 + i)^{-3} + \ldots + B(1 + i)^{-T} \quad (1)$$

which can be summed to

$$PVB = \frac{B[1-(1+i)^{-T}]}{i} \qquad (2)$$

PVB is bigger than C if

$$\frac{B[1-(1+i)^{-T}]}{i} > C \qquad (3)$$

Another way is to compare the "annualized" values of B and C. B is already an annual value, but C needs to be converted. What annual value of x, incurred each year during the next T years, would have a present value of C? This means calculating the value of x that solves the following equation:

$$C = x(1 + i)^{-1} + x(1 + i)^{-2} + x(1 + i)^{-3} + \ldots + x(1 + i)^{-T} \quad (4)$$

Summing the right-hand series and solving for x yields

$$x = \frac{iC}{1-(1+i)^{-T}} \qquad (5)$$

B is bigger than x (the annualized costs) if

$$B > \frac{iC}{1-(1+i)^{-T}} \qquad (6)$$

Equations (3) and (6) are clearly asking the same question. We will use both alternatives at different times, either comparing present values of benefits and costs or comparing annualized values of benefits and costs.

Cost per Expected Death Avoided

Another way of looking at equation 1-1 and the benefit–cost analysis surrounding it is to calculate, for any given policy, the expected cost per life saved (or, what is the same thing, the expected cost per death avoided). That is simply the total cost of the policy divided by the expected deaths avoided by the policy. Whether the policy passes its benefit–cost test depends upon whether the cost per life saved is smaller or greater than the value we put on expected lives (V_L)—that is, $C/(\Delta pN) \gtrless V_L$—and a little algebraic manipulation will quickly convince you that the two different-sounding tests, this inequality and the inequality in equation 1-1, are actually identical.

areas free of life-threatening air pollution or crime? How much discount due to lesser safety is there in used-car prices?

The figures vary from study to study, but the majority yield estimates between \$1 million and \$5 million per expected life saved (Dillingham 1985; Fisher et al. 1989; Viscusi 1996). Throughout this book, whenever we think quantitatively about life-saving devices or policies, we will consider this range of values for V_L.

Lest you think that this approach to the saving of expected human life is useful "only in theory," let's have two looks at studies of actual government policies. The first study is of the U.S. EPA's implementation of the asbestos ban (Van Houtven and Cropper 1996). In deciding which asbestos products to ban, and which not to ban, the EPA estimated the expected number of cancer cases that would occur if the product were not banned and the cost of replacing (or giving up) the product if it were banned. Figure 1-2 shows the cost per expected cancer case avoided (the vertical line) and the number of expected cancer cases avoided (the horizontal line). The products represented by squares were banned, and the products represented by triangles were not banned. The less costly asbestos products were much more likely to be banned. With one big exception, and a couple of small exceptions, products costing less than \$60 million per cancer case avoided were banned and those costing more were not banned. (You may find it interesting to speculate

Figure 1-2

Asbestos Products Banned or Not Banned by the U.S. EPA

Note: The original EPA calculations have been converted to 1997 dollars.

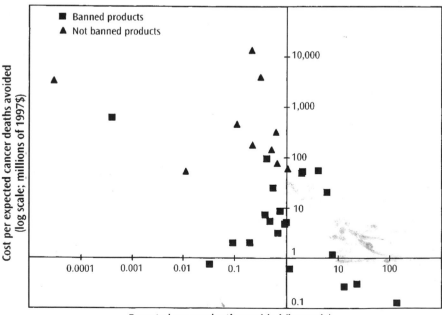

about the anomaly, where the asbestos ban cost more than $600 million per death avoided. It was in automatic transmission components, whose replacement, the EPA estimated, would cost only $200,000, but which would prevent only 0.0004 deaths.) Although $60 million is much higher than most estimates of the value of life policymakers use, this study does show the EPA's implicit awareness of cost as a factor.

A second study examined the anomalies in cost-effectiveness of different life-saving policies (Tengs and Graham 1996). They examined 187 federal life-saving policies and found that they cost $21 billion and saved 56,700 lives. That's an average cost of only $376,000 per life saved. That sounds low in comparison with other value-of-life estimates. But these authors went on to ask another question: If we stopped spending on the policies that cost more than $7.5 million per life saved, and spent that money on the policies that cost less than $7.5 million per life saved, how many more lives could we save? The answer turns out to be an additional 60,200 lives—we could have saved more than twice as many expected lives! When we spend on policies that cost a lot per life saved, within a given government budget, we actually *lose* lives.

Note

1. There is also the more direct effect that for every dollar of construction expenditure in the United States, there averages 4 cents worth of injury and fatality costs. So if a public expenditure constructs something that saves less than 4 cents worth of injury and death per dollar of expenditure, it will actually, in a net sense, *lose* lives (Viscusi 1996). This is just a statistical variant of the more sensational stories about rescuers dying while trying to save an endangered mountaineer, so that the net loss of life is higher with a rescue attempt than without. Even the Coast Guard, which long prided itself on its selfless "we have to go out, but we don't have to come back," has quietly stopped using that motto (Berry 1964).

PART 1

Solid Waste Creation, Collection, and Disposal

Business Waste

We smell chicken manure where they smell money.

—Jerry Taylor, neighbor to Perdue Farms in Laurel, Delaware, which processes
600,000 tons of chicken "litter" a year (quoted in Clines 1999)

A huge amount of business waste is generated in the United States. We can only guess at exactly how much, because much of it is either ignored, as with mining and agricultural wastes, or taken care of by the business that generated it, without fanfare or statistics. Official guesses say that industrial waste (not counting mining and agricultural waste) runs to 250 million tons a year, somewhat more than municipal solid waste (Kesler 1994).

Most business waste is closely regulated—*now*. But that has not always been true, as we shall see in Chapter 15. There are exceptions, however, and in this chapter we will look at two sectors in which relatively unregulated waste causes serious external costs: mining and agriculture. Why they have gone unregulated, and how they should or could be regulated, are the questions we will discuss in the first half of the chapter.

All businesses, when they sell their products, pass on to their customers not only the product but also the duty to dispose of the packaging and, when its useful life has ended, the product itself. The consumer, of course, pays for the product and its package, but not (usually) for their disposal costs, which are covered by hidden subsidies from the municipal waste disposal system. How we might make producers aware of the disposal costs of their packaging, and thereby induce them to use more economical packaging, are the topics of the second half of the chapter.

Unwanted Industrial By-Products

Manufacturers essentially use inputs to produce outputs. They make more profit by selling more outputs, and they make less profit by buying more inputs. Clearly, inputs that do not become outputs are unwanted. Not only are such unwanted industrial by-products a waste of money when the inputs are purchased, but the unwanted material also costs money to dispose of. Unlike households, manufacturing and commercial firms rarely have their waste picked up and disposed of at zero marginal private cost. In short, manufacturers have a strong profit incentive either to reduce by-products that require costly disposal or to find ways to reuse, recycle, or sell them. In New York City, for example, where the municipal solid waste system collects only 18% of the waste paper for recycling, private-sector haulers, who must pay high prices at landfills, recycle 89% of their collected waste paper (Hershkowitz 1997).

As a result, manufacturers are continually seeking ways to produce their products with less input, as well as less wasted input. Every consumer (over a certain age) has noticed that most manufactured goods steadily get smaller as well as better. For example, during the past three decades, a typical refrigerator that holds 12 cubic feet has come down in weight from 295 to 144 pounds, in outer dimensions from 31 to 25 cubic feet, and in energy consumption from 136 to 45 kilowatt-hours per month (Franklin Associates 1997).

Life-Cycle Analysis

A currently fashionable way to get at the full and complete environmental impact of a product is through life-cycle analysis, which looks at the entire life of the product, from the birth of its raw materials, through its factory production and household use, and on to its death in a landfill or incinerator. Let us look at an example of this, the disposable coffee cup debate: Paper or plastic (polystyrene)? The private cost to the consumer is about the same. Life-cycle analysis tries to answer the question, which type of cup is *environmentally* less hostile?

The paper cup means cutting down trees, a renewable resource, whereas the plastic cup is made from petroleum, a nonrenewable resource. The production of the paper cup consumes 36 times as much electricity, generates 580 times as much wastewater, and results in greater emissions of most kinds of air pollutants. But the manufacture of the plastic cup uses 3 times as much cooling water and causes pentane gas emissions. (Pentane carries the following hazard warning: "Danger! Extremely flammable liquid and vapor. Vapor may cause flash fire. Harmful or fatal if swallowed. Harmful if inhaled. Affects central nervous system. Causes irritation to skin, eyes and respiratory tract"; http://www.jtbaker.com/msds/p0737.htm).

The plastic cup is more homogeneous and hence more easily recycled (the paper cup has a plastic coating). Both burn well, but the plastic cup produces twice the energy. In a landfill, the paper cup degrades, giving off carbon dioxide and methane. But the plastic cup takes up nearly twice as much landfill space (though it weighs less), and it doesn't degrade at all (Franklin Associates 1990; Hocking 1991; Kamberg 1991; Passell 1991).

So where does all this lead us? To trying to add up apples and oranges (e.g., trees and petroleum, pentane and methane). At its best, life-cycle analysis is the beginning of an analysis of social cost-effectiveness because it tries to list all the possible environmental external costs that are incurred in the production, consumption, and disposal of a product. But it stops short of putting money values on the various kinds of external damages so that they can be added up and compared. This is a tough and uncertain step, to be sure, but it is an essential one if the analysis is to be of any help in policymaking (Arnold 1995; Menell 1995; Portney 1993).

One other issue with life-cycle analysis is the so-called boundary problem. You certainly want to examine the environmental problems caused by raw materials, but do you then examine the environmental impact of the inputs that produce the raw materials? And then the inputs that produce those inputs? And so on (Arnold 1995; Portney 1993). Maybe the ultimate boundary problem is hinted at in the title of a recent article: "Never Mind Paper vs. Plastic Bags, How Did You Get to the Grocery Store?" (Meadows 1999).

Manufacturers are helped in their efforts to reduce inputs and unwanted by-products by the fact that most factory-floor waste—called "preconsumer waste"—is easily collected and quite homogeneous, and hence is more cheaply reused or recycled. Each factory can become specialized in handling its own waste because there are usually only a few different types of waste in any particular production process. For example, though postconsumer plastic is still little recycled in the United States, 95% of all preconsumer plastic scrap was already being recycled a decade ago (OTA 1989b).

Does this suggest that manufacturers do not contribute to the waste problem in the United States? No, for two reasons. First, manufacturers can often dispose of their unwanted by-products at prices that do not adequately reflect the costs of that disposal. Landfills, and other waste disposal processes, can be priced below their marginal social cost. When this happens, manufacturers are given inadequate incentive to reduce their wastes. And second, much of what ultimately becomes manufacturing waste is passed on to the consumer in the form of packaging, which becomes waste almost immediately after its purchase; and in the form of the product itself, which becomes waste when its useful life is over. If neither the consumer nor the producer is charged a price for this disposal—a price that reflects the marginal social cost of disposal—then neither the consumer nor the manufacturer has a sufficient incentive to prefer smaller, lighter, and more compact packages and products.

In this chapter, we will discuss each of these market failures. The first, that waste can be disposed of without charge, happens in many U.S. businesses, but the problem is quantitatively most serious in two industries: mining and agriculture. We look at each of these sectors in the next two sections. Then we return to the other problem, that manufacturers who produce consumer products can pass on the cost of disposing of their packaging waste to the consumer—or even further, to the municipality in which the package is discarded.

Mining Waste

Extraction waste, mostly from mining, amounts to 2.5 *billion* tons a year, an order of magnitude larger than either industrial waste or municipal solid waste (Kesler 1994). Although some of this mining waste is backfilled into the mines from which it was taken, most of it ends up as slurries, fines, spoils, tailings, solutions, slags, muds, drosses, residues, and ash, all of which carry acids and metals that can be washed or leached into nearby surface waters and groundwater (Rampacek 1982).[1] Nor is water the only external cost. The Surface Mining Control and Reclamation Act of 1977—which dealt almost entirely with coal mines—listed the potential external costs of open-pit mining:

> … many surface mining operations result in disturbances of surface areas that burden and adversely affect commerce and the public welfare by destroying or diminishing the utility of land for commercial, industrial, residential, recreational, agricultural, and forestry purposes, by causing erosion and landslides, by contributing to floods, by polluting the water, by destroying fish and wildlife habitats, by impairing natural beauty, by damaging the property of citizens, by creating hazards dangerous to life and property, by

degrading the quality of life in local communities, and by counteracting governmental programs and efforts to conserve soil, water, and other natural resources. (http://www.osmre.gov/smcra/101.html)

How important are the social costs of these mining wastes? Roughly 5% of the most contaminated places in the United States (to be discussed in Chapter 15) have been created by mines, and these are not just ancient mines with old mining practices—more than half of these sites are still active or have been closed only within the last quarter-century (U.S. EPA 1998d). The U.S. EPA estimates that it will cost $20 billion just to clean up these 60 worst sites, and billions of dollars more for something like 20,000 other, less contaminated but still hazardous sites.

Why are mines permitted to treat their wastes so casually? The answer goes back more than a century. Following the Civil War, in an effort to encourage growth in the western part of the United States, the General Mining Law of 1872 offered public land to miners at very low royalty rates and with very little required environmental protection—no or few standards for prudent mine operations, mine site cleanup, reclamation, or restoration. Though amended, this law still prevails (for a readable summary of the law, see http://imcg.wr.usgs.gov/usbmak/anat1. html). Indeed, when today's laws about the handling and treatment of waste were passed in the late 1970s, the Bevill Amendment specifically exempted most mining waste.

The 1872 law has done three bad things. First, it has permitted the excessive environmental degradation of mining areas. Some such degradation is necessary and even desirable— after all, the minerals are valuable, and there is no way to extract them without some scarring. But the law permitted miners to ignore too many external costs. Second, because mines on private land are subject to greater royalties and environmental restrictions, the law has prevented cost-effective mineral extraction, because miners prefer to extract ore from public sources with low private costs—even if these mines have higher social costs. And third, because mining on public land is more

---

Cyanide, Gold, and Wildlife

All mining causes environmental problems, but gold mining has compounded the problems for a century by using a technology called *cyanide leaching*. The days of nuggets and panning are long gone. Today's gold consists of microscopic flakes, which are recovered through chemical processes involving a cyanide solution—one *ounce* of pure gold can be profitably mined from many *tons* of ore. The cyanide that cannot be reused is channeled to ponds, where air, sun, and natural biodegradation by microbes break down the cyanide residue (OTA 1992; Clark 1997).

While it awaits this breakdown, cyanide is of course still one of the most toxic substances known. And these ponds are often vast, sometimes 200 or more acres. Although there is no evidence that an accidental discharge from one of these ponds has ever caused a human death, there is much evidence of animal deaths, when wildlife come to drink or when scavenging animals eat cyanide-poisoned corpses. Cyanide mining waste has killed 10,000 trout at the Richmond mine in the Black Hills of South Dakota; 900 birds were killed when they drank water at an Echo Bay mine in Nevada; and hundreds of bats have died at gold ponds in Arizona, California, and Nevada.

Attempts to prevent such deaths have included noise-making devices and colored pennants, but nothing has been as effective as covering the ponds with nets. Unfortunately, many mill-tailings ponds are too large to cover or are too difficult to reach. Even for covered ponds, wind, snow, and ice take their toll on the nets. Netting is privately costly; so are inspecting and reporting. The life of a deer, bird, or bat is neither easily valued by society nor privately profitable to the mine owner. It could be worse. Until a century ago in the United States, mercury, which is far more dangerous (to humans), was used to separate the gold, and it is still used in many developing countries (UNIDO 1997; Schuman 2001).

---

than marginal, the overmining of public land lowers prices of virgin-source minerals. This, in turn, causes two inefficiencies: Minerals, with artificially low prices, are overproduced; and virgin minerals, with artificially low prices, are overproduced relative to recycled minerals. (We will return to the choice between virgin and recycled materials in Part 2.)

Now that the U.S. West has been settled—and from the viewpoint of water availability perhaps excessively settled—we might well ask why we continue to encourage excessive and careless mining on public lands. Part of the answer is no doubt political clout.[2] Some Western politicians seem to have little else on their minds in Washington than perpetuating the subsidized use of federal lands. But a large part of the answer is cost. Requiring mine owners to treat their large volumes of waste in the same way as other industries are required to treat their much smaller volumes of waste would have a huge impact on the costs of mines. If U.S. minerals were protected from foreign competition, this would greatly increase the U.S. price of most minerals. If they were not protected, it could mean the end of mining, and thousands of mining communities, in the United States. Where the nonenvironmental consequences are great, the environmental consequences often get ignored (U.S. EPA 1985).

This is an aspect of a general problem in U.S. waste policy. We refuse to do a benefit–cost analysis if lives are involved. Rather, if lives are threatened, *no* expense is considered too great to protect those lives—in principle. But if the expense is in fact perceived as too great, the only logically consistent alternative is to assume that no lives are threatened and therefore that no life-saving expense is called for. We will see throughout this book examples of this all-or-nothing choice, which derives from our refusal to publicly "value" lives and hence a refusal to make in-between choices.

What we should be doing is estimating how much damage mining waste is doing and how much abatement of that damage is called for, balancing cleanup costs against life-saving benefits. If this optimal abatement of mining pollution bankrupts some mines, then we are better off without them (but we then should have transition policies to retrain or relocate suddenly unemployed workers).

Agricultural Waste

Agriculture is the source of nearly half the water pollution of the United States (Lyon and Farrow 1995). Animal manure—more than 100 million tons a year—can contribute a variety of pollutants to nearby water if not properly handled, and the application of fertilizers and other chemicals to the soil also leads to runoff of pollutants into nearby surface waters and leaching of pollutants into the groundwater. Historically, agricultural wastes have received favorable treatment from government at all levels, partly because farmers were considered poor, and partly because most farms were in sparsely settled areas so that the pollution affected few people.

Manure is usually poured into a "lagoon"—that is, a depression in the earth—where it undergoes anaerobic decomposition. At the end of the process, a valuable soil nutrient has been created, but during the process, groundwater leaching and surface runoff are risked, and the odors can be powerful. Because the alternatives to anaerobic decomposition are usually capital-intensive and expensive, this way of handling of manure probably made sense when farms were small, neighbors were few, animals were few, and the land-to-animal ratio was high.

In the past quarter-century, however, livestock production has undergone tremendous change. Concentrated animal feeding operations (CAFOs) have taken over the production of poultry, swine, and cattle. Modern techniques offer large economies of scale, and corporate farms have taken advantage of them. With swine, for example, packed cheek by jowl, sows produce 22 offspring a year, and the piglets are grown to their full 250 (low-fat) pounds within five months. In the past decade, the number of swine farms in the United States has dropped by 72%, and the largest 3% of the remaining farms (those with an inventory of at least 1,000 head) now produce 60% of pork sales (Copeland and Zinn 1998). This means much more animal

More on Pig and Poultry Manure

During the 1990s, North Carolina's swine population quadrupled, chiefly at the expense of other states, because it left its CAFOs almost entirely free of environmental regulation (Warrick and Stith 1995; Copeland and Zinn 1998; Marks and Knuffke 1998; Kilborn 1999; Innes 2000). Most of this new hog production was located in the state's southeastern counties, in areas with shallow water tables, many wells, and many rivers. The by-product is 10 million tons of hog manure a year.

External costs abound. First of all, there are odors and pests. Next, leachate and runoff contaminate surrounding surface water and groundwater. And finally, animal waste lagoons spill regularly—in 1995, one eight-acre lagoon in Onslow County breached and, within an hour, poured more than 20 million gallons of feces, urine, and water into the New River, killing an estimated 10 million fish. But the worst was yet to come—Hurricane Floyd in 1999, accompanied by heavy rain, spilled not only manure but the drowned hogs themselves into North Carolina's waterways. Seepage into the groundwater was not far behind—wells near CAFOs tested at three times the usual rate for fecal bacteria. (Fecal bacteria can cause significant disease in humans, such as diarrhea, dysentery, cholera, and typhoid fever. Some of these bacteria can also cause infection in open wounds. Untreated fecal material also adds excess organic material to the water. The decay of this material depletes the water of oxygen, which may endanger or kill fish and other aquatic life.) The North Carolina Pork Council has asked Congress for $1 billion to replace the storm-damaged facilities, *exactly* as they were previously constructed and operated—even though alternative technology is now available (Thompson 2000).

Meanwhile, at the federal level, there is a move afoot to expand tax credits for non-fossil-fuel energy production to the burning of animal manure, and some states are considering adding to this federal subsidy (Morris and Nelson 1999; Adams 1999; Snyder 2001). One proposed federal law is called the Poultry Electric Energy Power Act, or PEEP. Such electricity generation would remove the external costs of fertilizer production by making lagoons obsolete and by using enclosed anaerobic digesters to capture the methane, but it would add new external costs by generating air pollution and destroying the desirable nitrogen and organic content of the manure (and hence destroying its usefulness as a fertilizer). Subsidizing the electricity disposal route is a curious way, and probably a very wrong way, to try to solve these market failures. That energy generation and manure production both cause external costs, and hence are underpriced, requires taxing both, not subsidizing one.

One more point before we move on. Delaware is subsidizing plants that turn poultry manure into fertilizer without endangering groundwater (Clines 1999). If, for some reason, we cannot tax an activity that causes external costs, it may be necessary to subsidize an alternative activity, but this is far from ideal. Aside from its budgetary impact, it encourages the production of chickens and the collection of poultry manure from far away.

waste on each animal farm and much less land per animal on which to spread the manure.

Yet the various federal water quality laws of the past three decades have generally either exempted or ignored most animal wastewater pollution—the U.S. EPA only prohibits CAFOs from directly discharging such waste into navigable waters (except during heavy rainstorms). Control is left to the states, which in turn do little.[3] Because industrial water pollution and municipal sewage (i.e., human manure) treatment plants are very tightly controlled, the total lack of control over animal manure is evidence of a lack of concern for cost-effectiveness in U.S. water pollution policies. Surely the marginal social cost of the first bit of abatement in agriculture is lower than the marginal social cost of the last bit of abatement in factories and (human) sewage treatment. Putting that in less technical words, the looseness of agricultural pollution controls in itself probably explains why nearly half of U.S. waterways still fail to meet legislated water standards.

In short, there is no good economic reason for failing to attack the external costs of agriculture with the same enthusiasm with which we attack the external costs of industries and municipalities. But even if we tried to do so, it would be very difficult to attack agricultural water pollution, because much of it, unlike animal manure, is *nonpoint-source*—meaning that it is difficult for the government to know which farmers caused which pollution. In California, for example, where 80,000 farms use a fourth of all the pesticide in the United States, 46 of the state's 58 counties register detectable levels of pesticides in their drinking water (Martin 1999).

Because it is hard to know who is doing the damage, both regulation and Pigovian taxation are hard to implement without huge monitoring expenditures. These monitoring problems are of two types. First, every farm's application of every polluting input must be tracked. And second, the impact of each input on the resulting surface water or groundwater pollution must be estimated. Just taxing, say, fertilizer use would be inappropriate, because some farms are so situated that their fertilizer never reaches water, whereas other farms' fertilizer use may be a major cause of water pollution.

In theory, it is possible to achieve a first-best input use through the application of taxation (Segerson 1988). The government must estimate the optimal amount of, say, river pollution (call this optimal pollution, P_{opt}) and the marginal damage done by pollution levels in excess of that optimum (call this marginal damage, D_{opt}). Then each farmer is taxed an amount equal to $D_{opt}(P_{act} - P_{opt})$, where P_{act} is the actually observed pollution level of the river. If each farmer knows how the use of fertilizer affects the river's pollution, then each farmer will consider this D_{opt} whenever considering an additional dose of fertilizer. The external cost is internalized by the tax.

There is just one problem. The amounts of taxation (or subsidies) involved could be huge, especially if there are many farms involved. If, despite the neat theory, some farmers misjudge and slightly overuse fertilizer in any given year, a bankrupting level of taxation may be imposed upon all of them. It may be efficient, but is it fair for *each* farmer's tax to equal the damages done by *all* farmers?

If the chemical wastes of agriculture are to be taxed to deter their excessive use, and we cannot tell which waste caused which pollution, it may be necessary to tax the inputs themselves. This is second best, of course, because each input will be taxed at the same rate even though inputs on different farms cause different

Figure 2-1

Voluntary Farm
Environmental
Programs

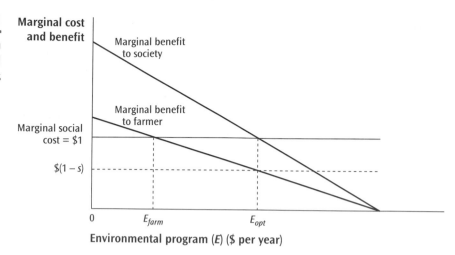

amounts of pollution. Even if we are willing to take this second-best route, there is still the question of *which* input to tax. Although the fertilizer or insecticide "causes" the pollution, the pollution would not occur if the chemical input had not washed into the nearby surface water or underlying groundwater. So, in a very real sense, the irrigation water is just as much a cause of the pollution as the chemical itself. Indeed, taxing the irrigation water has two big advantages. First, it is easier to note where the water is used than the fertilizer, so a water tax could be levied only where water pollution is a problem, whereas a fertilizer tax would need to be levied over a larger region. And second, irrigation water is (usually) already metered, so monitoring and taxing it would be easy, especially if it is provided by a public agency, in which case the tax could be simply included in the price.[4]

Because of these difficulties, economic as well as political, most agricultural environmental programs end up being voluntary. That such programs are inadequate can be easily seen. Consider Figure 2-1. On the horizontal axis, line up a gamut of possible environmentally friendly programs, from the biggest environmental benefit per dollar to the lowest benefit per dollar, and measure them in dollars of social cost. The marginal social cost of a dollar's worth of program is therefore always $1, but as we move to the right, we are undertaking expenditure dollars that yield fewer and fewer environmental benefits. In Figure 2-1, we draw two marginal benefit curves. The lower curve is for the farmer and family—farmers benefit most of all from living amid a healthful, esthetically pleasing environment. The upper curve is for society as a whole—everybody who comes into contact with the water, including the farmer and family. It is higher because it takes into account the willingness to pay of all the neighboring nonfarmer families that would enjoy the improved environment at the farm.

Without any government encouragement, the farmer would choose only E_{farm} dollars of environmental programs, for only that much environmental protection yields more benefits to the farmer and family than it costs. Society as a whole would prefer E_{opt} dollars' worth. One way to encourage the farmer to choose E_{opt} is for the government to subsidize, by an amount s, each dollar of the farmer's environmental program. This induces the farmer to choose the optimal program, but at the possibly high cost of sE_{opt}. If government budgets are tight, less than s will be offered, and less than E_{opt} will get done.

This abstract Figure 2-1 does, in fact, come close to describing U.S. voluntary agricultural environmental programs, as embodied in the Environmental Quality Incentive Program, begun in 1996. The federal government pays up to 75% of any approved environmental beneficial practices that farmers undertake. The government spent $167 million on this program in 1999. That's about $0.17 per farm acre in the United States.

Mining and agriculture are mostly located in sparsely populated areas and hence (directly) affect relatively few people. But their waste can be a major problem for those few who are affected. We turn now to packaging waste. This is more a city problem. Although its per capita impact may be smaller, packaging waste affects many more people.

Packaging Waste

To talk about waste from packaging, we must first get rid of the widespread notion that all packaging is a total waste. Packaging serves many purposes—such as preservation, sanitation, security (from theft and tampering), safety, and consumer convenience.

Nowhere is this more clear than with food packaging, which accounts for more than half of all packaging. There is a highly *negative* correlation between the amount of packaging and amount of food waste (Alter 1989). The World Health Organization maintains that in the developing countries of the world, between a third and a half of all food decays before it reaches consumers—recall the discussion of the cross-country composition of solid waste in Chapter 1—whereas in the United States, only 2–3% of food so decays. Most researchers agree that better food packaging (and refrigeration) made an important contribution to the dramatic decline in the incidence of stomach cancer in the United States during the past century (Palli 1997).

Because the cost of producing the packaging is almost always passed on through higher prices to consumers, the packaging must be worth it to consumers, or else we would still be buying Uneeda biscuits from an open barrel. The producer passes on the problem of disposing of the packaging to the consumer—and, as we shall see in the next chapter, the household that actually throws away the package also usually passes on the cost. Consumers end up paying the cost of producing the packaging but not the cost of disposal. The social problem is to make producers and/or consumers aware of the costs they impose on society when they use bulky or complex packages.

Two ways have been suggested to "internalize" the disposal costs into the manufacturer's decisions about packaging. The first is to make the manufacturer directly responsible for the collection and disposal of any package used—often called "producer take-back responsibility." The second is to tax manufacturers for their packages by an amount equal to the marginal social cost of its disposal—often called an "advance disposal fee." We look at each of these approaches in the next few pages. To anticipate the outcome of this examination, the first approach will be seen to be necessarily inefficient, whereas the second may lead to an efficient outcome.

Producer Take-Back Responsibility

Take-back responsibility means that manufacturers are required to assume responsibility for any product or package that they put into the marketplace. In strictest terms, this means that they must *physically* take it back from the consumer and dispose of it themselves—or reuse or recycle it—at their own expense, or hire someone else to do it. As Reid Lifset puts it, such take-back responsibility "transforms the transaction between the producer and the consumer from one of purchase to one closer to leasing" (Lifset 1998). This is intended to make producers consider the cost of disposal when they decide how much packaging and what kind of packaging to use. It certainly does that. The problem is that it is a very expensive way to do it.

It is expensive because it forgoes the advantages of economies of scale in waste collection. The cheapest way to collect waste is to have a single truck pass by each house and pick up all that household's waste as it passes. Even having two trucks pass by, one to pick up waste for disposal and another to pick up waste for recycling, raises the cost a great deal—as we shall see when we look closely at recycling (in Part 2). But take-back responsibility increases the costs even more because, in theory, one truck passes to pick up empty Campbell soup cans, another to pick up empty Crest toothpaste tubes, yet another to pick up empty Reebok shoe-boxes, and on and on. Of course, in reality each manufacturer would not actually pick up or take back each discarded package. But that is beside the point—the law requires that *someone* make the special effort to collect all packages and take them back to the original producers or to proper disposal sites.

In fact, where take-back responsibility has been implemented, producers quickly realize the enormous expense of collecting their own packaging, and they band together to pick up waste collectively. This reduces the cost, but not by very much in comparison with a well-run municipal solid waste collection system, because the manufacturers' collection system at best no more than supplements an already operative municipal system and effectively duplicates it for no good reason. (The manufacturers' collection cannot replace the municipal collection because

Should All Packaging Be Recycled?

One way around the problem of nonrecyclable packaging is simply to ban the use of nonrecyclable packaging, as did the Environmentally Acceptable Packaging Ordinance of Minneapolis–Saint Paul (Fishbein and Gelb 1992). But should all packaging be recycled? The answer is almost certainly no. An example shows why.

Beverages can be packaged not only in glass, plastic, and metal containers but also in aseptic boxes that are very unfriendly to recycling, being a composite of aluminum, polyethylene, and paper bonded together with adhesives. Because of their unrecyclability, environmentalists frown on them, and Maine even banned their use for a while. But aseptic containers weigh less per ounce of beverage than bottles or cans and hence, when discarded, take up less landfill space. This reduced original demand on resources *may* make the "one-way" aseptic box superior (in the sense of the present value of net resource costs) to the initially heavier but recyclable traditional container (Fierman 1991).

The proof of this superiority involves messy algebra, but the common sense is obvious—aseptic boxes will be less costly relative to metal cans: (a) the smaller the can recycling rate; (b) the smaller their volume (or weight) relative to metal cans; (c) the smaller the marginal social cost of landfills relative to the can's recycling costs; and/or (d) the smaller the social value of a recycled metal can relative to its recycling cost.

The reduced resource demand "may" make the aseptic box superior, but does it do so in fact? That answer must wait until Chapter 10 (specifically to the box in that chapter titled "Aseptic Boxes versus Aluminum Cans Revisited"). The point here is simply that unrecyclable packaging *may be* less socially costly than recyclable packaging. Not all packaging should be recycled. Accordingly, not all packaging should be recyclable.

there is much nonpackaging waste still to be handled.) Even if we consider recycling, the take-back system is redundant, duplicating both the municipal trash collection service and the municipal recyclables collection service, because packaging waste is at best only partially recyclable.

Furthermore, there is no incentive under take-back responsibility for households to participate in the return of packaging. Indeed, if it requires extra effort by the household, there is an incentive not to participate. We shall see (in Chapters 6 and 14) that there are products for which special disposal is important and for which take-back responsibility may be appropriate, but we shall also see there that household incentives can and must be attended to.

In short, the idea of producer take-back responsibility is correct—to make manufacturers consider the disposal costs of their packaging when they design their packages—but the requirement that they physically take back the package leads to enormous, unnecessary costs. Before turning to an alternate system that internalizes packaging costs without the awkward side effects of high-cost collection, it is interesting to look briefly at the actual operation of one such producer take-back system, the German "Green Dot" program.

The German Green Dot Program

In 1991, Germany made the manufacturer of each product responsible for collecting and recycling its packaging. In principle, the consumer was to return each package to the retailer, who would return it to the manufacturer, who would recycle it (Shea and Struve 1992; Fishbein 1994; Scarlett 1994; Reynolds 1995; OECD 1998)—or sort of recycle it. (Definitional problems intrude; e.g., incineration initially was not permitted as a means of "recycling," but it later became acceptable for some materials, particularly plastics.) The potential cost of many small and independent recycling systems was so staggering that German manufacturers quickly came up with an alternative. They formed a company, Duales System Deutschland (DSD), to collect all the packaging waste of its members and arrange for its recycling. The member firms paid a fee on each package they put into the economy and then placed a green dot on the package to indicate to consumers that the package could be returned to any of the specially marked yellow DSD collection bins.

Effectively, the Green Dot program has added a new collection system to the already existing municipal waste collection and recycling systems. This duplication has been, by general agreement, very costly. Moreover, the volume of packaging waste has declined little—apparently, either packaging was not excessive to begin with or the fee structure failed to provide an incentive for manufacturers to reduce it. Mostly, the program has simply shifted waste and recyclables from one set of bins to another. And the new bins are much more expensive ways of collecting recyclable materials than the traditional municipal programs—some estimates say more than twice as expensive per ton of recyclables (Brisson 1993; Fishbein 1994). One study found that the Green Dot program cost nearly $500 per ton of recyclables (Boerner and Chilton 1994). Another found that the costs per ton "approach the costs of handling a ton of hazardous waste" (OECD 1998, 33).

The fee levied on participating manufacturers was initially based on their (self-reported, and often underreported) volume (Louis 1993). Thus, the fee ignored differences in the weight per volume and in the recyclability of the material, and hence it gave no incentive to the manufacturers to produce lighter or more easily recycled packaging (OECD 1998).

Where households must pay a fee for municipal waste pickup, they tend to use the DSD bins for all their waste—and the German government admits that nearly half the contents of the DSD bins is not packaging waste and much of it is not even recyclable (Gehring 1993). And those who contract with DSD to collect from their yellow bins have no incentive to monitor their use (or abuse) because they were being paid by the ton. Finally, the firms to which DSD sells the collected material (or pays to have it taken) must guarantee to recycle it, but there is evidence of corner cutting here—ranging from warehousing it, to sending it surreptitiously to landfills, to exporting it. Such exports, moreover, collapse the prices of recyclables throughout Europe, and hence Germany's augmented recycling tends to undermine efforts to increase recycling in neighboring countries (*Economist* 1993a, 1993b; Genillard 1994).

Is there no good news about Green Dot? Yes, two pieces. First, German manufacturers have found many ways to reduce or simplify their packaging—knowledge that the rest of the world may soon value. And second, the volume of German recycling has increased immensely. But the fact remains that Green Dot recycling has been very costly—and unnecessarily so, for the same desirable outcome can be achieved without excessive costs by means of an advance disposal fee.

Advance Disposal Fees

Once a package has served its purpose, something must be done with it—which costs resources that are not provided by the manufacturer who produced the package. The costs are borne by others, either as external costs (e.g., packaging that becomes litter), or as implicit subsidies (e.g., municipal waste collection programs that are financed out of general fund taxation). To make manufacturers aware of these costs, it is necessary to estimate them and tax each package according to the marginal social cost it imposes at the end of its use. With an advance disposal fee (ADF), the manufacturer pays for the later handling and disposal of the package, but unlike take-back responsibility, the manufacturer does not physically undertake that handling and disposal (Pearce and Turner 1992). (As you might guess, this situation gets more complicated when we introduce recycling. But we will wait until Chapter 10 for that.) The actual disposal process can continue to be the traditional lower-cost municipal waste collection system.

Advance disposal fees would not have major impacts on the price of most products. Estimates of marginal social disposal costs typically run 1–2% of product prices (Pearce and Turner 1992; Little 1992; Ackerman 1997). But ADFs would generate a lot of government revenue—a 2% tax would in effect be a significant increase in state sales taxes—and this, curiously, presents a major problem (the states' median sales tax is 5%; see http://www.taxadmin.org/fta/rate/sales.html). Politicians desperate for greater revenue might embrace high ADFs for the wrong reason, just to raise revenue. And politicians strongly opposed to increased taxation might reject reasonable ADFs for just that reason, despite their desirable social purpose. Of course, ADFs could always be adopted along with a comparable cut in the traditional sales tax; then, the only effect would be to raise the prices of packaging-intensive products and lower the prices of little-packaged products.

Few U.S. states now use ADFs—except for very specific products such as tires or car batteries—but something very close to such a fee is being implemented in France. Although supposedly its own version of the German Green Dot program, France's Eco-Emballage system comes much closer to being an ADF. In 1992, French manufacturers were given various options with respect to their packaging, but most have chosen to work with Eco-Emballage, a semipublic firm that collects a fee from its member companies and uses this revenue to assist municipal recycling programs and municipal waste-to-energy incineration facilities. Although the recycling of packaging is far from guaranteed under the French system, it does make use of existing municipal trash and recycling programs, with the result that the advance disposal fees are in most cases a fifth to a tenth as high as the fees charged under Green Dot (Ackerman 1997).

Why Not Packaging Regulation?

Often, when some activity imposes environmental costs on others, it is possible to regulate that activity either through a price mechanism, such as an ADF on packages, or through some quantitative mechanism, such as regulating the size, shape, or composition of packaging. In the previous section, we leapt at the ADF approach. Why not consider the quantitative regulation of packaging?

One answer is administrative complexity. It is not too difficult to determine the ADFs to be paid by various kinds of packaging, but it would require a huge bureaucracy to tell each manufacturer what kind and amount of packaging should be used. But there are theoretical reasons as well for preferring an ADF.

To simplify, let us assume that all packaging consists of a single material, each pound of which costs c to collect and dispose of. Two shippers use this packaging, one of marbles and one of eggs. Each can reduce the amount of packaging used, but the egg shipper can reduce it only at high cost (e.g., customer dissatisfaction, messy cleanups, and the need to replace damaged eggs). The marble shipper can

---

Florida's Advance Disposal Fee

In the early 1990s, Florida introduced an ADF, charging a small tax on the cans, bottles, jars, cartons, and newspapers sold in the state. The revenues were used to fund such programs as recycling initiatives and landfill upgrading. But any material that was at least 50% recycled was exempt from this charge, and many firms altered their packaging, or fostered its recycling, to qualify for the exemption. Despite the fact that the tax rate was doubled in its second year (from 1 to 2 cents per container), the tax revenues fell by half (Ackerman 1997; Scarlett et al. 1997).

The details of the Florida law point up some of the difficulties of ADFs. Cartons and containers were taxed 2 cents each, whereas newspapers were taxed 50 cents a ton. This implies that containers cost more than $1,000 a ton to collect and dispose of or recycle, but old newspapers cost less than $1 a ton. The ADF needs to be more carefully tailored to the actual disposal cost if it is going to give the right incentives to manufacturers. And exempting materials from any tax at all if they achieve a 50% recycling rate makes it an either–or tax system that further removes the tax from an accurate proxy for either the disposal or the recycling costs (Alexander 1993). (You might be wondering whether the ADF fee ought to be lower—though perhaps not zero—if the product is extensively recycled. The answer is no, or rather not exactly, but the explanation is complex and will have to wait until we discuss recycling policies in Chapter 10 and recyclables markets in Chapter 12.)

The Florida policy can be interpreted as a great success, where 1 or 2 cents a container significantly affected firms' packaging decisions. Or it can be interpreted as irrelevant, where the changes were made, not to save a few cents, but to avoid the public-relations stigma of being labeled an environmentally uncaring firm. In any case, Florida packaging changed a lot during the two-year term of the law, and stayed changed when it lapsed.

easily reduce the packaging. To regulate by quantity, we would need two special studies, one of egg shipping and one of marble shipping, to determine the optimal quantity of packaging for each. In each study, we would need to discover the marginal cost of packaging reduction, equate that to the marginal cost of collecting and disposing of packaging (c), and promulgate a regulation requiring that optimal volume of packaging.

An ADF of c per pound of packaging achieves the same result without the special studies of eggs and marbles. Each shipper now considers c to be a part of the marginal private cost of packaging. The marble shipper will increase profits by cutting back on packaging a lot, whereas the egg shipper will find it profitable to pay a large ADF rather than ship broken eggs to customers. The shippers themselves figure out what is their optimal amount of packaging!

One more point: The ADF will end up affecting the relative prices of marbles and eggs. The relative price of marbles will fall, as it should, because marbles will be shipped with little socially costly packaging. The relative price of eggs will rise, as it should, because eggs are packaging-intensive. (For a fuller version of this theory of uncertain marginal costs and benefits of pollution abatement, see Baumol and Oates 1988, chap. 5.)

The real world, of course, is more complex than this story of eggs and marbles. But the conceptual difference between price and quantity regulation of packaging applies just as well—indeed, more so, because the real world consists of thousands of products and packages, each of which would require special study before optimal quantitative regulations could be issued. General George Patton once said, "Don't tell people how to do things—tell them what to do and they will surprise you with their ingenuity" (Patton 1947). ADFs essentially follow Patton's advice. Tell manufacturers how much they can save by not using a pound of packaging—they can save the ADF on each pound not generated—and let them surprise us with their ingenious ways of optimally reducing packaging, while they think they are just avoiding taxes.

Final Thoughts

Clearly, making environmental waste disposal laws by sector makes little sense. If a farm or mine is miles from anywhere—and especially far from people and water susceptible to pollution—then it makes sense to place fewer restrictions on its waste disposal techniques. But the weaker laws should apply because of the distance, not because of the kind of production. Cost-effective pollution reduction requires that the marginal cost of reduction be the same for all polluters—farm, mine, or whatever. For mines and farms that are near water and people, more careful waste disposal is appropriate, just as careful (in the equal-marginal-cost sense) as for the factory producing the same kind of waste next door.

In the latter part of this chapter, we have been looking at packaging, because that is the biggest remaining problem with manufacturing waste. But the products themselves also will need to be disposed of some day. Eventually, ADFs should be levied on the products themselves, especially on short-lived ones. (For long-lived products, e.g., houses, the present value of disposal costs is low because of discounting. One dollar, appearing 50 years from now, discounted at 4%, has a present value of only $0.14 today, i.e., $1.04^{-50} = 0.14$.) Such fees on the products themselves would encourage firms to offer to reacquire their worn-out products, perhaps even paying customers to return them, to get their ADFs lowered. (The correct ADF would be the disposal cost times the fraction of the products that would eventually be disposed of at public expense.) Now, even without such fees, many companies acquire and "remanufacture" old products and parts into new ones—for example, photocopier toner cartridges, postage meters, office furniture, auto parts, and "disposable" cameras (Deutsch 1998; Duff 2001d). We are most unlikely to get back to the refillable, reusable bottles ubiquitous before 1950—on this, we will say more in Chapter 6—but we might be surprised to find out how much remanufacturing is possible once it is stimulated by ADFs.

The main achievement of ADFs, however, is that they "internalize" the disposal cost of packaging and give a profit incentive to manufacturers to reduce the amount of packaging they employ. Leasing, and hence recapturing the product after its lease-life, would mean a zero ADF for the manufacturer. Companies such as Xerox and Pitney-Bowes have found that remanufacturing from recaptured leased products has cut their overall manufacturing costs—they lease without the incentive of an ADF.

How big should ADFs be? We have a ways to go before we can talk about that. For one thing, we haven't brought in recycling yet. But first we must recognize that households also participate in the creation of solid waste. In the next chapter, we will examine household waste activities and how policies may shape them.

Notes

1. In the rest of the world, only about half the mines are surface, open-pit mines, but in the United States, 90% of the mines are, with their attendant surface scarring and surface waste (Hartman 1987). Why the difference? The character and depth of the ore matters a lot, but economics also matters. Underground mining is more labor-intensive, and hence more likely to be practiced where wages are low, whereas open-pit mining is more capital-intensive, and hence more likely to be practiced where wages are high. Typically, an open-pit mine removes not only the ore but the overburden (the worthless rock on top of the ore), and the overburden may be 5 to 10 times as great in volume as the ore. The ratio of mining waste to marketable output depends on the value of the metal, running at times even higher than 350,000 to 1 for gold (Rhyner et al. 1995).

2. The clout goes much further than just avoiding Pigovian taxation to correct external costs (Humphries and Vincent 2001). For example, most minerals receive a percentage depletion allowance, which means that some fraction of the value of the mineral output is counted as depreciation for tax purposes—even if the discovery of the mineral deposit cost absolutely nothing and hence there is nothing to depreciate. If U.S. output of a mineral is more than marginal, this leads to global inefficiency as well as U.S. inefficiency, because there will be an overmining of U.S. ores and an overlow global price. Of course, if other countries have even less environmentally protective regulation of their mining than does the United States, tightening our laws will lead to overmining abroad and may further underprotect the global environment from mining activities.

3. Michigan, for example, promotes voluntary measures, provides financial and technical assistance to help CAFOs implement generally accepted management practices, exempts CAFOs that use such practices from "nuisance suits" (i.e., suits brought by CAFO neighbors seeking compensation for odor or well contamination), and offers tax breaks and low-interest loans to CAFOs that use such practices (www.ncsl.org/statefed/cafocht.htm).

4. Whether a water or fertilizer tax would be better in fact depends upon various production parameters. But one study of lettuce in the Salinas Valley of California, used estimated production functions to derive the welfare costs per hectare of three kinds of taxes, each designed to reduce water pollution by 20%: the cost-minimizing mix of water and fertilizer taxes (welfare cost of $6.67), a water tax alone ($6.83), and a fertilizer tax alone ($90.16) (Helfand and House 1995). The "welfare cost" was calculated as the difference between (a) the profits of all the farms when no taxes were levied and (b) the profits when taxes were levied plus the tax revenues collected. An input tax lowers the profits because less output is produced after the tax is levied and because the input mix of taxed and untaxed inputs is distorted by the tax.

Household Waste Collection

The problem of refuse has … become more complex than formerly, and this complexity may not yet have reached its limit.
—Rudolph Hering, 1913, president, American Public Health Association
(quoted in Melosi 1981, 96)

For the most part, Americans pay nothing for their trash collection and disposal in the *marginal private cost* (MPC) sense. In the vast majority of cities and towns where some kind of collection is provided, its cost is covered either from general fund revenues or from a flat charge on residents, where by flat charge is meant a charge for the service that is not related to the amount of trash that is put out for pickup (Jenkins 1993). It makes no difference with respect to household trash-generation incentives whether the general fund or a special time-based trash charge is used; the end result is the same—the MPC of putting out an additional unit of waste is zero (Kemper and Quigley 1976). And if a special time-based trash charge is used, it makes no difference whether the total revenue collected from households covers the total cost of household trash collection; the end result is still the same—the MPC of putting out an additional unit of waste is zero.

Americans respond to this "free" service by generating large amounts of municipal solid waste (see Table 3-1). Per capita solid waste nearly doubled between 1960 and 1990, and the weight has not declined much since then, despite the great increases in recycling during the past decade. Not

Table 3-1

U.S. Municipal
Solid Waste

Year	Waste generated (pounds per person a day)
1960	2.68
1970	3.25
1980	3.66
1990	4.50
1994	4.50
1995	4.40
1996	4.32
1997	4.43
1998	4.46

Note: These data are now collected annually for the U.S. EPA by Franklin Associates; no official figures were available before 1960, and only census-year estimates were made for the period 1960–1990.

Source: Franklin Associates 2000.

all of this MSW comes from household residences. Small businesses also use the MSW collection service, and about 40% of the total collection is estimated to come from such firms, although a number of arbitrary assignments must be made to make this estimate (e.g., tires and major appliance disposal is assumed to be made by small firms, though they are often just providing a service for households). In any case, this adds up to a lot of trash, more than 200 million tons a year. U.S. cities collect it and take it somewhere for further handling or disposal, almost all for free.

Waste Collection—Costly but Unpriced

But there *are* marginal social costs to the pickup and disposal of solid waste. There are private costs—wages, fuel, machinery, land, but in the vast majority of U.S. municipalities, these costs are paid with general revenues. Solid waste collection also imposes a number of external costs—reduced highway safety, added street noise, increased traffic congestion, more air pollution. (Solid waste *disposal* also imposes external costs, which we shall cover in the next few chapters.)

When we fail to price solid waste disposal, we create two kinds of market failure: When price does not cover marginal private cost, we introduce a largely unintended subsidy to waste generation and thereby induce households to generate too much waste. Or when price does not cover the external costs (i.e., the excess of marginal social cost over marginal private cost), we foist waste collection costs onto third parties and thereby further induce households to generate too much waste.

A simple diagram may make this point clearer. Figure 3-1 shows the demand for municipal solid waste collection and the marginal private and social costs of collecting it. (Each cost is assumed to be constant for simplicity.) Currently, the price of such collection (P_0) is zero, and households react to that price by creating a large volume of waste (W_0). If the waste collection price were raised to cover MPC (to P_1), less waste would be produced (W_1). But the price that covers all social costs (P_{opt}) would lead to even less waste (W_{opt}). Indeed, W_{opt} is the optimal amount of waste—for any W in excess of W_{opt}, households are not willing to pay as much as the social cost. (Recall that the demand curve is also a schedule of WTP of households for waste collection and dis-

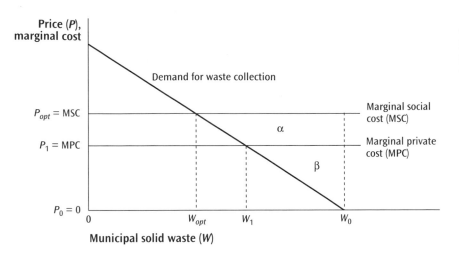

Figure 3-1

Deadweight Loss of
Underpricing Waste—
Theory

posal.) Moreover, pricing waste at zero instead of P_{opt} creates a deadweight loss. The deadweight loss is the sum of the two areas marked α and β, which show the total amount by which MSC exceeds WTP when the price is lowered from P_{opt} to zero.

It is possible to put some numbers on these concepts. It requires three pieces of empirical information: the actual current volume of MSW disposal, the MSC of waste collection and disposal, and the price elasticity of demand for MSW disposal. Because size, income, and waste generation will vary across municipalities in different parts of the United States, we will use rough averages. The current volume of MSW disposal is about 4.4 pounds per person a day; the MSC of collecting and disposing of waste is thought to be about 5 cents a pound (i.e., $100 a ton) (Repetto et al. 1992; Stevens 1994); and the many estimates of the price elasticity of demand for MSW cluster closely around –0.2 (Goddard 1994; Wertz 1976; Stevens 1977; Jenkins 1993; Skumatz 1990; Strathman et al. 1995; Reschovsky and Stone 1994; Miranda et al. 1994). (When we come to pricing when recycling is possible in Chapter 10, we shall find that the appropriate trash charge is probably less than the MSC of trash collection and disposal, but the following estimates do give us an idea of the magnitude of the welfare losses due to inefficient pricing.)

These numbers let us give empirical relevance to Figure 3-1; this is done in Figure 3-2. Note first that at a price elasticity of –0.2, raising the price of waste disposal from 0 to 10 cents a pound would reduce the daily per capita volume of MSW from 4.4 to 2.9 pounds.[1] Where would the other 1.5 pounds go each day? We are going to be thinking a lot about that, but just to anticipate for now, it would go in some combination to the reduced purchase of waste-producing products and packages, to the longer lives of waste-producing goods, to the recycling of what was previously waste, and—alas—to illegal disposal.

The welfare gain from changing from zero pricing to optimal pricing would be the triangle consisting of $\alpha + \beta$, which is conceptually comparable to the similarly marked areas in Figure 3-1. But this area can now be measured—it is just the area of the triangle. The welfare gain from optimal pricing would be a little under 4 cents per person a day, or $14 per average American a year—that's $3.5 billion a year for the whole United States. (Using an elasticity of 0.1, we find the welfare loss to be about 2 cents per person a day, or $7 per person a year, or a total of $1.8 billion a

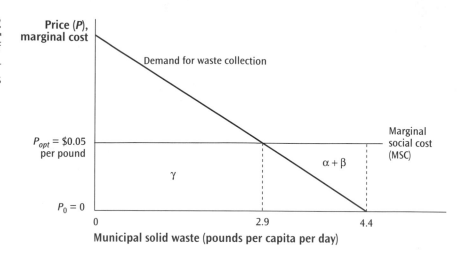

Figure 3-2

Deadweight Loss of Underpricing Waste— Empirical Aspects

year for the entire United States.) Although this estimate of the deadweight loss from zero-pricing waste is somewhat higher than most other estimates, the relevant point is that zero-pricing trash is socially expensive.[2]

Before leaving this exercise, we should also note that the waste disposal fee means that the average American would pay an additional amount each day equal to the area marked γ in Figure 3-2. This amounts to new solid waste fees of just above $50 per average person a year, but this is not necessarily a new cost to people, because it would be partly offset by reduced external costs in waste collection and it could be more than *fully* offset by a reduction in whatever taxes or fees that are currently needed to finance solid waste collection and disposal. Below, when we have considered illegal disposal and recycling, we shall find that the proper pricing of trash disposal is more complicated than it seems here, but this exercise makes the general point that *any* trash-pricing system—where the price is no greater than the MSC of trash collection and disposal—can reduce deadweight loss and save total cost.

In short, Americans put out too much trash for MSW pickup and disposal, simply because it doesn't cost anything at the margin. A welfare gain of more than a billion dollars a year is waiting to be achieved. And only pricing waste can achieve it. Command-and-control approaches, long favored by bureaucrats, really cannot get at the problem. Try to think of some—perhaps a limit on the weight of waste that each family could put out? Do we want to deny high-income families the right to exercise their highly income-elastic demand for waste? Or more realistically, do we want to drive well-off families to using a separate (and duplicative) private-sector pickup? Do we really want to enlarge the bureaucracy so that it could issue extra waste permits to large (or otherwise "waste-needy") families? Here, a price-incentive solution is not only more efficient—it may be the only way to reduce U.S. household waste.

Who Should Be Charged for Solid Waste?

In the last chapter, I suggested that the disposal cost should be paid by the firm that actually manufactures the product (as an advance disposal fee), and in this chapter

The Messy Diaper Debate

The debate between disposable and reusable diapers has been raging off and on for the past third of a century—though disposables now seem to have won out with most parents. Whether one's concern is trees, water pollution, cash cost, biodegradability, energy, or landfill usage, the issue is not a minor one—a baby undergoes as many as 10,000 diaper changes (i.e., one ton of diapers) before it is toilet trained. The debate goes on and on because of the proliferation of red-herring arguments by both sides (Little 1990; Tryens 1990). The issue is really quite simple. Let us examine each of the interested parties.

■ *Consumers* (i.e., the parents who pay for the diapers and who, we assume, also speak for their babies). They vote with their pocketbooks, either for disposables at $20 a week or for a diaper service that costs a little more than half as much money but much more time, because cloth diapers need to be changed nearly twice as often. (For brevity, we will ignore the people who buy diapers and wash them themselves; this choice also depends upon the parents' relative valuation of their time and their money.) By their choice, they reveal the diapering style that makes them better off; and this, after all, is what economic efficiency is all about, making people (now and in the future) better off.

■ *Producers*. Both the manufacturers of disposables and the provider of reusable diaper services are in business for profit, and by their very survival, they reveal that they do not in fact consistently lose money. What consumers pay at least covers the private costs they incur in providing their product or service.

■ *Others*. Diapers would be like broccoli, no cause for public policy or debate, were it not for external costs. The question is, simply, does the production or consumption of diapers incur social costs that are not paid by either the consumers or producers? Is energy underpriced? (Cloth diapers use more.) Are trees underpriced? (Disposable diapers use more.) Is water underpriced? (Cloth diapers use more.) Is waste disposal underpriced? (Disposable diapers use more.) The answer to the first three questions is "probably yes," and the answer to the fourth question is "definitely yes." Most of the debate about diapers would disappear if those four markets were priced correctly.

Most of the debate would disappear—but not all. Disposable diapers pose a special danger to groundwater that is not posed by most other household trash because disposable diapers contain human feces. How large is this special external cost? As far as I know, nobody has ever made any effort to measure it. For illustrative purposes, suppose this extra external cost of disposable diapers is $50 a ton. Divide that $50 by 10,000 diapers. That's half a cent per diaper. The simple solution is to levy a Pigovian tax of half a cent per disposable diaper to cover the extra external cost (either at the time of production or at the time of sale), and then let consumers buy whichever they like. What one should not do, as many suggest we should, is to ban disposable diapers outright or forbid their use in child care centers or nursing homes—such a ban would effectively be an infinite tax, far in excess of their external costs.

One final point: Currently, there is no federal tax on disposable diapers; and of the states, only North Carolina taxes them (at 1% of the price). If the total number of diapers used by a baby can be considered independent of the pricing scheme, subsidizing reusable diapers is almost the exact equivalent of taxing disposable diapers. Indeed, in Seattle and in several cities in Austria and Germany, city solid waste departments provide up to $100 in coupons good for the purchase of reusable diapers, on the grounds that much more than that will be saved in landfill disposal costs (McConnell 1998).

Garbage Disposers

Domestic in-sink food waste disposers (FWDs) have always received mixed press on the environmental front (Kemper and Quigley 1976; Bailey 1997). Some U.S. cities ban them; others require them (in new homes). Basically, the question is whether food waste is socially cheaper to dispose of through the sewage system or through traditional MSW collection and landfilling. In principle, it is not hard to do a social benefit–cost analysis of the decision to buy and use an FWD.

The first thing to consider is the private decision to buy and use an FWD. The FWD costs money to buy, install, and operate, so presumably the convenience to the household makes this purchase worthwhile. The problem is that other costs and benefits of operating FWDs may or may not be considered by the household in making its FWD purchase decision.

■ *Water.* Ownership and use of an FWD typically means that the household uses one more gallon of water per person each day. Where water is subsidized, the FWD buyer's decision is distorted. By how much? If water is underpriced by a third of a cent per gallon, this amounts to $1.22 per person each year.

■ *Sewers.* The introduction of FWDs may cause increases in suspended solids and oil and grease in the sewer collection system. One study estimated that this amounted to $0.18 per FWD user a year (NYC, DEP 1997). Moreover, the FWD roughly doubles the solids content of a household's sewage, and this added sewage flow must be treated. The cost of this treatment varies greatly, depending upon how extensive it is, and that depends on past and future regulatory actions. One study guessed that the cost on this account could be anything from $1.00 to $10.23 per FWD user per year (NYC, DEP 1997). Even if sewage is charged for—usually not through metering sewage but by assuming it to be proportional to water use—the additional cleaning, backup, and treatment costs are not levied on FWD owners. (Note that these water and sewage costs *are* borne by people with wells and septic systems, which is the chief reason that they rarely install FWDs.)

■ *Solid waste collection.* The FWD reduces the amount of food waste to be collected by the municipality by about half. Because food waste averages about a third of a pound per person each day, this will mean a collection cost savings of $1.52 per year (assuming a marginal social cost of collection of $50 a ton). But in most U.S. cities, the FWD buyer would not recognize this social benefit because there would be no marginal price on solid waste disposal.

■ *Landfilling.* Ultimately, the food waste must be landfilled. But food waste that emerges from the sewage system—sludge—can either be positively valued as fertilizer or be expensively treated as hazardous waste because of its toxic compounds, including heavy metals. Sludge can be treated to remove the toxics, but that of course adds to the treatment costs. It is, of course, arguable that from a social viewpoint, it matters not by what path these toxics and metals are released to the ground—household "sludge" would be similarly expensive if it were administratively feasible to charge for it.

In short, there are many ways in which the private decision to buy an FWD reacts to wrong social prices, possibly leading to an incorrect social decision. But the error can be in either direction and will probably not amount to much in dollars of net benefits or costs. FWDs, like diapers, probably fall under the category of "don't sweat the small stuff."

I am arguing that the disposal cost should be paid by the household that actually disposes of the product. But this, as you have probably already realized, would be double taxing the disposal. One or the other must be taxed, not both. To anticipate the argument below, it doesn't matter (much) which party is charged—until we consider recycling (in Part 2).

Consider a simple ballpoint pen. It could be sold without any package at all, or it could be marketed in a fancy plastic, cardboard, and paper package that costs $1 to dispose of. We could tax the pen manufacturer $1 if he uses the fancy package; and this $1 advance disposal tax would almost surely get passed on to the consumer in the form of a higher price for such packaged pens; a pen in a fancy package would end up costing the consumer $1 more than an unpackaged pen. Or we could directly tax the consumer $1 when he puts a pen package in the solid waste bin. Either way, the rational, informed (or simply observant) consumer will realize that a packaged pen will end up costing $1 more than an unpackaged pen, and the consumer will choose to buy a packaged pen only if the extra convenience of the package is worth the extra $1.

As long as we are dealing with such a simple production–consumption process—that is, purchase the good, use it, put it in the trash, collect it, and landfill it—it makes no difference whether the product carries an advance disposal fee (levied on its manufacturer) or a waste disposal charge (levied on the consumer household). The outcome is very similar to the tax-incidence theory worked out in every introductory economics course in the country, where it is shown that it doesn't matter whether you levy an excise tax on the supplier or on the demander—the end result on the price of the product is the same.

At the risk of belaboring the obvious, this is a good time to introduce a diagram that we shall find useful in more complex matters below, when we start to consider incineration, illegal disposal, and recycling. There are two equally appropriate means to handle the internalizing of the MSC of waste disposal in, say, a landfill (shown in Figures 3-3 and 3-4).

This is the basic result so far—it doesn't matter at which level the waste tax is levied. (The charge could also be levied 50–50, half on the producer and half on the consumer—or indeed in any other proportions; to anticipate, we shall find good reason below for levying some advance disposal fee on manufacturers *and* some trash fee on households.) There is no particular virtue to the producer pays principle, even though it is much heralded by environmentalists. It takes two to make trash, a producer to produce and sell it and a consumer to buy, consume, and pitch it, and it is as impossible to assign blame to one or the other as it is to decide which blade of the scissors cuts the paper. Anyway, guilt is irrelevant here. The consumer ends up paying for the trash disposal no matter where the charge is levied. Having made the point that the two approaches are basically identical, we should nevertheless look at some ways in which they do in fact differ, especially as these differences will become more important below, where we will consider more complexities in the waste disposal process:

- Collection and landfill costs differ in various parts of the country, as labor wages, population density, and land prices vary. An advance disposal fee cannot predict where the product (or its packaging) will end up being collected and landfilled, but a trash collection charge on the household can consider these differences.

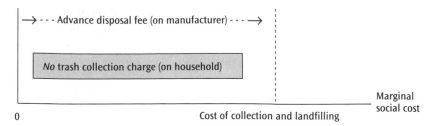

Figure 3-3

Means 1:
Charge Producer for
Future Waste Disposal
When Product Is Sold

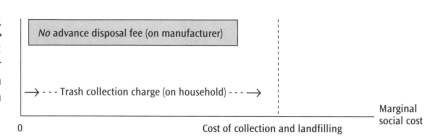

Figure 3-4

Means 2:
Charge Household for
Waste Disposal When
Product Becomes Trash

- An advance disposal fee raises the product price to the consumer by the same amount, no matter how long the consumer keeps the product. So it offers the consumer no inducement to keep and reuse the product. But a trash collection charge does offer such an inducement, because household reuse postpones that charge. Reuse should be encouraged—a postponed call on landfill resources is a reduced use of resources in a present-value sense.

- An advance disposal fee can vary according to the cost of collecting and disposing of the product, but a trash collection charge—unless it is prohibitively costly to operate—must be uniform across products, so much a pound or barrel or bag. A uniform trash collection charge will be too low on materials that are expensive to dispose of, and too high on ones that are cheap.

- There are fewer producers than there are consumers; as a result, the administrative and monitoring costs of an advance disposal fee may be significantly lower than those of a trash collection charge system (Fullerton and Kinnaman 1996).

- A trash collection charge invites illegal dumping of trash, where an advance disposal fee cannot be evaded by the household. We will think more closely in Chapter 6 about whether this issue is important, and if it is, what to do about it.

So much for theory. (Actually, so much for theory *for now*; when we come to illegal disposal in Chapter 6 and recycling in Part 2, we shall find more ways in which it matters where the disposal tax is levied.) It is time to look at actual trash collection charges—or "pay-as-you-throw" or "unit pricing," as these trash collection charges are sometimes called in the United States.

Actual Trash Collection Charges

In recent years, the practice of charging a marginal cost for solid waste disposal has rapidly increased in communities across the United States. Before 1986, only 126 municipalities marginal-priced trash (www.epa.gov/payt/comminfo.htm). Today, something like 6,000 municipalities do (including, by law, every town and city in Minnesota).[3]

Basically, there are three ways in which a city can implement a system of trash collection charges (Miranda et al. 1996):

- *Priced tags,* where the household must buy specially marked tags and attach them to its usual plastic trash bags. This system has many advantages. The tags themselves are cheaply produced and, since they are small in size, they can be sent through the mail or sold in small stores. But tags can fall off or be stolen, and tags do not ensure uniformity of bag size.
- *Priced bags,* where the household must buy specially marked bags and use only them to set out its trash for pickup. This system is also low cost; if bags are already used for trash, only the printing represents a new social cost. But bags can tear or be torn by animals, and a somewhat more cumbersome distribution system is needed for garbage bags than for tags.
- *Subscription cans,* where each household must contract for a certain number and size of trash cans and stay within that limit with each pickup. The disadvantages are obvious—the cans mean a significant initial investment, the variety of contracts necessitates a complex monitoring and billing system, and there is no incentive for a household to reduce its trash below the contracted can limit (Nestor and Podolsky 1998).[4] But the advantages are many. Cans are neater and greatly reduce the litter attendant to pickup days. And multiple efficiency and equity goals can be addressed through complex pricing schemes.

All of these systems are volume-based in the sense that the household ends up paying by the cubic foot or by the gallon, not the weight, of its trash. Actual prices in bag or tag systems range widely, from $0.02 to $0.10 a gallon (Miranda and Aldy 1998). The problem with such volume-based pricing is the "Seattle stomp" (named after the city where it was first observed), whereby households buy and use home compactors to reduce the volume and hence the cost of their trash. This is socially inefficient, because the compactor trucks that pick up the trash do that same job over again, better and cheaper, so every home compactor purchased solely to save trash fees is a complete *social* waste of resources. Home compactors also place external costs onto the backs of people who must lift the stomped bags or cans.

Furthermore, stomping messes up the optimality of the pricing system. The marginal social cost of trash depends upon both its volume and its weight, and pricing by one or the other is efficient only if, on average over time and households, there is a fixed and immutable proportionality between the two. When some households stomp some or all of the time, and others don't, that proportionality is destroyed, and incorrect signals about trash generation, reduction, and reuse are inevitably being sent to some households. If the price assumes a stomp, then non-stomping households are being overcharged; and if the price assumes no stomp, stomping households are being undercharged.

The way out of this problem is of course weight-based charges, and this has been tried in various cities (McLellan 1994; Skumatz et al. 1994; Andersen 1998). But weight-based charges mean weighing, and this requires scales on the collection trucks, it slows collection (by 10% in Seattle), and it necessitates a complicated billing system (rather than the simple tags, bags, or can count).

Tag, bag, and can-number versions of trash collection charges are all basically fixed-price systems (although can-number systems permit some flexibility in this respect). Efficiency—setting the price equal to the marginal social cost of collection and disposal—suggests a price that declines with volume, because there is a cost to simply driving the collection truck by a house that has no trash and because picking up twice as much does not cost twice as much, but the inefficiency of constant price is probably small (Johnson and Carlson 1991).

Equity, however, means worrying about the impact of trash collection charges on the poor, who lose their "free" trash service. This worry diminishes once we remember that most trash collections are now financed from the municipal general fund, which comes mostly from property taxes, and these taxes should go down—as should the rents of the poor—partially offsetting the new trash charges.[5] It is only partially offset because the property tax is more progressive than a trash fee would be—because most studies conclude that the income elasticity of demand for housing is greater than the income elasticity of trash generation (Kemper and Quigley 1976; Fullerton and Kinnaman 1996; Miranda et al. 1996). If, however, equity worries persist, the municipality can overcome this by providing, say, the first tag (or bag or can) for free to all.[6]

One reader of a preliminary draft of this chapter noted that most publicly provided goods with economies of scale—for example, gas, electricity, and water—charge by the volume of consumption and concluded that the fact that solid waste collection companies do not regularly seek to also charge by volume indicates that the administrative problems of trash collection charges are so great as to outweigh the benefits of efficient marginal pricing. When given the choice, however, as with business collections, many private trash haulers do charge by volume. Traditionally, with household waste collection, the municipal-

The Lansing, Michigan, Experience with Trash Collection Charges

Lansing, the capital of Michigan, is a city of more than 100,000 people. Since 1975, the city has offered its residents a weekly fee-per-bag waste collection system. Multiunit complexes and those that do not want to subscribe must arrange service with a private waste hauler. Private haulers charge less than the city (for households that generate more than 250 gallons a month), and the city ends up servicing about half of its single-family households. For city service, households purchase special 30-gallon green plastic bags—sold in most supermarkets and convenience stores—for $1.50 ($1.00 before 1992). Five cents a gallon is comparable to the prices charged in most other cities with trash collection charges. In the year after the fee increase, there was no decline in the number of households choosing the city's service, but the amount of solid waste collected went down by nearly 40%.

Since 1991, the city has also collected recyclables and yard waste, which are put out on the same day in separate containers. This service is mandatory. Households put out one bin with glass, plastic, and metal; they put out paper products in paper bags or bundles; and they put out yard waste in special paper bags. The recycled materials are collected at no variable charge, but there is an annual $57.50 charge per household for recycling and yard waste collection, which is assessed along with city property taxes. The green-bag fees cover the cost of the trash collection, and the $57.50 covers the cost of the recycling and yard waste collection and treatment (net of revenues received). The city estimates that about 30% of its total solid waste (excluding yard waste) is recycled and that about 4% of its total solid waste is illegally dumped or burned. The 30% figure is on a par with other cities that charge for trash collection and recycle; the 4% figure is much higher than others—or at least higher than what other cities believe or admit (Miranda and Aldy 1998; Bauer and Miranda 1996).

ity decides on the pricing scheme, and this has historically evolved into time-based (rather than volume-based) pricing or no explicit pricing at all. When the trash collection is put out to bid, private collectors are rarely asked how they would like to price, only at what cost will they do the job. Even if they were asked, most would probably not urge trash collection charges simply because it would be a new pricing scheme—one in which they have no experience at the household level, and hence one in which they risk large losses in the early stages from winning bids that turn out to be too low (what is called in the bidding literature "the winner's curse").

Indeed, time-based trash pricing is *only* sustainable as long as private, competitive trash collection is *prevented* by the municipality. To see this, consider a town that allowed private haulers to enter and compete for trash customers. Suppose, to keep the example simple, that it costs $1 a bag to pick up trash and that the average household puts out 8 bags a month. Two trash haulers enter this market: one offers to pick up trash for $1 a bag, and the other for $8 a month. Each expects to cover costs. Families that discard fewer than 8 bags a month choose the $1-a-bag hauler, and families that put out more than 8 bags choose the $8-a-month one. The $1-a-bag hauler breaks even, but the $8-a-month one loses money and has to raise the price, say to $11 a month. Now families that dispose of 8 to 10 bags a month will switch to the $1-a-bag hauler. Again, the $1-a-bag hauler breaks even, but the $11-a-month one loses money and has to raise the price still further. Eventually, the hauler who charges per month raises the price so high that all customers switch. In sum, the hauler who charges per month cannot survive if there is a competing hauler who charges per bag. Pricing per bag (or tag or can) is the way a free market would charge for trash collection. Only tight municipal control of trash collection has made time-based pricing possible in the past.

The Logistics of Collection

For rural areas or small towns, especially in developing countries, household trash collection is usually optional. Any health or aesthetic costs of inadequate disposal are largely borne by the household itself. As cities grow, however, income, density, and waste also grow. Ill-disposed waste begins to generate ever larger external costs. At some point, these external costs become so great that city dwellers begin to require the proper disposal of waste. Some households do it themselves, making a weekly trip to "the dump," but competitive private-sector collectors also begin to appear; some even work for free to get garbage to use for animal feed. Eventually, the illogic of having several collectors combing through the same routes becomes so apparent—at least in most U.S. municipalities—that a monopoly collector, either public or private, is chosen.[7]

Although this stereotypical history of urban waste disposal seems accurate, there is in fact mixed evidence for its empirical basis. The original study of MSW found no evidence of economies of scale (Hirsch 1965), and later more sophisticated studies have provided mixed evidence (Kemper and Quigley 1976; Kitchen 1976; Collins and Downes 1977; Stevens 1978; Tickner and McDavid 1986; Dubin and Navarro 1988; Tawil 1996). Nevertheless, MSW collection probably is a natural monopoly. Multiplying the gasoline and time costs of driving by houses must increase the cost per house. There is evidence that twice-weekly pickup and back-yard pickup are each 20–50% more expensive, and most U.S. cities have given up each of these luxuries during the past few decades (Tickner and McDavid 1986; Stevens 1989; Ezzet 1997). Also long gone in the United States—owing to larger cities, greater health concerns, and more distant pig farms—is the separate collection of edible garbage from other trash.

Not only are pig farms more and more distant from cities, so also are landfills—as we shall see in the next chapter. Landfills in the United States have recently become larger and less common, so even small towns often face ever higher transport costs in trash disposal. The standard compactor truck almost universally used for trash collection in the United States is a particularly poor choice for transport. When it is merely driving, all its expensive compaction mechanism is idle, it is getting poor gasoline mileage, and all but one of its workers (where there are still multiworker crews) are idle.

The solution to this problem is transfer stations, where the trash is transferred to large, more fuel-efficient semi-trailers—or trains or barges—for the haul to the landfill. The transfer station offers an additional cost-saving advantage if the landfill costs are based on volume (rather than weight), because the trash can be further compacted when it is transferred from the collection vehicle to the semi.

This cost differential is shown in Figure 3-5. The actual trash collection and landfilling costs, C_0, and the dashed line there show that the cost of hauling rises proportionately to the distance to the landfill. Suppose a transfer station is located at distance D_0 and the trash is there transferred to a semi at a cost of $(C_2 - C_1)$; more precisely, $(C_2 - C_1)$ is the *net* cost of the transfer—the cost of the trans-

---

Incentives in Trash Collection

Every city has at some time pondered the question whether trash should be collected by a city solid waste department or by a private hauler. If free trash collection is provided to households, the issue is simply which of the two systems costs the municipality less. Economic research has not helped settle this debate. About half the studies find that contracting trash collection out to private haulers is less expensive than public collection; about half find the reverse (Viscusi et al. 1992).

Once cities move to trash collection charges, new issues arise in the choice. A competitive, unregulated private hauler will have to charge average private cost to avoid losses. It should charge MSC, which may be higher or lower than APC. MSC is above MPC because of external costs, but MPC may be below APC because of declining marginal costs in trash collection. Getting the private hauler to charge the right price is a bureaucratic burden that may suggest that public collection be chosen.

But public collection is not without its own incentive problems (Bailey 1995b). In some cities, collection workers get overtime pay to finish their routes if they cannot finish them during regular working hours. It is not surprising that collectors there are slow. Elsewhere, if collection workers finish early, they can go home and still collect a full day's pay. It is again not surprising that collectors there become "speed balls" who cut corners on noise, safety, and vehicle maintenance.

---

fer *minus* any landfill savings due to the extra compaction. The cost per mile is thereafter lowered (the solid line). At a distance D^*, the total trash cost per ton becomes lower with a transfer station. Empirically, for most cities this distance D^* is 15–20 miles (Schaper and Brockway 1993; Rhyner et al. 1995).

As trash becomes ever cheaper to transport, and the differential in the cost of landfills in different regions becomes ever greater, trash is being transported ever greater distances—even interstate (to be discussed in Chapter 7). A half-century ago, transfer stations were rare in the United States. Today there are thousands.

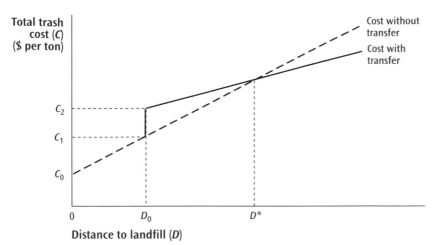

Figure 3-5

Cost Differentials with a Transfer Station

Trash Collection in the Developing World

Trash is handled very differently in developing countries. To begin with, households are much more involved in the solid waste collection process than in the United States. This is especially true of the poorest households, which are generally neglected with regard to urban services by their city governments and which often are housed in such dense and streetless areas of the city that modern compactor trucks could not offer standard curbside pickup if they tried. Although poor cities generate less waste—barely a pound per person a day—the trash is more putrescent, making it a more pressing public health problem. But poor cities do not have sizable resources to attack the problem. How do they handle it?

In Jakarta, each small residential area informally chooses a "village" leader, who in turn decides what kind of waste pickup is to be supplied and how much each person will pay. Typically, in the poorest areas, the method of choice is a handcart, operated by a low-paid resident who regularly services perhaps a hundred families, picking up their trash and delivering it to a nearby depot to await pickup by a city truck. The advantages of the system are that the trash does, indeed, get removed from the immediate area and that the leader can set a fee for each household that reflects the household's ability to pay. The disadvantage is that, in the poorest areas, feasible fees are so small that service deteriorates; when this happens, people begin to refuse to pay, and service deteriorates further—equilibrium may mean no service at all (Porter 1996).

In many African cities—such as Accra, Ghana; Harare, Zimbabwe; and Gaborone, Botswana—the government offers two kinds of waste service, the standard house-to-house curbside pickup (at a price) and neighborhood trash bins (called "skips") that are periodically emptied. This latter service is free but requires that households carry their own trash to the nearest skip. Households, or residential areas as a whole, choose the service they want (and can afford)—where time is scarce, more money is paid and where money is scarce, more time is paid (Fischer and Porter 1993; Porter 1997). For cash-starved city budgets, the advantage is that the money price can be set high enough so that the rich end up subsidizing the waste collection of the poor—and the health externalities may even make the system of different prices efficient.

In Curitiba, Brazil, where some shantytowns are so densely packed that garbage trucks cannot enter, the government hit on a novel, and successful, way to get the garbage out. The garbage trucks gather on the outskirts and exchange food, bus tokens, and school notebooks for sacks of garbage that the residents themselves carry out (Brooke 1992; Rabinovitch 1996). It may seem perverse to subsidize trash, but when people are living in dense proximity, trash can create all kinds of disease—both directly and indirectly from the scavenging of rats, dogs, and birds—and subsidizing a positive externality is appropriate (Benenson 1990).

Trash pickup and its transport to a landfill is done differently in most cities in the developing world. And it should be. Where capital is scarce and labor cheap, we would expect more labor-intensive methods to be employed. Generally, they are. Most cities use open trucks, each with a crew of several men, who pass the waste up into the truck with shovels, baskets, or cans (Porter 1996, 1997). Still, productive inefficiency of two kinds can intrude.

The first kind is excessive capital intensity. City waste administrators often try to hasten the day when the city will have "modern" waste collection by importing large mechanical compactor trucks, which are operated by one or at most two workers. And they are often encouraged in this temptation by foreign aid administrators, who want to help with waste collection *and* want to sell the kind of equipment produced in their own countries. Not only does this throw away the advantage of cheap labor, but it introduces new problems. Waste in developing countries primarily consists of food scraps, not paper, and it does not compact much, if at all; moreover, it is wetter and causes regular breakdown of the expensive-to-repair compaction mechanism.

The second kind of productive inefficiency is excessive labor intensity. In many developing-country cities, officials attempt to solve unemployment problems through overallocations to the labor-intensive parts of the city budget. If this meant better waste pickup and better health as a result, this might not be altogether bad. Waste pickup, however, requires expensive trucks, so excessive street cleaning is tried instead—after all, a street-cleaner needs little more than a broom and a barrel on wheels. But manicured streets, especially when provided for free in the richer parts of the city, are hardly a high priority use of the city's budget. Recycling also is handled very differently in developing countries, but we leave that part of the story for later (see the box in Chapter 8 titled "Recycling in Cities of the Developing World").

Final Thoughts

There just isn't any good reason why governments should subsidize the collection and disposal of trash in the United States. Perhaps there is good reason in the cities of developing countries, where the poor cannot afford to pay for collection and where, if their trash is not collected, serious epidemics will periodically break out that affect everyone. But that argument does not even remotely apply to U.S. cities.

Although we are still a long way from a final decision on trash fees, we should not leave this chapter without firmly fixing in our minds that the purpose of trash fees is not to raise money for city budgets or to punish those who are guilty of creating trash. The purpose of trash fees is to *internalize* the collection and disposal costs, so that manufacturers will produce products and packages that contain less trash and so that consumers will resist buying trash-laden products and packages.

Notes

1. The equation for the arc elasticity of demand in this price range is: $-0.2 = \{(x - 4.4)\ /\ [(4.4 + x) \div 2]\} \div \{(0.05 - 0.00)\ /\ [(0.05 + 0.00) \div 2]\}$, where x is the volume of trash put out at a price of $0.05 a pound. Solving this equation yields $x = 2.9$ pounds per person a day. If we consider a price elasticity of -0.1, the average daily trash per person falls from 4.4 to 3.6 pounds.

2. For example, Van Houtven and Morris (1999) estimate the deadweight loss avoided by per-bag pricing in a pilot program in Marietta, Georgia, to be about $25 a year *per household*. But Van Houtven and Morris consider a slightly lower price elasticity (0.16), a smaller initial waste disposal (5.8 pounds a day *per household*), and a less than optimal price (4 cents a pound, with MSC estimated at 6.25 cents a pound). Some estimates of this deadweight loss run higher than that of the text. For example, Morris and Holthausen (1994) consider three simultaneous changes—a trash collection charge, initiating curbside pickup for recycling, and going from semiweekly to weekly trash collection—and they estimate the average gain to be $117 a year *per household*.

3. Although most programs are so new that few generalizations have yet emerged, there is extensive documentation of these experiments in trash collection charges. See the U.S. EPA website: http://www. epa.gov/epaoswer/non-hw/payt/research.htm.

4. This lumpiness in the pricing structure can cause a perverse incentive—as long as the contracted containers are not full, the marginal price to the household for additional trash is zero. Indeed, there is mixed empirical evidence on whether total waste generation is larger or smaller under can subscription programs (Nestor and Podolsky 1998; Van Houtven and Morris 1999). This lumpiness can be prevented to some extent by adopting smaller container sizes (Miranda and Aldy 1998; Rathje 2000).

5. Property taxes are deductible on federal income tax returns, whereas refuse collection charges would not be. This point is relevant for particular cities making this choice, though it should not be relevant for a national choice between the two.

6. In cities that do this, there is a temptation to compensate for this by raising the price of subsequent trash units above their marginal social cost. This temptation should be resisted. $P > MSC$ sends the wrong signal to households, and it may encourage private, cream-skimming duplication in the trash system. Indeed, any $P > MPC$ may encourage households and small businesses to switch, if permitted, to private trash collection where the price will be $P = APC$. A tax on private collection may be needed—ideally, a tax equal to the external costs (i.e., equal to the difference between MSC and MPC).

7. In most cities, the commercial hauling market, which collects for businesses too big to settle for weekly pickup in small containers, usually consists of many competitors. Is there an inconsistency here? No, because the commercial haulers pick up larger volumes at each stop and hence make much more frequent trips to the disposal site, so there are fewer economies to monopolizing a route.

Solid Waste Landfills

Nowhere in the world is there such a waste of material as in this country.
—Austin Bierbower, "American Wastefulness," *Overland Monthly*, April 1907

As villages become cities, there comes a time when municipal planning replaces individual waste disposal decisions. So solid waste "dumps" are as old as cities themselves. In the United States, cities began to assume responsibility for waste collection and dumping by the end of the nineteenth century (McBean et al. 1995). The first dumps were little more than open pits, with frequent spontaneous combustion—even intentional fires to reduce volume—and attendant smoke, odors, noise, rats, and seagulls. Typically, until the late 1970s, each city and large town operated its own small landfill, charging a modest "tipping fee" for commercial and industrial users and for the trash of small towns and villages on its periphery.

Things have changed during the past 20 years. Today there are few landfills, and those few are mostly big and privately operated. Along with the change in numbers and size, the real tipping fee has doubled, up from a national average (in 1997 dollars) of about $12 a ton in 1985 to just over $30 in 2000 (Repa 2000).

What triggered these dramatic changes? Basically, our increased concern for the effects of dumps on our health and the environment has led to new regulations for the opening, operation, closing, and postclosure monitoring of "sanitary landfills." There have always been economies of scale in landfill activities, but until recently these were not great enough to warrant the additional transport costs to distant large landfills (Gallagher 1994). The new regulations of the 1980s increased these economies of scale, in three ways. First, many of these regulations imposed the same additional costs on landfills almost regardless of their size, which meant that the cost per ton of such regulations was much higher for small landfills. Second, many of the new regulations required special expertise—political, administrative, engineering, and legal expertise—that small landfills could not acquire at reasonable cost. Third, many of the new regulations applied only, or more strictly, to *new* landfills, which made the expansion of existing landfills a less costly alternative.

Landfill Scarcity, Size, and Numbers

The numbers indicate the magnitude of these changes. At the start of the 1970s, there were 20,000 landfills in the United States, but by the end of the 1980s only 6,000, and by 1998 barely 2,000 (U.S. EPA 1988; Repa 2000). Small landfills closed. Big landfills grew in number and size. By the end of the 1980s, a few hundred landfills handled half of all the municipal solid waste generated in the United States (U.S. EPA 1991b). The cost figures explain why. The EPA estimated that a landfill handling less than 25 tons a day costs more than $40 a ton (converted to 1997 dollars), whereas a landfill with capacity greater than 400 tons a day costs less than $10 a ton (U.S. EPA 1988, 1991b).

The ownership of landfill capacity was also changing. Small cities quickly recognized their inability to handle the new and complex regulations at a reasonable cost. Even many large cities began to close their landfills—today, only 38 of the 100 largest U.S. cities own their own landfills (Ezzet 1997). Meanwhile, large private specialized waste management companies, such as Browning-Ferris Industries (BFI) and Waste Management Incorporated (WMI), were acquiring massive amounts of landfill capacity (Bailey 1992). By the early 1990s, these two firms alone owned more than 2 billion cubic yards of permitted landfill capacity. How much is 2 billion cubic yards? That's three years worth of all the U.S. MSW at current discard rates. By the mid-1990s, the permitted daily volume of BFI and WMI could already handle two-thirds of the nation's MSW (Kravetz 1998a; Bailey 1998).[1]

Private Cost and Price of Landfills

Conceptually, thinking about the marginal private cost of a landfill is easy. When one extra unit of waste is placed in a landfill, two kinds of costs are incurred. First, the variable cost of interring the waste—digging, pushing, lining, cover-

The World's Biggest Dump

For the past half-century, the Fresh Kills landfill on Staten Island has been absorbing much of New York City's municipal solid waste, 13,000 tons a day. It has grown to 3,000 acres and 500 feet high, and more than 100 million cubic yards of waste have been collected there (Walsh 1989; Magnuson 1996; *Big Apple Garbage Sentinel* 1998). By the time it stopped receiving trash in March 2001, it had passed the Great Wall of China as the largest human-made "structure" on the planet (Rathje 1989). The closing of Fresh Kills, scheduled to be completed by 2002, is becoming costly—the network of wells, pipes, underground walls, and treatment stations will cost more than $1 billion.

For the past decade, New York City has been trying to postpone the closing, and its attendant costs, by discouraging private haulers of commercial waste from using Fresh Kills. The tipping fee was raised from $36 to $80 a ton in 1988, and again to $150 a ton in 1992 (Gold 1990; Cairncross 1993). This worked: The commercial half of the city's waste is now landfilled elsewhere—but not without the social cost of nearly 100 private transfer stations, where the trash is moved to 18-wheelers, railroad cars ("pee-yew choo-choos"), or barges ("gar-barges") for export to ever more distant landfills. It is not surprising that the new transfer stations sprung up where land was relatively cheaper—and where zoning laws permit such facilities—and that happens to be where New York City's poor live (McCrory 1998).

To avoid this burden on its poor, New York has recently started to ship its trash directly out of the city, to transfer stations in New Jersey, and thence to landfills hundreds of miles away. Not only is this expensive for the city—transport and landfill (not counting collection) will cost some $90 a ton—but the people near the transfer stations and waste destinations are more and more unhappy at becoming part of New York City's new waste disposal process (Konheim and Ketcham 1991; Walby 1999; Martin and Revkin 1999; Stewart 2000a).

Collecting, transferring, and transporting New York City's trash is not going to be cheap. The 2002 sanitation budget is nearly $1 billion for an estimated 11,400 tons a day, an average cost of $240 a ton (Gynn 2001).

ing, and so on. Second, when that unit is buried, it moves up the date (albeit slightly) when the landfill must be closed, postclosure monitoring must begin, and a new landfill permitted and opened.

This inclusion of the costs of an infinite succession of closed and opened landfills makes the calculation of the MPC complicated, and we will skip the detailed algebra (Ready and Ready 1995). But we can take a shortcut that comes close to getting the right figure for the MPC. Consider a landfill that is closed today and a new one simultaneously opened; all these closure and opening costs amount to F dollars. The new landfill is intended to last T years and receive Q units of waste a year. In other words, $(1/T)$th of the landfill is to be used up each year. Suppose that in each year, this used-up $(1/T)$th of the landfill were replaced, yielding an average replacement cost per *year* of F/T—note that our simplification here means ignoring economies of scale, because such a tiny replacement would surely cost much more than F/T. The average replacement cost per *ton* would then be $F/(QT)$, because Q units of waste are landfilled each year. There is also an interest cost (or more correctly, an opportunity cost) on the capital invested. At interest rate i, this is iF/Q per unit of waste. And finally, there is a variable cost V in operations and maintenance per unit of waste; by assumption, V is both the average and the marginal cost of interring waste. The final formula for the average private cost and marginal private cost of putting waste in the landfill is therefore

$$ \text{APC} = \text{MPC} = V + \left(\frac{F}{Q}\right)\left(\frac{1}{T}+i\right) \tag{4-1} $$

Note that we must count both the depreciation cost (or, more accurately, the replacement cost) and the interest cost.

Numerical calculations of this MPC of solid waste will vary greatly, depending on local wage rates and land prices as well as the size of the landfill. But two recent, careful studies are worth reviewing (and a third is described in more detail in Appendix A on page 65). See Table 4-1. Using equation (4-1) and the data of these studies yields estimates for the APC (in 1997 dollars) of $29.33 and $24.88 a ton, respectively (Gallagher 1994; Walsh 1990a, 1990b).

Landfill tipping fees are broadly comparable to these cost-per-ton estimates, averaging $36 a ton for the 100 largest cities in the United States (Ezzet 1997).[2] But

		Value assumed	
Table 4-1	*Variable*	*Gallagher*	*Walsh*
Two Estimates of the Average Private Cost of a Landfill	Waste flow, Q (thousands of tons a year)	300	365
	Landfill life, T (years)	20	20
	Interest rate, i (annual percentage)	4	4
	Variable cost, V (dollars per ton)	9.33	13.75
	Fixed cost, F (present value, in millions of dollars per landfill)	66.68	45.12

Note: The fixed cost above consists of all the opening and postclosure costs, with the postclosure cost discounted to present value.

Sources: Gallagher 1994; Walsh 1990a, 1990b.

this average conceals a great deal of variation, from $9 a ton in Denver to $97 in Spokane. Statewide averages also vary widely, from $8 a ton in New Mexico to $75 in New Jersey (Ackerman 1997). And internationally, there is even greater variance—where land is scarce and expensive, as in Germany or Japan, tipping fees run as high as $300–400 a ton (Hawken 1993).

As was mentioned above, tipping fees have about doubled in real terms during the past 20 years, indicating that the economies of scale achieved by the new and expanded large landfills have not fully offset the many forces leading to higher costs and concomitant higher tipping fees: stricter state and federal regulations, the need for landfills to provide compensation to landfill neighbors, the closing of most small landfills, the increased monopoly power of the single nearby landfill in many regions, and state taxes on landfill tonnage (Glebs 1988). We turn now to the reasons for the stricter regulations.

The External Costs of Landfills

Dumps have always been unpleasant places with their smell, litter, vermin, and smoke. But, aside from trying not to live near them, people worried little about dumps until recently. In the past few decades, however, the threat to groundwater from landfill leaching has been recognized more and more, and old solid waste dumps have been discovered to be hazardous to health and to the environment. Indeed, more than a fifth of the serious Superfund sites—see Chapter 15—are old solid waste dumps (OTA 1989b). Landfills also emit methane, which can explode and which is a greenhouse gas—about a third of U.S. emissions of methane are from landfills (U.S. EPA 1998c). And even the best of modern landfills still have problems with odor, noise, and litter. Adding these diverse external costs up in dollar terms typically yields a rough cost of $45–75 per ton of solid waste (Repetto et al. 1992).

More careful efforts to measure the external costs imposed on nearby residents often seek to discern differences in their property values (adjusted for other differences in housing and neighborhood characteristics). In principle, consider two houses that are identical in every way, except that one is near an obnoxious landfill. The price of the house near the landfill should be lower by the capitalized value of the willingness to pay of the *least* bothered potential buyer. Some of these property-value studies find that nearby landfills do depress housing prices. One such study found house values lower by as much as 15% for houses within a tenth of a mile of a landfill (Skaburskis 1989). Other studies have found that the price of housing would be a few percent higher if that housing were located a mile away from a landfill (Havlicek 1985; Nelson et al. 1992; Reichert et al. 1992). Still other studies find *no* effect at all from landfill proximity (Pettit and Johnson 1987; Bleich et al. 1991; Garrod and Willis 1998). There are even some pluses for landfill neighbors—their roads become wider, are paved, and get plowed in the winter.

It is easy to see why it is hard to discern a landfill effect on property values. There are many more important determinants of house prices that may swamp any landfill effect. Houses are not sold very often, so land price studies often rely on assessed values, which in most cities are only loosely related to actual market values. Indeed, if assessors seek, consciously or unconsciously, to compensate landowners for their locational misfortune, assessed values will be low around landfills, but not for market reasons or because of external costs.

Beneath the surface problems of noise, dust, litter, and so on lurks the threat landfills pose to groundwater. Traditionally, water was ignored, and many old landfills were sited in swampy areas because that land had little opportunity cost and hence was cheap for the city to buy. It is now recognized, however, that landfill water must be carefully monitored and collected (for recirculation or disposal as sewage), lest it leach the landfill's dangerous metals and chemicals into the groundwater beneath it. Most of our contemporary policy about landfills, as we will shortly see, is concerned with preventing such leaching.

Indeed, one of the ironies of today's approach to landfills is that they are kept much drier—so much so in some cases that the garbage in them rots very slowly. Some people consider this a good thing—because biodegradation generates potentially harmful methane gas or leachate (Rathje 1989). But others worry that the caps and liners installed to keep them dry will wear out before the garbage in them finishes biodegrading. Were this to happen, our effort to reduce the near-term threat to our groundwater would be solved only by increasing the long-term threat (Tenenbaum 1998). Fortunately, the importance of faster biodegradation of landfills is being recognized more and more—partly to reduce the danger of liner failure, partly to enhance the profitability of making electricity out of landfill gas, and partly to reduce the time needed before the land can be again used for normal activities (Hull 2000; Sullivan and Stege 2000). This faster biodegradation in these "bioreactor landfills" is achieved by recirculating the landfill leachate and even adding new water to the landfill.

The final external cost of landfills is their contribution to greenhouse gases and global warming. As a result of rising planetary temperatures, we are already experiencing a reduction of the northern hemisphere's snow cover, a decrease in Arctic sea ice, a rise in sea levels, and more frequent extreme rainfall events (Houghton et al. 1996). Landfills add to this potentially catastrophic problem through their emissions of methane. When solid waste is put in a landfill, it is decomposed by anaerobic bacteria, whose by-product is landfill gas (LFG), about half methane and half carbon dioxide, both of which are greenhouse gases. This LFG also contributes to local air pollution—methane can produce explosions, and some of the other gases are known or suspected carcinogens.

External costs mean that marginal social cost exceeds marginal private cost (MSC > MPC) and that government policies are appro-

The United Kingdom's Landfill Tax

In the United Kingdom, a landfill tax of about $20 a ton is levied on solid waste. This has been passed on into higher tipping fees and has induced manufacturing and commercial firms to reduce their waste or find alternative ways to dispose of it. But it is not surprising that studies show no effect on households, which rarely face trash collection charges at the curb—the U.K. recycling rate remains under 10%. Furthermore, the volume of household solid waste seems to have risen, possibly because small businesses put more of their waste out as household waste, because business waste disposal now costs more.

The tax was intended also to stimulate recycling, and so landfills were permitted to claim a tax credit of up to 90% of any contribution they made to an approved local environmental project. Landfills did indeed take advantage of this credit, even though such contributions still cost them 10% of the total amount, to gain public-relations points with their neighbors. But two thirds of their tax-credit contributions went to improving a public park, public building, or other public amenity in the vicinity of the landfill. Again to no one's surprise, other kinds of contributions—especially those that might have encouraged recycling or other alternatives to landfills—were not often chosen by the landfill operators for their contributions (Morris et al. 1998).

priately considered (see Appendix A to Chapter 1). Reliance on Coasian negotiation or courtroom thrusts to solve landfill externalities would clearly be unhelpful. And Pigovian taxation, although conceptually promising, would require the careful monitoring of each of the landfill's many externalities.

Taxes based on landfill tonnage badly miss the Pigovian mark. Although such taxes may hasten the closing of small, high-cost landfills and raise tipping fees at the remaining large landfills, landfill taxes have no direct impact on household waste generation unless there are trash collection charges at the curbside—although they may have indirect impact if the tax revenues are used to finance the expansion of local recycling programs. But in any case, landfill tonnage per se does not cause any of the external costs associated with landfills. And a landfill tonnage tax does nothing to induce landfill operators to reduce their litter, their leachate, their methane, and so on.

Nevertheless, nearly half the states levy some kind of landfill tax, though they do so not so much as an attack on landfill externalities as to raise revenue for recycling subsidies. In the United States, such taxes are almost always modest. But this is not so in Europe; there, landfill taxes are widespread and sometimes quite high. In Denmark, for example, the landfill tax is almost $50 a ton (Andersen 1998).

Partly because market incentives in landfill safeguarding are not easy to conceive, and partly because the U.S. EPA has always preferred direct command-and-control approaches to environmental regulation, the EPA has mostly selected direct physical safeguarding regulations for landfills. But before turning to these policies, let us look at the theoretical issue: How much safeguarding is optimal?

Consider the groundwater leachate problem. If hazardous substances leach into the groundwater, those living nearby who use wells for drinking water are in danger of health problems—for now, we will simplify the theory by calling all the various health problems "expected cancer deaths." (In fact, cancer is the most serious, but not the most frequent, problem caused by leaking landfills; smelly and dirty well water and diarrhea are much more common, and it is just for theoretical simplicity that we use "expected cancer deaths" as a catchall for these problems.) A landfill, as shown in Figure 4-1, if not safeguarded at all against leachate, will cause C_{max} expected cancer deaths. These can be avoided by taking precautions—plastic liners to prevent leaks, monitoring wells to quickly discover leaks, and remediating observed leaks to prevent further damage—but avoiding cancer deaths is ever more expensive as more and more precautions are taken. This is shown as the rising marginal cost of avoiding expected cancers in Figure 4-1. In our policy choices, we have decided that avoiding a death should be undertaken if it can be done for less than V_L—in other words, we act as if we value expected lives at V_L (see Appendix B to Chapter 1). Thus, V_L is the marginal benefit of taking precautions enough to save an additional expected life. Precautions should be taken as long as the MC is less than this benefit, V_L. This is shown in Figure 4-1 as C_{opt}. (As Figure 4-1 is drawn, C_{opt} is reached before C_{max}. If, however, the marginal cost of avoiding expected cancers is still less than V_L at C_{max}, then the optimal safeguarding gets rid of all expected cancer deaths.)

If regulations are to be promulgated, those activities that save lives at an MC less than V_L should be initiated, and C_{opt} expected cancers should be avoided. Notice that C_{opt} is less than C_{max}—even with optimal regulation, expected deaths will occur. And notice also that society loses if there are *either* more *or* fewer deaths than C_{opt}.

Figure 4-1

Optimal Safeguarding
of a Landfill

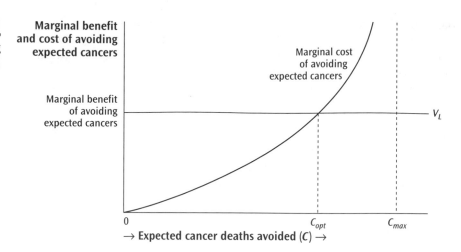

Excessive regulation uses resources that could have been better spent elsewhere; and inadequate regulation fails to use resources that could be better spent here.

Landfill Policies

Not until 1976 did the federal government assume a role in landfill management and standards, but landfill regulation continued to be left almost entirely to the states until 1991. States varied widely on what they required of their landfills—some required leachate management, some required control of explosive methane concentrations, and some specified minimum distances from human-made structures, surface water, or groundwater. Subtitle D of the Resource Conservation and Recovery Act of 1976 did little more than ban open dumping, and the Hazardous and Solid Waste Amendments of 1984 only required the U.S. EPA to report on health and landfills.[3] This EPA report was duly produced in 1988, calling for greater federal control over location criteria, design criteria, groundwater monitoring, corrective action, closure procedures, postclosure care, and financial responsibility. These new regulations—The Solid Waste Disposal Facility Criteria—were promulgated in 1991 (Rasmussen 1998). In some states, these new regulations represented new restrictions on landfills, but in many, the new federal rules were less strict than the rules already enforced by the state.

As of 1994, solid waste landfill operators must demonstrate their ability to pay the costs of closure, postclosure care, and cleanup of any releases during the postclosure period.[4] New and expanding landfills must be designed for groundwater protection. Landfills must not be sited in environmentally hazardous locations, must prevent hazardous waste from being accepted, must cover with six inches of soil at the end of each day, must control insects and rodents, must monitor for methane gas leakage, must not burn, must not accept liquid waste, and must limit public access.

Nearly half of the municipal landfills are small, serving communities of fewer than 10,000 people, and RCRA regulations let the states exempt some of these from some requirements. Landfills become eligible for exemption if they receive less than 20 tons a day, show no signs of groundwater contamination, and either are far from an alternative landfill or get less than 25 inches of rainfall a year (Rasmussen 1998). This list of exemption criteria is interesting. Little rainfall and no groundwater contamination make sense, but they would also make sense as criteria for bigger landfills. The small tonnage and alternate landfill distance make little sense from the viewpoint of landfill externalities—they really only make sense as a means of keeping down the costs of small and remote landfills for noneconomic reasons. That may have seemed politically necessary to the U.S. EPA, but from an economic viewpoint, if the benefits are worth the costs, it should not matter who incurs the costs. Federal subsidies can always be offered to shift the burden of these costs.

How much postclosure care should a landfill receive? Most states now require 20 to 30 years. Because some of the landfill contents will remain toxic for much longer than that, 20 to 30 years is clearly a compromise between zero and forever. Let's look conceptually at this decision. There are four kinds of postclosure costs to landfills: the costs of the remediation of observed leaks; the remaining expected environmental damage that still occurs after optimal remediation of observed leaks; the costs of monitoring the landfill so that leaks will be quickly observed; and the environmental damage done by unobserved, and hence not optimally remediated, leaks. Each of these costs is very uncertain. It may be beyond the ability of science to provide the information needed to choose optimums. Even if the scientific information could be provided, it would be costly. Because these costs will differ for every closed landfill, the very process of such expensive data-gathering may fail a benefit–cost test. Arbitrary regulations about monitoring and remediation may be the best we can do.

Many states ban certain items from being placed in a landfill. There are two reasons for this: certain items are hazardous and contribute critically to the potential for leachates to damage groundwater, to which we will return in Chapter 14; and certain items are believed to be better recycled than landfilled, which we discuss here. Typical of these are bans on landfilling compostable yard waste, tires, and white goods (e.g., stoves and refrigerators). Where households are not charged correct prices for their disposal choices, such bans may fulfill a better-than-nothing purpose. But complete bans are heavy-handed instruments, implying that it is optimal to recycle *every* single unit of the banned category.

> ## Landfill Regulation in New Jersey
>
> New Jersey was the first state to begin regulating landfills as public utilities (in 1971), and a brief review of its rocky history indicates some of the pitfalls of regulation (Kleindorfer 1988). New Jersey not only imposed strict environmental regulations on its landfills but also fixed the landfill operator's portion of the tipping fees very low. The total tipping fee, however, was increased greatly by the addition of various taxes. This put three-pronged pressure on the state's landfills—costs up, (net of tax) prices down, and demand down. As a result, its landfills were closing even before the U.S. EPA regulations hit, and they had neither the accumulated past profits nor the ability to borrow to finance the new required safeguards (U.S. EPA 1987). There were 331 landfills operating in New Jersey in 1972, but only 13 in 1988, by which time half of the state's MSW was being exported to other states for disposal.

Some states tax all landfilling (as in the United Kingdom; see the box titled "The United Kingdom's Landfill Tax"). Again, we should notice that this will influence businesses, because a landfill tax will probably be passed forward into tipping fees, but households will not be much affected unless the tax is passed forward into their marginal trash-disposal costs. Minnesota is currently considering not just a landfill tax but a total ban on any use of landfills—effectively an infinite landfill tax (Johnson 1999). The purpose, of course, is to encourage recycling, but such a ban equally encourages incineration, export, or illegal dumping. In any case, the optimal tax on landfilling is unlikely to be infinity.

EPA regulations concerning the control of landfill gas require LFG collection systems on landfills operating after 1993—but not on those that stop receiving solid waste before then, even though such closed landfills are also generating methane and carbon dioxide and will continue to do so for another quarter-century. Collecting the methane and just flaring it helps—it converts it to carbon dioxide, a less harmful greenhouse gas, and it removes the possibility of explosion. But methane collection is expensive (see Appendix B on page 67) and has not been required of "old" landfills, partly for economic reasons (old landfills produce less methane) and partly because it might prove difficult to locate the responsible party.

Since 1978, the Public Utility Regulatory Policies Act has obligated electric utilities to buy electricity from outside generators at a price equal to the utilities' own marginal cost.[5] With this guaranteed market and having to collect its LFG, many operating (and some closed) landfills have elected to use the collected methane to generate electricity.[6] The needed equipment is expensive, so only large landfills can do it profitably—see Appendix B for the results when a small and closed landfill tries. Of course, if the trend toward more rapid biodegradation of landfills continues, this will accelerate the production of LFG and may make its use to produce electricity more profitable.

Federal subsidies of LFG energy (which have been offered in the past (see Appendix B) and are again being contemplated) will also encourage LFG energy production (Duff 2001b). Even if fossil-fuel energy production generates higher external costs than non-fossil-fuel production, subsidies are not the best way to approach the problem. Both energy sources should be taxed, with fossil-fuel energy being taxed at a higher rate (see Appendix A to Chapter 1). Subsidizing LFG energy recovery instead of taxing it leads to the anomaly that environmentalists, who usually embrace nontraditional (i.e., non-fossil-fuel) energy production, vehemently oppose LFG energy subsidies for fear that they will cheapen landfilling and thereby discourage recycling (Duff 2001e).

The EPA's Analysis of Landfill Regulation

Knowing that there is *any* possibility of groundwater contamination *and* that there is *any* possibility of it affecting someone's drinking water is enough if you are looking for zero risk. If, however, your goal is more modest, it is necessary to make some quantitative estimates of these risks and how proposed policy changes the risks, and then to compare them with the costs of the policy. Indeed, since 1981 when President Ronald Reagan promulgated Executive Order 12291, all major regulations must be accompanied by a regulatory impact analysis. This the U.S. EPA provided for the 1991 landfill regulations (U.S. EPA 1991a, 1991b).

The EPA began by noting that 54% of the MSW landfills in the United States have no wells within one mile. Of the remaining 46%, if no new rules were issued, 12% would pose a risk greater than $1*10^{-5}$ of causing a cancer over a lifetime of maximum exposure, another 26% pose a risk between $1*10^{-5}$ and $1*10^{-6}$, and the other

62% pose a risk of less than $1*10^{-6}$. Let's explore those numbers a bit. Assume that greater than $1*10^{-x}$ means exactly $1*10^{-(x-1)}$ and that a lifetime risk of $1*10^{-(x-1)}$ means an annual risk of $1*10^{-(x+1)}$. Then, the average U.S. landfill, without new rules, would have generated a probability of causing a cancer death each year of $0.46(0.12*10^{-6} + 0.26*10^{-7} + 0.62*10^{-8})$. That's 0.00000007, less than 1 chance in 10 million of causing a cancer death each year. To give intuitive meaning to that figure, consider that the probability that you will be killed on the highway by another driver is that high every nine hours.

Does this mean that we should be doing nothing about the leaching of our landfills? Hardly. To begin with, landfill leaching does lots of bad things besides cause drinking-water cancer deaths. But let's pretend that cancer deaths are the only externality. And various landfills are different; some cause more cancer deaths than others. But let's pretend that they are all alike. Still, the 0.00000007 figure tells us nothing about whether the money we were already spending on safeguarding landfills was too little or too much. To answer that, we would need to know how many cancers would have been caused if we had not done what we were already doing. If we were already spending x dollars a year safeguarding our landfills and preventing y cancer deaths each year, then it was costing us x/y dollars per cancer death avoided—and we can compare that figure with the cost of saving lives elsewhere (V_L) by other means and then decide if that degree of landfill safeguarding was a sound expenditure of our limited resources. The 0.00000007 probability of the previous paragraph does not tell us anything about either x or y.

What the 0.00000007 probability *does* tell us is something about whether we should safeguard our landfills yet more carefully. Reducing the probability of 0.00000007 to zero would mean saving about 18 expected American lives a year. If these 18 cancer deaths were the only externality and all landfills were equally dangerous and equally safeguarded, should we be doing more safeguarding? How much should we be willing to spend to completely remove the risk of leaching and the resulting well-water cancer deaths in the United States? Again, think of public policy that consistently values lives saved at $1–5 million ($V_L$)—that is, embraces policies that save lives at a cost of less than $1 million per life saved, rejects policies that save lives at a cost of more than $5 million, and thinks very carefully about the policies in between. Then, to completely remove the remaining cancer risk from all U.S. landfills, we should be willing to spend no more than $18–90 million a year (18 deaths times $1–5 million equals $18–90 million).

The actual EPA rule promulgated for landfills was estimated by the EPA to cost about $450 million a year (converted to 1997 dollars)—$25 *million* per cancer death prevented, an order of magnitude higher than we are usually willing to spend to save expected lives. And of course, *no* landfill regulation can reduce expected groundwater-caused cancer deaths to *zero*. This simple, straightforward calculation strongly suggests that Americans had better things to do with their resources than further increase landfill protection.

The detailed EPA estimates agreed that not all the 18 well-water cancer deaths could be avoided. The EPA calculated that the new "final rule" for landfills would in fact save 2.4 American lives over the next three centuries (U.S. EPA 1991a, 1991b). Unfortunately, it is difficult to compare the *annualized* costs during the next few *decades* with the *total* lives saved during the next few *centuries*. But the magic of present value comes to the rescue. Assuming the annualized landfill cost estimates pertain to each of the next 30 years (i.e., 20 of operation and 10 of postclosure monitoring), the present value of $450 million a year is $8.46 billion. (These present values are calculated using a 3% discount rate, the rate the EPA used in its analyses.) The 2.4 lives saved during the next 300 years are the equivalent of 0.008 expected lives saved a year, or a present value of 0.267 expected lives. By the EPA's own analysis, the cost of the new landfill rules is (a present value of) $32 *billion* per (a present value of) each expected life saved. Even without present valuing the lives saved, the cost runs to $3.5 *billion* per expected life saved. This is *three orders of magnitude larger* than the value of life that most U.S. policies are being judged by.

An aside on the methods used here: Finding the present value of a number of lives may strike most people as curious, if not macabre. But if we do not do such a calculation, we are essentially assuming that a life saved at some distant time is just as important to policy decisions as a life saved next year. Lawyers and economists agree! The courts once overturned an EPA regulation because it calculated the present value of dollar costs per *not*-present-valued lives saved (Augustyniak 1998). And the person in the street agrees, according to surveys. For example, when given the choice between a policy that saves 100 lives this year or a policy that saves 200 lives 25 years from now, 70% of those surveyed preferred to save the 100 lives now (Cropper et al. 1994). Note that the present value of 200 lives in 25 years is greater than 100 lives now only if you use a discount rate smaller than 2.8%.

What does this review of the EPA analysis tell us? One possibility is that the EPA thinks there is much more to clean groundwater than just saving lives. The EPA analysis explicitly quantifies only the benefits of expected cancers avoided, but it mentions and may be implicitly considering such additional benefits of clean groundwater as the impacts on the ecosystem (e.g., wetlands), public confidence in the safety of landfills (thus making it easier to site them in the future), the value of clean groundwater to people who don't use groundwater but who feel

Landfills in the Developing World

A moment's reflection tells us that getting highly putrescent waste out of densely populated cities in the developing world is much more important, from a public health viewpoint, than interring it carefully in its ultimate rural burial place. So we should not be surprised to find that landfill regulations are much less strict in developing countries. Indeed, they probably should remain less strict for some time to come, with landfills in those countries being plagued by frequent fires, much dust and litter, little or no soil covering, liquid waste in open trenches, and no groundwater monitoring (Porter 1996, 1997).

The U.S. news media like to highlight the resulting landfill disasters in developing countries, such as the "waste-slide" that buried hundreds of villagers living below a mountainous dump in Manila (Mydans 2000). Such disasters lead foreign visitors and advisers to poor countries to sternly urge that "sanitary landfill practices should be adopted" (World Bank 1989, 1993). Before urging state-of-the-art landfills on countries like the Philippines, we should remember that most of the municipal landfills in the United States were "still dumps" until recently, presumably because we did not collectively think their improvement to be a very high priority until we were richer.

better knowing that it is clean, fewer diseases and chronic health problems as well as fewer deaths, fewer explosions, and less esthetic nuisance (Rasmussen 1998; U.S. EPA 1991b). (Also mentioned as a possible benefit by the EPA analysis is the impact of clean groundwater on property values, but property values are at most just a way of measuring other benefits, not a separate category of benefit.) If the only benefit of clean groundwater were deaths avoided, the EPA's own analysis shows that could be more cheaply achieved by replacing all contaminated well water with bottled or piped water. But all this raises a question: Why did the EPA dwell at such length on one kind of benefit if, in the end, it did not consider it to be a particularly important benefit?

A second possibility is that the EPA does not believe that the call for a regulatory impact analysis is a call for a complete benefit–cost analysis. What is produced is a lengthy but token treatment of benefits, and a lengthy and careful treatment of costs, especially the cost impact on those who will be most heavily affected by the new regulations.

Let us look once more at the regulatory impact analysis of the new landfill regulations (U.S. EPA 1991a, 1991b). There, many pages are devoted to calculating the cost per ton of solid waste, the cost per household, and the cost per municipality of the new rules. Not just the average cost but also the distribution of costs, and thus the percentage of seriously affected tons, households, and municipalities can be identified. This makes sense because the federal government may want to soften the impact of the new rules on the most heavily affected landfills (or those used by poorer people). But the softening of the impact should be done by some sort of side payment, not by any weakening of the rules. Cost-effectiveness requires that the marginal cost of saving an expected life through landfill safeguarding should be equal across all landfills, not more extensive just because the landfills are big or are in wealthy areas.

Final Thoughts

Nobody (who remembers them) wants to go back to dumps, but the U.S. EPA itself offers evidence that we are now doing more than we should be doing to safeguard our landfills. The 1991 landfill rules, according to the EPA, will save only 2.4 expected U.S. lives during the next three centuries, and at a cost of a half-billion dollars a year. The cost per life is astronomical. I cannot resist mulling the question, how many lives could we save if we put that half-billion dollars into something more promising?

APPENDIX A

What Did the Ann Arbor Landfill Cost?

Ninety-Ninety Rule of Project Schedules: The first ninety percent of the task takes ninety percent of the time, and the last ten percent takes the other ninety percent.
—Anonymous

The cost of a landfill would seem a fairly easy thing to estimate, especially when the landfill is already closed—as is the one in Ann Arbor, Michigan. But many of the costs of this landfill are distant, probabilistic, imputed, unpriced, or hard to put dollar values on. All these problems make it more difficult to estimate the Ann Arbor landfill's true economic cost, even now, a decade after it ceased receiving trash.

Although the City of Ann Arbor had undertaken some solid waste operations before 1959, it was in that year that the city opened its first landfill and began the publicly funded collection and disposal of the garbage and trash of its citizenry. A third of a century later, in 1992, the landfill proved unable to meet the ever more stringent regulatory constraints on its operation and was closed. What did it cost? The long answer to this question is contained in Bitar and Porter (1991); Higashi (1990) and Pyen (1998) provide narrative histories of Ann Arbor's past and future landfill troubles. But the bottom line of this Appendix is that—even though the landfill is essentially filled and closed—its final social cost is not yet known.

Its "on-budget" cost, for the period 1959–1990, was about $80 million—or $6 per cubic yard of solid waste (converted into 1997 dollars). (The city did not keep track of waste disposal by weight, so estimates of the cost per ton of solid waste require an estimate of the average density of the waste. Because the city moved during this period from open trucks to compacting trucks, and the composition of its MSW was steadily changing, a single estimate is even harder to make; it might be about $25 a ton.) But speculation about future postclosure costs, monitoring costs, remediation costs, and unremediated environmental costs might well suggest a doubling of this figure.

The calculation of the Ann Arbor landfill cost is made easier than the theory of the text suggests by the fact that it was Ann Arbor's first and last public landfill. So waste deposited in it did not hasten the day when the landfill would have to be closed and a new landfill opened. This means that most of the private costs of the landfill were simply the budget costs. The major exception is the cost of the land it covered.

In principle, the cost of the land when it was originally purchased by the city represented the capitalized value of the future rent stream of the best alternative use of that land. Thus, the original purchase price (adjusted to 1997 dollars) should capture the present value of the social cost of the forgone alternative resource stream from 1959 into the distant future. With hindsight, however, we now know that real land prices on the outskirts of Ann Arbor have escalated dramatically in the intervening years. Using this information about recent prices of land around the landfill lets us make better estimates of the land costs, not only during the landfill's operation but also into the future—during which time the land is settling and usable only for low-weight, low-value purposes (e.g., parks and softball fields). Incorporating the steadily rising opportunity cost of the land into the cost estimate adds another $18 million, bringing the total cost to almost $100 million, or just above $7 a cubic yard.

But landfill costs do not end when the landfill is closed. There are three remaining kinds of costs: the actual budget costs of sealing, covering, and monitoring the landfill for many years into the future; the expected budget costs of remediating any breaks in the integrity of the landfill to prevent environmental damage to the adjoining air, surface water, or groundwater; and the expected environmental damage that will occur anyway, despite the attempted remediation efforts. Of these three costs, only the first is easy to estimate.

During the last decade of operation of the Ann Arbor landfill, wells had been showing that the groundwater under it had become contaminated. When the landfill was closed, the city was required to build a slurry wall to deflect future groundwater flow away from the landfill leachate and to install four hydraulic pumps to remove existing contaminated groundwater and discharge it into the city's sewer system. The city is also under contract to monitor and maintain these systems for the next 30 years. The present value of all this cost is estimated to be $20–25 million (Pyen 1998; Bershatsky 1996). This brings the landfill cost to $9 a cubic yard.

Despite these precautions, environmental leaks may still occur and either do damage or require additional remedial expenditures. Indeed, even without new leakage, current state and U.S. EPA regulations about closing and closed landfills may become stricter and require new expenditures. There is no way to forecast these future damages and expenditures. These highly uncertain, distant, and possibly huge costs are the horse in Mishan's horse-and-rabbit stew. In short, the rabbit of Ann Arbor's landfill cost about $9 a cubic yard. Only during the next century will the horse part of the cost become clear. This cost uncertainty applies to all landfills.

APPENDIX B

A Social Benefit–Cost Analysis of Ann Arbor's Landfill Gas Recovery

Zeal without knowledge is the sister of folly.

—John Davies

Because Ann Arbor's landfill was closed before 1993, federal regulations require the city to do little to control its landfill gas emissions. Nevertheless, the city decided to recover this LFG and generate electricity. The city contracted with Biomass Energy System to extract the landfill methane, and with Michigan CoGen to turn it into electricity and sell that electricity to Detroit Edison (southeastern Michigan's electric utility). The electricity began to flow early in 1998. Because hundreds of other U.S. landfills have already made a similar decision, it is interesting to explore the economics of landfill gas recovery and energy production through a social benefit–cost analysis (Asarch and Cort 1995; and discussions with the involved firms).

The Ann Arbor landfill has been closed for five years, and the quantity of methane a landfill produces declines dramatically with the age of the interred waste. Moreover, the acreage and accumulated waste of the landfill are near the minimum considered feasible for energy recovery. So it was always expected that this project would at best prove no more than marginally profitable.

The social costs of the project are of two kinds: the initial costs of the gas collection system, the generators, the Detroit Edison interconnection, and the necessary permits ($2.50 million); and the annual ongoing costs of collecting the gas and generating the electricity ($0.15 million a year for a present value of $1.22 million). (The present value of a unit flow each year for 10 years, discounted at 4% a year, is $[1 - (1.04)^{-10}] \div 0.04$, which equals 8.11; multiplied by $0.15 million, that makes a present value of $1.22 million.) The total present value of these two costs is $3.72 million.

The obvious social benefits are of three kinds: the electricity-generating costs avoided for each year of the next 10 years by Detroit Edison, valued at their marginal cost of generation ($0.12 million a year for a present value of $0.95 million); the value of an installed, 10-year-old, but still useful gas collection system at the end of the project life (worth $0.80 million which, when discounted back to the present, is $0.54 million); and the value of 10-year-old but still useful electricity generators at the end of the project life (worth $0.87 million which, when discounted back to the present, is $0.59 million). The present value of these benefits is $2.08 million.

(Two explanatory points: First, the Michigan CoGen generators can produce 14 million kilowatt-hours a year, but the landfill only yields sufficient methane to generate slightly less than 6 million kilowatt-hours. Detroit Edison's "avoided cost" is $0.020 a kilowatt-hour. This means the total avoided cost is $0.12 million a year. Second, the gas collection system cost $1.20 million when new and is expected to last 30 years. Straight-line depreciation suggests that its value after 10 years will be $0.80 million, which after discounting to the present is $0.54 million.)

The costs ($3.72 million) far exceed the benefits ($2.08 million)—someone must be losing in this process. So far, we have looked only at *what* the benefits and costs are. Perhaps it is time to look at *who* receives the benefits and *who* pays the costs. Five agents in this project gain benefits and incur costs:

■ *The City of Ann Arbor.* The city receives some payments in royalties and taxes from the generators, and it has some administrative costs each year, but these are small enough to be omitted here. The city's chief gain is the gas collection system at the end of the project. The present value to the city is *plus* $0.54 million.

■ *Biomass Energy System.* Biomass incurred $1.20 million in initial capital costs and incurs $0.06 million a year in ongoing operation and maintenance costs (for a present value of $0.49 million). It sells the gas to Michigan CoGen for $0.20 million a year (for a present value of $1.62 million) and receives a federal tax credit of $0.07 million a year (for a present value of $0.55 million). The overall present value to Biomass is *plus* $0.48 million.

■ *Michigan CoGen.* CoGen incurred $1.30 million in initial capital costs and incurs $0.09 million a year in ongoing operation and maintenance costs (for a present value of $0.73 million). It also pays Biomass $0.20 million a year for its gas (for a present value of $1.62 million). In turn, it sells the electricity to Detroit Edison for $0.34 million a year (for a present value of $2.75 million)—this is nearly three times Detroit Edison's avoided cost ($0.0575 versus $0.020 per kilowatt-hour). Finally, CoGen still will have useful generators at the end of the project, worth a present value of $0.59 million. The overall present value to CoGen is *minus* $0.31 million. Is CoGen losing money on this project? Not really; it has contracted for a bundle of projects with Detroit Edison, and it presumably makes up elsewhere for any losses on the Ann Arbor project.

■ *Detroit Edison.* This electricity generator is saving $0.12 million a year in generating costs but is buying that electricity at $0.34 million a year. To make Detroit Edison willing to perform this public service, the State of Michigan's electricity regulators have permitted Detroit Edison to raise its electricity prices sufficiently to cover the deficit (of $0.22 million a year). The present value to Detroit Edison's profit ends up at zero, plus any goodwill it garners for showing its environmental concern.

■ *The rest of society.* Somewhere, on the projects where Michigan CoGen makes up for any Ann Arbor losses, someone is paying too much—at least $0.31 million too much in present value. Millions of Michigan residents end up paying $0.22 million more each year for their electricity (for a present value of $1.80 million). (We ignore for simplicity the fact that the higher price would induce them to consume less electricity.) And hundreds of millions of Americans end up either paying more taxes or enjoying fewer other services each year to raise the $0.07 million in tax-credit subsidy for Biomass (for a present value of $0.55 million). The overall present value to the rest of society is *minus* $2.66 million.

The overall present value is, of course, the same *minus* $1.64 million, whether one does a "what" or a "who" benefit–cost analysis. But the "who" results make it clear why the city and the various involved businesses are either willing or anxious to undertake the project. For the millions of other Americans who lose in the process, each loses only a few pennies a year, almost not enough for anyone to notice.

Before concluding that this is just another bad project—with low benefits concentrated on the few and high costs spread over the many—we should notice the many less tangible external benefits that we have so far overlooked. The improved handling of the landfill methane will reduce the probability of explosion, will reduce odor around the landfill, will improve local air quality, and will reduce the number of dead birds at the landfill. It will also contribute to a reduction in the global warming problem in two ways: It converts the landfill methane into carbon dioxide, which is a much less serious greenhouse gas than methane; and it reduces the amount of coal that Detroit Edison needs to burn each year. (The earlier present value analysis considered the *private* costs avoided by Detroit Edison but not the *external* costs avoided.) This means that Detroit Edison will emit fewer tons of carbon dioxide each year during the life of this project. The present value of this avoided carbon dioxide tonnage during 10 years is about 0.17 million tons. Dividing the present value of the dollar loss by the present value of the carbon dioxide tonnage gained indicates that the project reduces greenhouse gases at a cost of about $10 a ton—a very favorable rate, in comparison with almost all currently available alternative means of reducing carbon dioxide.

> ### Greenhouses Too?
>
> Ann Arbor stops at energy. The nearby private Wayne Disposal landfill in Belleville, Michigan, used to collect the landfill gas—not only to generate electricity, but also to produce heat to operate an acre of greenhouses that hydroponically grow vegetables and herbs, which are sold throughout southeastern Michigan (Graff 1989; hydroponic growth occurs without soil, because the plants grow in water infused with nutrients). Generally, greenhouse hydroponic vegetable cultivation is too costly because of the needs for land and heat. But the land atop a settling landfill can't be used for much anyway, and, once the methane is collected, it adds little to the cost to produce both electricity and heat. Nevertheless, the revenues still did not cover the added costs, and the greenhouse hydroponic operations were eventually closed down.

Notes

1. Mergers and takeovers have rearranged the waste industry structure in the past few years. USA Waste Services and WMI have merged, and Allied Waste Industries and BFI have merged (Gynn 2000a, 2000b). Together, in 2000 these giants earned about three-fourths of the revenue of the 100 largest waste firms in the United States (Geiselman 2001b).

2. Tipping fees are sometimes quoted as dollars per ton and sometimes as dollars per cubic yard. The latter seems more sensible because it is volume, not weight, that fills up a landfill. If the price is quoted per cubic yard, this encourages compaction, and this sometimes matters— as, for example, in the decision whether to build a transfer station, where waste can be further compacted. It is not easy to convert prices per cubic yard into prices per ton because waste varies in density. Basically, MSW gets denser as it goes from household to landfill. Waste leaves

the house at about 200 pounds per cubic yard, is compacted to about 600 pounds per cubic yard in the compaction truck, and is ultimately compacted still further, to more than 1,000 pounds per cubic yard at a transfer station and the landfill itself (Vesilind 1997).

3. In a sense, Congress did a great deal about solid waste landfills during these years. It determined that, despite the hazardous waste that households accidentally or innocently include in their trash, MSW would not be treated as hazardous waste. During this period, non-hazardous (by definition) MSW continued to be largely unregulated under Subtitle D of the Resource Conservation and Recovery Act. Hazardous waste, conversely, was becoming liable to much stricter and more expensive regulation under Subtitle C. We will get to hazardous waste regulation in Chapter 14.

4. Providing "financial assurance" of ability to compensate postclosure damages is not easy. Proper insurance companies do not want the business, because actuarial estimates of expected costs are highly uncertain and there is no ceiling to the potential liability. Many landfills provide the required financial assurance cheaply by setting up a wholly owned subsidiary to provide it. Because the assets of this subsidiary most often consist of parent-company stock, it is questionable how much assurance they provide (Arner et al. 1999).

5. The Public Utility Regulatory Policies Act is scheduled to lapse shortly, but it may be replaced by federal or state requirements that utilities derive at least 15% of their electricity from renewable sources (for which landfill gas would qualify). This 15% target would be in the form of tradable permits, so utilities that achieved more than 15% could sell rights to others who achieved less. The price of the permit would effectively provide a subsidy to landfill gas recovery—better even than this act. There, the prices varied a lot across different forms of renewable energy. But the permit system would ensure cost-effectiveness across the different forms (Darmstadter 2000).

6. Landfills that do not want to collect methane can fulfill their legal obligations with few—probably too few—wells. So methane collection for electricity generation will require additional investment in the collection system. Once methane becomes desirable as a fuel, the landfill operator will want more water in the landfill to accelerate degradation and the concomitant methane production. A side benefit of this is that it reduces the length of time in which postclosure monitoring is needed.

Incineration of Solid Waste

An incinerator is just a landfill in the sky.

—Dave Dempsey, Michigan Environmental Council

An alternative to landfilling municipal solid waste is incinerating it, which greatly reduces the volume that remains to be landfilled. Humankind has a long history of burning its waste. As recently as 1960, almost a third of U.S. municipal solid waste was incinerated. The incinerators of 1960, of course, given the times, did not recapture energy and polluted at will. Increasing concern for this air pollution and the possible toxicity of the incinerator ash discouraged incineration in the 1960s and 1970s, but the dramatic fossil fuel price rises of the 1970s and fears of landfill price rises in the 1980s once again encouraged incineration—but a new kind of incineration, where the waste was converted not only to ash but also to energy (Ujihara and Gough 1989; Curlee et al. 1994; U.S. EPA 1995b). By the 1990s, this waste-to-energy incineration had stabilized at about 15% of U.S. MSW. *All* biomass incineration generates only about 3% of U.S. electricity output (DOE 1998b).

It is, however, a tenuous stability. To many environmentalists, "burn" is a four-letter word (as is "bury," but less so). States have been putting ever more costly restrictions on the operation of solid waste incinerators. And one state (West Virginia) actually bans incineration (Murphy and Rogoff 1994). But dismantling existing incinerators is seen by most cities as prohibitively expensive, and new incinerators will always be tempting where landfills are costly. It is no coincidence that more than 40% of the MSW is burned in the densely populated Northeast and less than 2% in the areas west of the Mississippi River (Glenn 1998).

Whether because the United States has more fear of incineration or because the United States has more land for landfills, we incinerate much less MSW than do most other industrial countries—Switzerland, Sweden, Denmark, Singapore, and Japan all burn more than half their waste (Bonomo and Higginson 1988; Alexander 1993). This greater reliance on incinerators elsewhere is not due to softer environmental regulations—Sweden, for example, has the most stringent dioxin emission targets in the world (OTA 1989b).

There will always be waste that cannot be recycled, and a choice will have to be made between incineration and landfilling. What makes the choice hard is that the two disposal techniques are so different in both their private costs and their external costs.

Comparing Incinerators with Landfills

An incinerator is not "just a landfill in the sky." Incinerators differ from landfills in many important economic ways. Leaving the differences in their external costs aside for a while, the private costs of incinerators and landfills display five significant differences that affect a community's choice between them as means of solid waste disposal: (a) Initial capital costs are much higher for incinerators than for landfills; (b) incinerators exhibit economies of scale to a much greater extent than do landfills; (c) land costs are a minuscule part of incinerator costs; (d) incinerators, once operative, are very costly if they do not receive a steady flow of burnable waste; and (e) incinerators can capture more energy more quickly from waste than can landfills, and hence their *net* operating costs, including energy revenues, are much lower than those of landfills. We will look at each of these characteristics in turn.

Capital

Estimates of the initial capital costs of incinerators vary greatly, from less than $100,000 per daily ton of capacity to about $200,000 per daily ton of capacity (U.S. EPA 1988; Denison and Ruston 1990; Alexander 1993; Keeler and Renkow 1994; Rhyner et al. 1995). Because incinerators in practice range from as small as 300 tons a day to as large as 3,000 tons a day, these cost ranges mean that the incinerator owner must initially lay out (or more realistically, borrow) $30–600 million. Indebtedness of this size can put serious stress on a municipality's budget situation for the entire life of the incinerator (or the life of its accompanying municipal bonds)— usually about 20 years. Even if the incinerator is privately owned and operated, it puts a strain on the municipal budget, in a way. The municipality is "indebted" in the sense that it usually has to promise to deliver sufficient burnable trash to the incinerator throughout its lifetime or pay a penalty for any shortfall—what is called a "put-or-pay contract" (Bailey 1993).

To understand the magnitude of these capital costs, it is useful to put them into costs per ton of the incinerated solid waste. Consider an incinerator in the middle of the above ranges, that costs $150,000 per capacity ton and burns 1,200 tons a day for 365 days a year and for 20 years. This comes to $30 per ton of waste burned.[1] Even before considering operating and maintenance costs, this incinerator is showing costs per ton comparable to landfill costs in many parts of the United States.

Scale

Large (mass-burn) incinerators are cheaper per ton than small (modular) incinerators, but not in the way that one might guess. The initial capital costs per ton are quite comparable—actually, the smaller modular facilities average slightly lower capital costs (U.S. EPA 1995b). But the operating and maintenance costs of the larger mass-burn facilities average about $10 per ton less than modular incinerators. Because the ash disposal cost and the electricity revenue are the same per ton for the two types of incinerators, this means that the larger facilities have lower net average total costs than the small ones.

It is not quite accurate to say that the initial capital costs (per ton of capacity) are comparable for different sizes of incinerators. There are initial costs beyond those of the facility itself, namely, the costs of getting permits to operate. The permit cost depends largely on the degree to which local residents oppose the facility, and that opposition is likely to be greater the larger the incinerator being contemplated; but one would still expect economies of scale in permitting, so that the permit cost per ton of capacity would be lower for larger incinerators. Thus, a number of factors suggest that incinerators are more appropriate for larger municipal areas than for smaller, although sparsely populated areas may be able to consider regional incinerators (Curlee et al. 1994). (For a formal analysis, see Keeler and Renkow 1994; for an illustration, see the Appendix on page 82.)

Land

Because incinerators require much less land space than do landfills, the cost of land makes a big difference to the relative costs of the two methods of solid waste disposal. Nowhere is this importance of land availability to the choice of incineration more clear than in Japan, where more than half the population lives in areas with densities of more than 10,000 per square mile, and more than twice as much municipal solid waste is incinerated as is landfilled (Hershkowitz and Salerni 1987). Land is expensive everywhere around heavily populated cities, and hence land is expensive in the Northeastern United States—recall the great differences in landfill tipping fees, ranging from less than $10 a ton in the rural parts of the Southwest to nearly $100 a ton in the urban parts of the Northeast.

Flow

Incinerators have a design capacity, and they do not operate with technical efficiency if they do not receive that flow—not on average, but at all times. All incinerators stockpile waste collected Monday through Friday so they can operate continuously on Saturday and Sunday, but few have space for long-term storage. If the design capacity is too small for the solid waste flow, the excess cannot be burned and must be disposed of in some other way. And if the design capacity is too large for the actual received waste flow, not only will the incinerator operate inefficiently but also the capital cost per (actual) ton of waste will be raised.

This inflexibility of incinerators puts them at a distinct disadvantage to landfills, which are flexible *par excellence*. If more waste is sent to a landfill, it just fills up more quickly and a new landfill will have to be opened sooner. And vice versa: If less waste is sent, it fills up more slowly and the landfill will last longer.

Energy

Fire is faster than decay in recapturing energy. Incinerators often recover their entire operating and maintenance costs through the sale of the resulting electricity (Keeler and Renkow 1994; U.S. EPA 1995b). Variable costs and electricity revenues each run $20–40 per ton of solid waste incinerated. This is very different from landfills, where the burning of the methane is at best a marginal net revenue producer (see Appendix B to Chapter 4).

Overall Private Cost Comparison

Because municipal solid waste must be collected in any case, we can begin an overall private cost comparison of incinerators and landfills at the incinerator or landfill.[2] Then we can summarize, qualitatively, the private cost advantages and disadvantages of the two disposal methods. The advantages of incinerators are three: They require much less land; they greatly reduce the volume of waste that needs to be disposed of in a landfill; and they roughly cover their operating costs through electricity sales. The disadvantages are two: huge initial capital cost and inability to adapt to fluctuating waste flows.

When these advantages and disadvantages are weighed quantitatively, incinerators are almost always found wanting. Most studies find the overall cost per ton of municipal solid waste is much higher in an incinerator than in a landfill. One careful, recent study calculated the total net collection and disposal cost, in a typical city of a half-million people, to be $95 per ton in a landfill and $147 per ton in an incinerator, roughly 50% higher (Franklin Associates 1994). Other U.S. studies have also found this large differential favoring landfills (U.S. EPA 1988). Similar cost differentials appear in European analyses (Cairncross 1993). If tipping fees can be taken as a clue to average total cost, the national average tipping fee of incinerators is roughly double the national average tipping fee of landfills (Repa 2000).

External Costs of Incinerators

The external costs of incinerators lie in the pollution incinerators produce—mainly, pollution from the gases that are emitted into the air and pollution from the residual ash that must be buried in the ground.

Air Pollution

When garbage and trash are burned, the combustion is incomplete. Some breaks down incompletely, some does not break down at all, and some of what does not break down goes up and out the flue (U.S. EPA 1987). Specifically, the flue gases of incinerators contain particulates that get into lungs and cause pulmonary problems and deaths; chlorinated dioxins and dibenzofurans (CDD and CDF), which are toxic and can cause cancer, reduced immunity, and birth defects; nitrogen oxides and carbon monoxide, which contribute to smog; sulfur dioxide and other acid gases, which cause corrosion and acid rain; and heavy metals such as arsenic, cadmium, lead, and mercury, which can cause serious human health problems.[3]

The air pollution of incinerators can be greatly reduced by optimization of the combustion and "scrubbing" to reduce air emissions. During the 1990s, the EPA forced much higher air pollution standards on incinerators, particularly on new and large incinerators (Berenyi and Rogoff 1999). Of course, the more air pollution is prevented, the higher the costs—state-of-the-art air pollution control can add up to $40 per ton of waste processed to the cost of the incinerator. Is it worth it? I can find no recent EPA studies of expected lives saved, but one EPA study in the 1980s suggested that the controls were *already excessive* then (see the box titled "The EPA Analysis of Incinerator Air Pollution Controls").

The air pollution of incinerators can also be increased by financial stress. When the landfills that the United States was supposedly "running out of" became cheaper in the late 1980s, this put pressure on incinerators to match their declines in tipping fees and/or operate below capacity. Desperate for new combustibles and new revenues, many incinerators turned from municipal solid waste to industrial waste. But industrial waste is much more likely to contain toxic chemicals; and if it is not monitored carefully, these chemicals will go up the stack or into the ash (Lipton 1998).

Ash Disposal

Not all trash burns. Glass and metals must be separated out, either before or after the incineration; and yard wastes are hard on incinerators for two reasons: The flow is seasonal, which prevents the smooth full-capacity operation that is needed for efficiency; and much yard waste is wet, which is worse than nothing for burnability. Even with glass, metals, and yard waste removed, incineration only reduces the volume of (the rest of) the household waste by 70–80%, leaving ash to be disposed of.

How does incinerator ash differ from the MSW from whence it came? No harmful pollutants have been added to it in the course of the incineration, so in a sense it is no different from MSW and should simply be landfilled. But the serious pollutants (e.g., cadmium, lead, and mercury) have not burned at all and hence are still there in the ash, with some four times the concentration as in the original MSW. Furthermore, the only reason that these toxic metals were not specially treated before incineration is the administrative difficulty and collection cost of separating them out in the MSW collection process. Once they are concentrated into incinerator ash, it is no longer administratively difficult to handle them more carefully.[4]

But it is still costly. When incinerator ash is considered hazardous and requires being sent to a hazardous waste disposal facility, the cost of its disposal goes up from about $30 a ton at a landfill or "monofill" to $200–500 a ton, depending on location and distance (Maillet 1990; Goddard 1994). (A monofill is a landfill that accepts only incinerator ash; its cost is usually a little higher than that of a landfill.) A cost of $200 per ton of ash means about $50 per ton of the original waste (because the ash is about a fourth of it by weight), which makes incineration totally uncompetitive with landfill disposal.

Should the ash be landfilled, monofilled, or treated as hazardous waste? It depends upon how much the public health exposure is improved by the greater dis-

The EPA Analysis of Incinerator Air Pollution Controls

The EPA analyzed the health effects of controlling the nation's existing (in 1986) 111 municipal waste combustion facilities so as to reduce their air emissions of organics by 95% and their emissions of particulate matter by 99.5% (U.S. EPA 1987). These existing incinerators were estimated to be causing something between 2 and 40 expected cancer deaths each year, almost all due to inhalation of chlorinated dioxins and dibenzofurans. With scrubbers and filters, the expected annual cancer deaths would be reduced by something between 1 and 35 expected deaths. The annualized cost of these controls on these existing facilities would be (converted to 1997 dollars) between $166 and $272 million.

These wide ranges yield even wider ranges for the cost per expected death avoided, all the way from less than $5 million (i.e., $166 million for 35 lives) to more than $200 million (i.e., $272 million for 1 life). The lower figure is in the range of acceptable life-saving policy; the high figure is nearly two orders of magnitude greater.

Figure 5-1

Optimal Disposal of Incinerator Ash

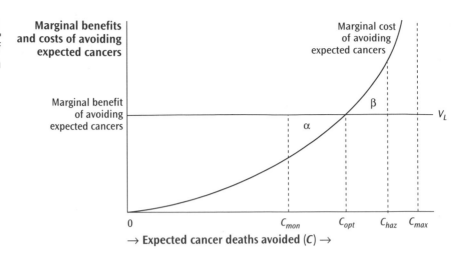

posal care, relative to the extra cost of the more careful disposal. Look at Figure 5-1. It simplifies the disposal issue by assuming that only expected cancers (C) are at issue. Each expected cancer avoided carries a social benefit of the value of life (V_L)—the maximum number of dollars we collectively agree on spending to save one expected life. Saving lives by ever more careful disposal carries increasing costs as the cheap and easy life-saving steps are taken first. In Figure 5-1, the optimal cancer avoidance is shown as C_{opt}, which is less than C_{max} (i.e., the number of expected cancers that would have occurred if no safeguards had been applied).

To better illustrate the debate over handling incinerator ash, I have drawn Figure 5-1 in such a way that neither the monofill ash disposal nor treating the ash as hazardous waste achieves the optimal expected cancer avoidance. Monofilling achieves too little cancer avoidance (C_{mon}), and hazardous waste treatment achieves too much cancer avoidance (C_{haz}). How can one achieve too much cancer avoidance? By using too many resources, resources that could be better used elsewhere. As Figure 5-1 is drawn, the area demarcated by the β represents the excess of marginal costs over benefits (V_L) of the excessive disposal care. The area marked with the α, conversely, shows the excess of potential benefits forgone over the costs because of too little care.

Again, as Figure 5-1 is drawn, the areas marked α and β are about the same size. In this case, it does not matter which disposal care is required—the same loss from the unattainable optimum is incurred in each case. The point of all this is that it is *not* "better to be on the safe side." Excessive care also imposes costs.

Everything in Figure 5-1 is hypothetical. In fact, one of the three kinds of disposal—ordinary MSW disposal, monofill disposal, or treatment as hazardous waste—is superior. But there seem to be no studies estimating the resource cost per expected life saved by the various ash disposal techniques—much posturing, but little evidence. All we know for sure is that there is a big difference in the costs of the two disposal techniques; it would be nice if we had a clue about the difference in the number of expected deaths prevented. The problem is compounded by the fact that the optimal handling of ash at one incinerator is not necessarily the optimal handling elsewhere, where water proximity and population density are different.

Until recently, the method of disposal of solid waste incinerator ash was left to the states, and they responded in many ways. In a 1993 survey of 23 states with municipal solid waste incinerators (and which had 104 of the 134 such incinerators in the United States), all but 3 required ash testing, all but 2 required an ash management plan, and all but 1 had some transport restrictions—such as covered or watertight trucks being used (Murphy and Rogoff 1994). But only 7 states required monofill disposal, and none required the ash to be treated as hazardous waste.[5]

In 1994, however, the U.S. Supreme Court decided that incinerator ash was not exempt from hazardous waste treatment if it did indeed contain hazardous wastes (DeWitt and Butler 1994). The reasoning was legal, not economic (Abbott 1995). Indeed, if we continued that logic, we would have to treat MSW as hazardous waste, and the resources that would be wasted in doing that are unthinkably large.

Neighborliness

Waste-to-energy facilities are a sort of combination of electricity generating station and solid waste landfill; but with respect to their desirability as neighbors, incinerators seem to be considered worse than either by those who live near them. There are few studies of the effects on property values of proximity to a solid waste incinerator, but the anecdotal evidence suggests that, from the neighbor's perspective, they add together the worst externalities of each.

Lower property values are a measure of the capitalized value of the external costs imposed on neighbors—lower property values are *not* an external cost in themselves. To the extent that we have already measured the external costs due to the air pollution and the toxic ash,

Incineration in Europe

Throughout Northern and Western Europe, there is heavy reliance on waste incineration—up to 80% of the residential waste is disposed of in this way. The obvious reason for this difference in disposal preferences between Europe and the United States is the relative scarcity of open, cheap land in Europe. But there is a second reason. In Europe, it is not uncommon to find common heating arrangements for entire districts, and this provides a ready market for steam generated by incinerators. Producing steam is more energy efficient and less capital intensive than producing electricity, and this contributes greatly to the profitability of European incineration. Indeed, few European incinerators generate electricity, partly because few European countries allow non-utility-generated electricity to be sold into the grid. In European incineration, there is typically state-of-the-art air pollution equipment, and ash is more calmly handled—going into road construction or into ordinary municipal landfills.

Incinerators in Eastern Europe, conversely, are usually older and lacking adequate—or in some cases any—environmental controls. Furthermore, the wetness of municipal wastes means that many of these incinerators cannot operate without supplementary fuel. Given the increasing concern for air pollution and the decreasing willingness to subsidize public enterprises there, incineration is losing its attractiveness in Eastern Europe.

adding in property value declines would be double counting, to some extent. But the immediate neighbors of an incinerator bear little of the air pollution cost and (usually) none of the toxic ash cost, so any decline in property values around an incinerator probably measures other external costs, such as noise, aroma, dust, and litter.

When potential neighbors fight the construction of an incinerator, whether from their own fear of its external costs or from a fear of declining property values due to other people's fear of external costs, very real costs are added to the project—added negotiations, meetings, time, and legal fees. Even if the protesters volunteer their time, the proposed incinerator owners or operators are not volunteering their time, and the initial private cost of the incinerator is raised.

Global Warming

Incinerators that burn solid waste, like plants that burn fossil fuels, create not only local air pollution but also the carbon dioxide that causes global warming. It seems sensible to add this as an external cost of solid waste incineration. In fact, however, incineration causes less global warming than landfilling, because landfills also generate methane, which has 21 times the potential for global warming as does carbon dioxide (U.S. EPA 1998c). Furthermore, incineration generates electricity, and this reduces the need to generate electricity by means of traditional fossil fuels—after accounting for that, incinerators add only very slightly to global warming gases (ICF Incorporated 1997). So if the choice were between landfilling and incineration on global warming criteria alone, incineration would be the winner. Indeed, some have proposed that tax credits should be offered to electricity production that uses biomass as its fuel (Naquin 1998).

Will the Right Decision Be Made?

In principle, it is not hard to make the correct decision between landfills and incinerators as means of disposing of municipal solid waste. But there is a lot of room for slippage between the in-principle cup and the in-reality lip.

External costs have to be quantified to make the choice between incinerators and landfills. Moreover, the *kinds* of external costs differ between the two disposal techniques, which makes their quantification more critical. To see this, consider two techniques for achieving the same (immensely worthwhile) objective: Technique A costs α and puts out only a units of some pollutant; whereas Technique B costs less, only β, but puts out more, b units, of the identical pollutant. All we need to do to make our choice is to calculate $(\alpha - \beta)/(b - a)$ and to decide whether the marginal benefit of abatement of this pollutant is greater or smaller than $(\alpha - \beta)/(b - a)$. But when there are two or more different pollutants involved—such as groundwater pollution and air pollution—the task is much more complicated and may involve trade-offs not only between money cost and pollution but also between pollutants.

Private costs also may be difficult to estimate, for two reasons. The first is that the relevant private costs extend several decades into the future and hence must be extrapolated (or in some other way estimated). One only has to look back a decade or two to see how big the forecasting errors can be—and how bad the decisions that result. During the 1970s and 1980s, cities faced simultaneously higher

---

Waste to Energy without Incineration

Incineration is not the only way to turn municipal solid waste into energy, a fact that is hinted at by the presence of landfill gas recovery systems (in Chapter 4). But landfill gas recovery has usually been an afterthought. One can also set up a special plant that removes glass, plastic, and metals for recycling and then converts the remaining solid waste into fuel-grade ethanol and industrial-grade carbon dioxide.

Such a plant is under construction in Middletown, New York (Hayhurst 2000). It will be paid a $60 per ton tipping fee to accept the municipal waste of 20 surrounding communities. Being paid for inputs *and* being paid for outputs sounds like a no-brainer winner. (A few people worry about the sulfur and lead emissions, about as much as those of some 50 people's chimneys and cars.) Unfortunately, although biomass-to-energy processes are old hat, the process has never been tried on municipal solid waste. The plant has already received federal assistance (because it is a renewable energy source), and it is not yet clear that it will be profitable.

---

energy prices and higher landfill prices. As a result of assuming that these trends would continue, and perhaps intensify, many cities opted for incinerators. By 1988 in Connecticut, for example, 62% of its municipal solid waste was being incinerated, and there were plans to increase this to 75%—with the rest being recycled (Meade 1989). Today, with landfill and energy prices falling or no longer rising, and with ever more expensive incinerator regulation, many cities are considering closing their expensive incinerators.

The second reason why private costs get distorted is subsidies. If land is "already owned," if capital is available through municipal borrowing at below-market rates, if the resulting electricity must be purchased by utilities at above-market prices, or if businesses can be forced to send their wastes to a higher-cost disposal facility, then decisions may be distorted, and all of these subsidies have already affected landfill versus incinerator decisions at different times and places.[6] When the subsidies disappear or are removed, previously distorted decisions become more visible.[7] Incinerators, built because they were thought to be the cheaper disposal method, have generally turned out recently to be the more expensive, chiefly because landfill costs have not risen as much as had been anticipated. In states which have both landfills and incinerators, the tipping fee is now almost always higher at the incinerators (U.S. EPA 1995b; Magnuson 1997).

There is one final way in which incorrect pricing may affect decisions between incinerators and landfills. The absence of any marginal cost to the household of setting out trash for collection means that there is too much trash being put out. To the extent that there are greater economies of scale in incineration than in landfills, more trash means that it is more likely incineration will be selected.

All this also offers one more reason for preferring a trash collection charge on households to an advance disposal fee on manufacturers. The marginal social cost of landfilling and incineration will differ in different communities and different regions, so costs of trash collection and disposal will also differ. An advance disposal fee on manufacturers must be uniform—one cannot predict in what part of the country the product will be sold. But a refuse collection charge on households can be set at the appropriate local rate: Each city should set the charge equal to the marginal social cost of collection and disposal when the cheaper of incinerator and landfill is chosen as the disposal technique. For completeness, the pricing means of Chapter 3 (Figures 3-3 and 3-4) are repeated in Figure 5-2, with the appropriate corrections to allow for incineration.

Given all these difficulties in estimating the relative costs of landfilling and incineration, it is not surprising that very few efforts have been made at this esti-

Figure 5-2

Two Means of Pricing Trash Collection

When the North East Solid Waste Consortium of 23 suburban communities in Massachusetts was formed nearly 20 years ago, energy prices were soaring, landfills overflowing, and inflation running full tilt. The communities committed themselves to building an incinerator to burn trash. The plant was to generate electricity to pay for itself and perhaps even produce a profit for the communities. Reality proved otherwise. The consortium's disposal system is the most expensive in the state, in need of all the help it can get.

One major factor is that electricity revenues have plunged. The price is now about 2.5 cents per kilowatt-hour instead of the dime hoped for at the beginning. Also, tightening environmental rules have required major incinerator improvements. Burgeoning recycling programs have reduced the flow of burnable trash from towns, which means trash has to come from other sources that pay lower tipping fees. The private operator of the plant, Wheelabrator Technologies, has won arbitration awards guaranteeing its profitability.

The net of all these negative developments is that the 23 communities are paying $95 a ton for waste disposal, a figure that will rise to $160 a ton when additional mandated upgrading of the plant is completed over the next five years. Countermeasures—refinancing debt, intercepting hazardous wastes, attracting private trash flows—have kept this figure from rising even further.

Member communities cannot escape these costs by dropping out. They are committed by contract and would have to pay the system's costs even if they used other disposal systems, which typically run $55 to $65 a ton.

State assistance of $3 million a year has been included in the House version of the budget, now in conference committee. The Senate provided no funding. Precedents exist for such aid—the state made a smaller contribution to a similar plant in Saugus and provides funding for recycling programs. State as well as federal requirements have imposed costs not foreseeable when the original structure was established. Helping pay those costs is appropriate.

[The above is quoted in its entirety from an editorial (*Boston Globe* 1999). Reprinted courtesy of the *Boston Globe*.]

mation. And those few vary a great deal, not only because of data difficulties, but also because observers have biases and because technologies, costs, and regulations differ in different places and different times. Nevertheless, it is interesting to look at two such efforts. The first uses Dutch data for the early 1990s; see Table 5-1. Despite the relatively high Dutch prices for land, this study finds conclusively for landfilling.

The second study examines the existing literature for four other countries; see Table 5-2. Costs here ignore land costs, assume that no landfill gas is recovered, and are given (usually) as a range. For each of the countries, there is considerable overlap in the estimates of cost per ton for the two techniques. Although Table 5-2 does not permit us to reach an indisputable conclusion, it strongly hints that incineration is the less desirable approach from a viewpoint of social cost.

Final Thoughts

Waste-to-energy incineration of municipal solid waste has been losing favor in the 1990s in the United States. Although relative costs differ in different cities, incineration usually seems to be the higher-cost option. And incineration suffers from inflexibility and uncertainty.

The inflexibility resides in the need to operate an incinerator very near capacity. This leads to incinerator tipping fee dilemmas. If the flow of trash falls below the incinerator capacity, the average net (of electricity revenues) total (capital and variable) incinerator cost per ton rises. This urges an increase in the tipping fee. But a higher tipping fee further discourages the delivery of trash to the incinerator, raising average net total cost even further. One solution cities (and states) have sought to this dilemma is requiring local trash producers to deliver to the incinerator, even though its tipping fee is relatively high. We return to this proposed solution in Chapter 7. The other aspect of incinerator inflexibility is that it locks in predictions (or plans) about recycling.

Cost category	Landfilling	Incineration
Average capital and operating cost	49	155
Energy generation[a]	−6	−25
Materials recovery[b]	0	−3
Land use	25	0
Air and groundwater pollution	11	67
Energy externalities avoided[c]	−7	−32
Material externalities avoided[d]	0	−8
Net average social cost[e]	74	153

[a]Gas recovery is practiced at the landfill, but it generates much less energy than does incineration. Revenues (and other positive social benefits) are shown throughout this table as negative costs.
[b]Aluminum and iron are assumed to be recovered before incineration. Zero means less than 0.5.
[c]Avoided by not having to produce this electricity by burning fossil fuel.
[d]Avoided by not having to extract virgin aluminum and iron.
[e]Totals do not add because of rounding.

Source: Dijkgraf and Vollebergh 1997.

Table 5-1

Net Cost Estimates for Landfilling and Incineration in the Netherlands (1997 U.S. dollars per ton of waste)

Country and disposal	Private cost	External cost	Energy gain	Total cost[a]
Germany				
Landfilling	51	3–15	n.e.[b]	53–66
Incineration	104–192	5–14	58–106	52–100
Sweden				
Landfilling	16–24	3–15	n.e.	19–39
Incineration	57–65	7–15	35–42	29–37
United Kingdom				
Landfilling	8–51	3–15	n.e.	11–66
Incineration	84–96	24–33	63–77	46–62
United States				
Landfilling	15–57	3–15	n.e.	18–72
Incineration	69–137	11–20	49–66	31–91
Netherlands[c]				
Landfilling	49	36	13	74
Incineration	155	56	57	153

[a]Total cost is private cost plus external cost minus energy gain (due to avoided energy costs); totals may not add due to rounding.
[b]n.e. means not estimated.
[c]The data for the Netherlands in Table 5-1 are repeated here in the same format as the other four countries to ease comparison.

Sources: Miranda and Hale 1997; Dijkgraf and Vollebergh 1997.

Table 5-2

Net Cost Estimates for Landfilling and Incineration in the Netherlands and Four Other Countries (1997 U.S. dollars per ton of waste)

The uncertainty associated with incinerators is both technical and legal. There appears to be no consensus about the hazards of dioxins and furans, nor about the hazards of releasing heavy metals into the air (Curlee et al. 1994). There are also regulatory and legislative uncertainties about incineration, especially about their air and ash pollution and about the impact of possible minimum recycling mandates. Given the huge initial costs of incinerators, this uncertainty about future costs is currently decisive in choices between landfills and incinerators. Of the hundred or so companies that were once in the waste-to-energy business, by 1999 there were only three left (Geiselman 1999).

Many of the waste-to-energy uncertainties are not caused by the nature of incinerators but by the strong opposition of environmentalists. For reasons still unclear to me after many conversations, "burn" is to them a dirty word. If burning is chosen over recycling because of misguided waste-to-energy subsidies, the hostility is sensible. The alternative for much waste, however, is a landfill, which means—unless the landfill gas is recaptured and used for energy—that an energy source is being totally wasted. For the foreseeable future, the opportunity forgone by landfilling this waste is the opportunity to extract and burn less fossil fuel.

APPENDIX

Incinerators on the Scale of Two Cities, Detroit and Ann Arbor

Why does a window-dresser always turn the price tags so we can't read them?
—Kin Hubbard, *Abe Martin on Things in General*

Size matters for the choice between landfilling and incineration. Perhaps two sets of real-world-type numbers will drive home that point. Below are brief benefit–cost analyses of incineration in Detroit and Ann Arbor, Michigan. Detroit's numbers are actual—it has an operating incinerator; Ann Arbor's are hypothetical—and indicate (one of the reasons) why Ann Arbor does not have an incinerator.

Detroit

Detroit's incinerator, a refuse-derived fuel facility, has been operating for almost 10 years, servicing the city and its suburbs. Batteries, glass, and ferrous metals are removed beforehand, and more than 3,000 tons a day are then burned, producing electricity and steam, all of which is sold to Detroit Edison.

The capital cost was $380 million (converted into 1997 dollars), including an upgrade of its original air pollution equipment, because it was later found to be inadequate. Assuming a 30-year life and a 4% discount rate, this yields an annualized capital cost of $28 million; (1/30 + 0.04)($380 million) = $28 million. Operating the Detroit incinerator costs about $35 million a year. The facility also generates some 250 thousand tons of ash per year, which costs $40 a ton (at a monofill) to dispose of, for a total ash cost of $10 million.

If Detroit were sending this 1 million tons a year of solid waste to the landfill Ann Arbor uses, it would be costing nearly $30 million a year. The steam, electricity, and sale of recyclables net the facility roughly $38 million a year. (I was unable to discover if Detroit Edison paid a price above its avoided cost for the electricity.)

The net of these five figures is very close to zero—the total annual benefits (i.e., $30 million plus $38 million) and the costs (i.e., $28 million plus $35 million plus $10 million) differ by less than 10%. Clearly, a definitive benefit–cost evaluation would need to put the data together more carefully. But even after that, it would almost certainly be necessary to look at the external costs, of the incinerator itself and also of the monofill to which its ash is sent, and at the external benefits, of the reduced landfilling of solid waste and of the air pollution not generated by Detroit Edison as a result of being able to buy this steam and electricity. What these rough estimates do tell us is that incineration on a scale such as this can be a close call (Schock 1995).

Ann Arbor

Because Ann Arbor is a small city with an extensive recycling program, it would need a small incinerator were it to choose incineration over landfilling. Currently, this would mean burning 150 tons a day; with increased recycling keeping down the growth of the city's waste to be incinerated over the incinerator's life, a modular incinerator with capacity of 200 tons a day would suffice. The capital cost would be about $20 million, and hence the annualized capital cost would be about $1.5 million. The primary benefit of such incineration would be avoided landfill costs, which would save Ann Arbor about $1.3 million a year at their current landfill rate and tipping fee.

Such a facility would cost about $0.7 million a year to operate. The roughly 13 thousand tons of ash to be disposed of each year would add $0.4 million to the cost if it were sent to Ann Arbor's present landfill as ordinary solid waste—somewhat more if it had to be sent to Detroit's incinerator ash monofill, and much much more if it had to be disposed of as hazardous waste.

In the process, 20–30 million kilowatt-hours of electricity would be generated each year. Regardless of what Detroit Edison actually paid for this electricity, the social benefit would be Detroit Edison's avoided cost, about 2 cents per kilowatt-hour. The social value of the electricity would thus be about $0.5 million a year.

All these figures are very rough—Ann Arbor has never conducted a close examination of the potential for an incinerator—but the total annualized costs (i.e., $1.5 million plus $0.7 million plus at least $0.4 million) sufficiently exceed the total annualized benefits (i.e., $1.3 million plus $0.5 million) that one need not bother toting up the balance of external costs (Zisman 1995).

Summary

Table 5-3 summarizes the various benefits and costs of the two incinerators. Note the question marks in the table, which indicate that neither the external costs of incinerator air emissions and ash disposal have been quantified, nor have the external benefits that the resulting electricity creates by reducing Detroit Edison's electricity generation from coal. The net benefit, without externalities, is negative in both cases, but the never-implemented Ann Arbor incinerator would have cost roughly three times as much per ton as does the active Detroit incinerator.

Table 5-3	Benefit (+) or cost (−)	Detroit	Ann Arbor
Benefits and Costs of	Capital cost	−28	−1.5
Detroit and Ann Arbor	Operating cost	−35	−0.7
Incinerators	Ash disposal cost	−10	−0.4
(in millions of	Landfill cost avoided	+30	+1.3
1997 annual or	Energy and recyclables revenue	+38	+0.5
annualized dollars)	External benefits and costs	?	?
	Net benefit[a]	−5	−0.8
	Annual tonnage of waste[b]	1,100,000	64,000
	Net benefit per ton	−$4.50	−$12.50

[a]Net benefit, excluding any external costs or benefits.
[b]Annual tonnage of waste if operated at capacity at all times.

Notes

1. The initial cost of the incinerator is $150,000 per ton times 1,200 tons capacity, or $180 million. To compare this initial capital cost with an annual flow of waste, we must find the present value of the waste flow; 1,200 tons a day, during the next 20 years has a present value of nearly 6 million tons (using a 4% discount rate).

2. This assumes that the distance from the collection area is the same to the incinerator or the landfill; if one can be sited closer to the city, this transport cost advantage needs to be considered.

3. Landfills also emit many pollutants into the air, but these are largely unregulated—for one simple reason, lining the top of a landfill to catch air pollutant emissions is considered prohibitively expensive. Backyard burning of trash, still prevalent in many rural parts of the country, is also a serious source of dioxins and furans—more than 10,000 times as much as burning in a well-controlled municipal incinerator, according to the EPA—and yet regulation of household burning is left entirely to the states and localities (ENS 2000).

4. This is the same phenomenon that we saw with sewage sludge in the box in Chapter 3 titled "Garbage Disposals"—toxics are taxed on one disposal path and not on another, for administrative reasons alone. Notice the dilemma: taxing neither path leads to too much being produced by both paths in total, and taxing one but not the other distorts the choice of path. The second-best policy usually requires taxing the path that can be taxed but by less than the full external cost.

5. The only restriction imposed at that time by federal law was that incinerators that did not recover energy were not exempt from hazardous waste regulations (Ujihara and Gough 1989). This restriction—obviously intended to encourage waste-to-energy operations, was equally obviously a very second-best way to do it—whether one recovers energy in the incineration process has nothing to do with the toxicity of the resulting ash.

6. The Public Utilities Regulatory Policies Act of 1978 required electric utilities to buy electricity from outside sources, at a price equal to their own marginal cost (called their "avoided cost"). Some states then went further and required that the price paid to some sources be much higher than marginal cost—in New York, for example, the price utilities had to pay for incinerator-generated electricity was set at nearly double the utilities' cost (Maier 1989). (Also see Appendix B to Chapter 4, for an example where the price was set at nearly triple.)

7. In return for $2 million per year in cash, jobs, and college scholarships, the poor and decaying town of Robbins, Illinois, invited Foster Wheeler Corporation to construct and oper-

ate a 1,600-tons-per-day incinerator (Jeter 1998). The profitability of the incinerator depended on the Illinois Retail Rate Act of 1987, which required utilities to buy electricity at the same price that they charged their customers. Once the incinerator was under construction, the law was repealed—curiously, for *all* electricity generators *except* those using landfill gas as a fuel. The incinerator operated at a loss for a few years before going into bankruptcy (McMullen 2000b). Nothing during this time period changed to affect the *social* profitability of the incinerator, but the change in the law turned it from privately profitable to privately unprofitable.

6

Illegal Disposal and Litter

Litter does not present the most pressing of disposal problems, but it may well be the most difficult to solve.

—George R. Stewart, *Not So Rich As You Think*, 1967

Once we start talking about pricing waste disposal, as we have in the last few chapters, we must also start talking about illegal disposal. Of course, we get some illegal disposal even without pricing, basically litter that people create because it is "too much trouble" to carry it to the nearest proper trash container. But if we add a trash fee of $0.02–0.10 per gallon to this "trouble," we have to worry that illegal disposal will move from a minor eyesore and inconvenience to a major social problem.

Illegal disposal takes many forms. Littering is the obvious form, but there is also surreptitiously carrying one's household waste to a commercial dumpster or burning one's combustible trash in the fireplace or a backyard barrel. Some of these generate new waste externalities. Most of them raise the total social cost of trash disposal, because illegal disposal usually creates greater *total* social costs of collection and disposal than would a proper disposal in the trash at one's own curbside. Because illegal disposal is in its essence a way of making somebody else pay for one's own trash collection and disposal, all illegal disposal ultimately leads to costly government and private countermeasures.

The possibility of illegal disposal adds a new branch to the consumer's waste decision tree. Until now, to create less solid waste and thereby avoid trash collection charges, consumers could do two things: reduce their waste-producing purchases or postpone their trash creation by reusing products. We now consider a third possibility, disposing of the waste illegally to evade waste collection charges—what we call "midnight dumping" and the British colorfully call "fly-tipping."

Waste Pricing and Illegal Disposal

In short, illegal disposal is socially costly in the sense that it adds to the total social cost of the disposal of municipal solid waste. If illegal disposal becomes a serious concomitant of trash collection charges, society must look for ways to discourage it.

Does illegal disposal become a problem *in fact* when trash collection charges are introduced? The evidence is still too little and too mixed for a definitive answer. One detailed study, but of a small sample, found that more than a fourth of the weight reduction of municipal solid waste after trash collection charges appeared was due to illegal disposal (Fullerton and Kinnaman 1995). A later study by the same authors, with a sample of 114 cities that apply trash collection charges, found that the introduction of a trash collection charge of $1 per bag reduced trash collection by nearly half, but barely 10% of the reduced trash showed up as recycling, which suggested to them extensive illegal burning or dumping (Kinnaman and Fullerton 2000). Conversely, a survey of 212 communities in the United States that have introduced trash collection charges disclosed that only 19% of the communities experienced any increase in illegal disposal as a result of the pricing scheme (Miranda and LaPalme 1997). Indeed, most studies, other than those by Fullerton and Kinnaman, conclude that illegal disposal may occur initially but does not long remain a serious problem (Miranda et al. 1996).

In the next few paragraphs, however, let us explore illegal disposal and litter on the assumption that they are likely to become problems. This means that we must reconsider the waste collection charge systems suggested in Chapters 2 and 3.

Trash Collection Charges and Refunds

We now seek a system of waste charges (and refunds) that achieves two goals: It *does* induce consumers to reduce and reuse to avoid or postpone landfilling or incineration costs; and it *does not* induce the consumer to dispose of waste illegally. We will continue to postpone consideration of the possibility of recycling until later.

One thing is very clear. It is no longer a matter of indifference whether the cost of landfilling is imposed on the manufacturer of the product as an advance disposal fee or on the household as a trash collection charge. The latter offers an incentive to illegal disposal; the former does not. So the basic disposal cost must be in the form of an advance disposal fee on products.

How big should the advance disposal fee be? There are two possibilities. First, the advance disposal fee should equal the marginal cost of legal collection and landfill (or incinerator) disposal. This returns the illegal disposal incentive to the same level as it is now, a matter not of money but of convenience. The cost to the household of legal and illegal disposal of products, once bought and ready for disposal, is the same—zero. If it is more convenient to litter, one litters; if it is more convenient to use the trash can (or if one fears being punished for littering), one uses the proper disposal method. This advance disposal fee has no effect on the litter situation.

But there is a second possibility, in theory at least (Fullerton and Kinnaman 1995). The advance disposal fee could be set equal to the higher social cost of illegal disposal, which includes not only the higher costs of collecting illegally dumped trash but also the disposal and external costs. The difference between the social costs of illegal and legal disposal could be refunded to the household if it chooses legal disposal—in effect as a subsidy for using the socially more desirable disposal system. This pricing scheme is shown in Figure 6-1. ("Litter cost" is used in the figure and hereafter as a shorthand term for all kinds of illegal disposal.)

Figure 6-1

Trash Pricing When
Illegal Disposal
Is Serious

The notion of a refund for legally disposed trash, however, raises many practical and administrative questions. First, there is the problem of administering the system, with all its attendant costs. Each household's trash would have to be somehow measured and recorded as it is collected, with periodic payments being mailed out (or deducted from property tax payments). It is not impossible to do this, but it is cumbersome and costly.

The second problem with a trash refund lies in its incentive to "create" trash that has not borne an advance disposal fee. Soil, water, and yard clippings are the obvious things that would be almost costless to acquire and that would bulk up the waste amounts, and their attendant "refunds." Perhaps this could be prevented by careful inspection by the trash collectors, but again we are talking cumbersome and costly.

Not only is this system messy in fact, it is not even the simplest, best approach to illegal disposal and litter externalities. Recall our discussion of Pigovian taxes in Appendix A of Chapter 1. The correct tax is a tax on what causes the externality—that is, on the improper disposal itself. Taxing the product and then subsidizing its legal disposal is a roundabout way of doing that. Why not penalize illegal disposal directly?

Penalizing Illegal Disposal

People dispose of waste illegally because either the inconvenience or the money cost (or both) of legal disposal is high. People who "litter" their trash cause external costs, and the ideal policy response is a fine for such activity. The amount of the fine would, in Pigovian fashion, be related to estimates of the eyesore damage and pickup costs of such litter. (I use the word "eyesore" as a catchall for all the damages that littered trash causes before it is picked up—not only esthetic displeasure, but also hand and foot cuts, farm equipment damage, wildlife death, and so on.) Unfortunately, although most states do advertise high fines for littering and illegal dumping, very few of the litterers are actually apprehended and fined. Law enforcement officers seem to have more pressing things to do than to scent out and ticket litterbugs. Once we multiply the probability of being caught and convicted times the amount of the fine, we find the *expected private* cost of illegal disposal is close to zero, no matter what the level of fines is.

Of course, one could always make the expected fine high by offsetting a low probability of being caught and convicted with an extremely high fine—say, $1,000 or even $10,000. But such a penalty structure would not prove tolerable in the United States.[1] If litterers were rarely apprehended but then threatened with huge fines, the courts would overflow with contested cases, juries would refuse to convict, and judges would balk at sentencing. In the United States, as in other democracies, people feel that the punishment must fit the crime.

Another drawback of money fines, big or small, is that they are not easily attuned to incomes; if they are fixed amounts for all offenders, they are regressive in their incidence. (Regressive incidence means here that the fixed money fine represents a larger part of a poor person's income than of a rich person's income.) For littering, "community service" or other time penalties represent more nearly proportional taxation and hence are politically more acceptable.[2] But the correct Pigovian tax, which reflects the marginal external damage done, is a money tax, not a time tax.

Overall, in theory, there is something to be said for either approach to illegal disposal—subsidizing legal disposal or penalizing illegal disposal. But the two approaches are hardly identical. Subsidizing legal disposal lowers the cost of producing (and later disposing of) waste, so it encourages greater waste; but penalizing illegal disposal raises the cost of waste and discourages its production. Subsidizing legal disposal costs government dollars but not real resources; but penalizing illegal disposal eats up resources— investigators, lawyers, courts (and possibly jails). Both approaches are subject to diminishing returns of a sort. Do~~es~~

A Hint at the Theory

Trying to theorize about subsidizing legal disposal versus penalizing illegal disposal gets quickly complicated, but a hint at the approach may be interesting (Sullivan 1987). Assume, for simplicity, that the total quantity of waste is 1 (in some units) and is not affected by the subsidy or penalty. The cost of legal disposal is C, and the cost of illegal disposal is so high that it must be stopped. The probability of apprehending an illegal disposer is P, and again for simplicity $P = \alpha r$, where r is enforcement resources and α is a parameter (and of course we will consider only values of r less than $1/\alpha$). If an illegal disposer is caught, the fine is F (and is exogenously given). To stop illegal dumping, the cost of legal dumping must be made equal to the expected cost of illegal dumping (and we presume that then no one would perversely continue to dump). The expected cost to a waste-generator of illegal dumping is PF, the expected fine. The cost to the waste generator of legal dumping is $(1 - s)C$, where s is the rate of the government subsidy paid on legal disposal. This means that

$$PF = \alpha rF = (1 - s)C \tag{1}$$

The government budget for anti-illegal-dumping operations is fixed at B, so

$$B = r + sC \tag{2}$$

Again for simplicity, but realistically, we will assume that the enforcement agency does not get to keep the revenues from fines. Also note that B must be less than C, or else the government subsidy would more than cover the entire cost of legal dumping (because the volume of waste has been fixed at one). Equations (1) and (2) can be solved to get the optimal values of r and s:

$$r_{opt} = \frac{C - B}{\alpha F - 1} \tag{3}$$

and

$$s_{opt} = \frac{\left(\dfrac{\alpha BF}{C} - 1\right)}{\alpha F - 1} \tag{4}$$

where r_{opt} and s_{opt} are the optimal values of r and s (provided that the exogenous fine is big enough that $F > C/\alpha B$). From these optimal values, r_{opt} and s_{opt}, we learn four things: First, if the budget (B) were larger, the agency should move from enforcement activity (r) toward subsidy (s). Second, if the cost of legal dumping (C) were higher, the agency should rely less on subsidy and more on enforcement. Third, if the fine (F) on illegal disposers could be raised, the agency should switch away from enforcement toward subsidy. Fourth, if enforcement became more efficient at catching illegal dumping (i.e., if α became higher), then the agency should switch away from enforcement toward subsidy. Of course, *all* these results depend heavily on the assumptions and functions of the model.

bling enforcement resources will not usually double the number of litterers detected. And doubling the subsidy rate means that ever more legal disposers are receiving wasted infra-marginal rents. These diminishing returns suggest that, in theory, some of each approach might be optimal.

Despite all this theoretical thinking, the continued criminalization of illegal disposal is almost surely more workable than subsidizing legal disposal. Nevertheless, whenever trash collection charges are imposed or increased, antilitter-enforcement resources may have to be expanded. Useful ways to use additional resources include: cleaning up rapidly any areas that attract illegal dumping, because the presence of previous litter reduces people's guilt feelings about adding new litter; dedicating litter police in unmarked cars for surveillance of potential illegal dumping sites or for searching illegal trash for clues about the owner's identity—as the Coast Guard already does for Great Lakes dumping from boats; and encouraging snitches by providing a litter tip hotline or by sharing any resulting fines with the tipster—New York City gives half the fine to the tipster provided he or she testifies (U.S. EPA 1998b).

Illegal Disposal of Used Oil

Of the 1.5 billion gallons of used oil currently generated in the United States, 13% is believed to be disposed of illegally (Sigman 1998b; and http://www.epa.gov/epaoswer/non-hw/muncpl/factbook/internet/mswf/mat.htm#9). Although used oil, like all petroleum products, is toxic, the U.S. EPA has never so classified it lest that classification deter oil recovery. So regulation of used oil has been left to the states, and they have varied greatly in their approaches.

About two-thirds of the states have banned the disposal of used oil except by special treatment, which of course raises the cost of legal disposal, and estimates suggest that such bans increased the frequency of illegal dumping by 28% (Sigman 1998b). As further evidence of price responsiveness, this same study found that increases in the salvage value of used oil reduce the amount of illegal dumping. Finally, states with greater enforcement efforts for their disposal ban reported fewer incidents of illegal disposal.

These findings suggest two approaches to midnight dumping. First, an increased enforcement effort. Apparently, firms and individuals respond greatly to even the smallest probability of being caught in illegality, and few enforcement dollars can yield big effects (Harrington 1988). Or two, a mandatory deposit on oil—that is, a tax on the sale of new oil that is refunded if that oil is ultimately properly disposed of (more on this just below in the text). The study of the previous paragraph estimated that a deposit of 10 cents a gallon on oil would reduce illegal dumping by 10% (Sigman 1998b).

Mandatory Deposits to Prevent Illegal Disposal

For some products, either the temptation to illegal disposal or the external cost of illegal disposal is so great that we choose not to rely only on antilitter ordinances and penalties to ensure their proper disposal (Dinan 1993; Palmer and Walls 1997). One example of illegal disposal for reasons of convenience is the littering of beverage containers in public places, and several states have reacted to this by imposing mandatory deposits on such containers—a special fee that consumers pay at the time of purchase and that is rebated when the container is properly returned. Examples of mandatory deposits on products where the external cost of illegal disposal is great are also beginning to appear—notably, automobile and household batteries and used motor oil, all of which can cause serious damage to groundwater if casually landfilled.

How do mandatory deposits differ from an advance disposal fee coupled with a household trash refund? In two ways, one unimportant and the other critical. The unimportant way is in the *impact* of the advance disposal fee, whether the producer pays all or only part of it and whether the consumer pays none or some part of it.

The *incidence* is almost certainly the same no matter how the impact is divided.[3] The price of the product will almost always go up by about the amount of the fee that the producer must pay, so the consumer ends up really paying the whole fee, no matter who it is levied on.

The critical distinction is in the re-collection of the product at the end of its life. With the trash refund, the disposed product is collected through the regular municipal collection system; with a mandatory deposit, a special collection path is established. If the intention of the deposit is to keep the product out of the landfill, the special collection may be essential. But if the intention of the mandatory deposit is only to prevent litter (and/or to increase recycling), the special collection system is an unfortunate additional cost of deposits.[4]

Mandatory deposits to prevent littering make some sense for products that are so bulky as to require a special collection system in any case—for example, automobiles or refrigerators (Lee et al. 1992). Mandatory deposits also make sense where there is no regular trash collection system. On Mount Everest, for example, climbers' empty oxygen containers accumulated around the summit for decades until Nike financed a litter-removal program, which "refunded" sherpas a cash bonus for each container they carried down and out (Krakauer 1997). This succeeded in clearing the summit of old containers, and new containers now bear a sufficient government deposit to keep the summit area clear. Mandatory deposits make the least sense on tiny items that do not harm landfills—such as cigarette butts, which Maine briefly considered for a mandatory deposit of five cents per cigarette (McMullen 2001).

The theory of mandatory deposits is straightforward. The deposit should be set equal to the extra social cost of improper disposal over proper disposal.

When a Tax Is Equivalent to a Deposit–Refund System

Sometimes, a tax on products using virgin raw materials turns out to be the exact equivalent to a deposit-refund system on the product (Sigman 1996a). Consider, for example, an auto battery that uses lead. And assume that virgin lead and previously used lead are perfect substitutes in the production of such batteries. Suppose virgin lead costs nothing and it costs p to turn it into a battery. If battery producers operate in a competitive market, the battery price will also be p.

Now the government places a tax of t on auto batteries made with virgin lead. Because the cost of a battery has now risen to $(p + t)$, the price will also go up to $(p + t)$. The tax is shifted forward onto the buyers, and the purchase price of a battery has gone up by the amount of the tax. When the battery is worn out, the buyer can throw it away costlessly. But the buyer can also offer it back to the battery producers. What will these producers be willing to pay for the lead in this worn-out battery? Because there is no tax on auto batteries made with recycled lead, it will only cost them p to produce a battery that will sell for $(p + t)$. Competition will force them to pay t for the used battery.

In the end, the battery buyer is essentially paying p for the battery, putting down t as a deposit on it, and getting the t back if the battery is returned to the seller—the tax of t is exactly like a mandatory deposit of t. But notice, this neat result only emerges if new and used lead are perfect substitutes in production and batteries made with new or used lead are perfect substitutes in consumption.

Before leaving this example, let us look at the one way in which these two systems are very different. With the tax, the producers *want* to buy back the used battery, even though it costs them t and virgin lead is free. With the deposit system, the producers do *not want* to buy it back—they are forced by law to take back the battery and redeem the deposit of t, but they would prefer not to since virgin lead is available to them at no cost at all. If it is easy to shirk the duty to redeem deposits, the tax may be better.

One final note: Recall the phrase, "wildcat banking." In the nineteenth century, U.S. banks issued their own banknotes, supposedly backed by gold and unrestrictedly redeemable into gold by the holder of the banknote. Many banks succumbed to the profitable urge to issue many more notes than they held gold. To forestall banknote holders from trying to convert these back into nonexistent gold, the banks located their redemption offices inaccessibly, "where only the wildcats throve" (Luckett 1980).

Then, if a person disposes of the product improperly, that person pays the external cost of improper disposal by forgoing the deposit. The threat of a forgone deposit becomes a Pigovian tax equal to the marginal external cost.[5]

Another way of viewing the mandatory deposit is as a system of Pigovian taxes and subsidies. When the consumer buys the product, the deposit is a tax levied on the assumption that the residual material will be improperly disposed of. And then, should the buyer return the product for proper disposal, the redemption is a subsidy for rejecting the easier, but socially undesirable, option of improper disposal.

While the theory is straightforward, the actual workings of mandatory deposits are often complex. It is worth our time to look closely at an example of the use of mandatory deposits, one that has been applied in the United States for nearly three decades: mandatory deposits on beverage containers.

Mandatory Deposits on Beverage Containers

Historically, there were always deposits on beverage containers. But these deposits were not mandated by law. They were applied by the beverage producers in an effort to recapture the relatively expensive bottles for reuse. Until recently, bottles were much too valuable to throw away—they were heavy, durable, and reused 15–20 times before breaking, getting lost, or chipping to the point where they were finally discarded. Steel cans appeared during World War II to provide beer and soft drinks to troops overseas, and they moved into the consumer market soon after the war (Bingham et al. 1989). The trend to "no-deposit-no return" cans and bottles was rapid, driven by many forces. The "one-way" container not only was convenient to consumers; it provided marketing advantages for the aggressive oligopolists of the beer and soft drink industries. One-ways were becoming steadily cheaper and lighter, they reduced transport costs, and they obviated the need for labor-intensive sorting, washing, and inspecting used containers. By the 1980s, the returnable, refillable container with its deposit had all but disappeared (Franklin 1991; Saphire 1994).

A by-product of this disappearance of deposits was greatly increased beverage container litter in public places. Litter provides two kinds of external costs: the obvious flow cost in the sense that someone else must dispose of the illegal trash, but also a stock cost in the sense that the public must endure the esthetic, equipment damage, and health costs before the litter is picked up. A return to deposits, this time government-mandated deposits, was seen as a means of ending this widespread illegal disposal of beverage containers. Moreover, with the advent of the energy crisis in 1973, mandatory deposits were also envisioned as a means of encouraging the return of the reusable container, which uses less energy per beverage delivery than do single-use containers. Also in the early 1970s, worries began to surface that the planet was running out of resources, and mandatory beverage deposits were seen as a way to push bottle reuse. Mandatory deposit laws began to pass, first in Oregon in 1972 and eventually in nine other states.[6]

The experiences of these ten states provide clear evidence of the impact of mandatory deposits on beverages and their containers (Porter 1978, 1983b; Bingham et al. 1989; Criner et al. 1991; Franklin 1991; Saphire 1994; Ackerman 1997). For a detailed benefit–cost analysis of one state's (Michigan's) experience with mandatory deposits, see the Appendix on page 97.

■ *Return rates and litter rates.* With deposits to be redeemed, beverage containers get returned. Studies of the 5- and 10-cent deposit states consistently show that 85–90% of the beverage containers are redeemed (Michigan Consultants 1998).[7] And of course the by-product is a decline in container litter, by nearly 80%, and a decline in overall litter by nearly half. There is also a decline of a few percent in MSW. Mandatory deposits *do* achieve what they are primarily intended to achieve.

■ *Beverage prices, costs, and consumption.* Studies consistently find that beverage consumption, both of beer and of soft drinks, goes down by 5–10% as a result of the introduction of mandatory deposits. This decline is the rational response to the fact that the *total* price to the consumer goes up, where by total price is meant the sum of three components: the actual retail price; the expected forgone deposit (because even conscientious consumers lose or break some containers); and the inconvenience cost of returning empties. With mandatory deposits, all three components increase, although the actual retail price increase may be small, because recycling revenues and unclaimed deposits to some extent offset the increased cost of beverage delivery. Indeed, given the oligopolistic nature of these beverage industries, it is never certain how much of the price increase reflects cost increases.

■ *Return to reusable containers?* One of the anticipated side benefits of mandatory deposit laws was a reversal of the trend away from refillable glass containers. Indeed, the mandatory deposit laws were once often called "bottle bills" because of this expected return to refillable bottles. Once bottlers and brewers were forced to re-collect the containers, it was thought, they would choose to re-collect reusable rather than unreusable containers. Not so. Mandatory deposits may have stemmed the tide toward one-ways, but not much—refillable beer bottles, for example, account for 13% of the market in deposit-law states and 3% of the market in non-deposit-law ones (Saphire 1994). This failure should have been anticipated. Without mandatory deposits, bottlers and brewers apparently found that the cost of new containers (b) was less than the costs of collecting containers (c) and washing them (w; "washing them" is shorthand for washing, inspecting, and replacing some of them). With mandatory deposits and required collection, they apparently found that buying new containers (b) less the revenue from recycling the old container (r) was less than the cost of washing reusables (w). In symbols, before mandatory deposits, $b < c + w$; after, $b - r < w$, or $b < r + w$. Although the collection cost (c) is larger than the recycling value (r), the outcome suggests that the difference was not large enough to alter the decision on which type of container to use.[8]

■ *Recycling impact.* Although increased recycling was not the original intention of mandatory deposits—the intention was to get back to reusable containers—once wholesalers were forced to re-collect their one-way containers, they sought ways to avoid the cost of landfilling them. Recycling was not widespread in the 1970s, but markets for the recyclable aluminum and glass quickly sprang up. Today, the 10 deposit-law states supply almost half the glass bottles, aluminum cans, and plastic bottles that are recycled in the United States (http://www.container-recycling.org). Before enthusing about this serendipitous boost to recycling, however, we should remember that re-collection of containers by their producers is a *very* expensive way to recycle—recall the German "Green Dot" program discussed in Chapter 2.[9]

The bottom line is that mandatory deposits are an effective antilitter program, but a costly recycling program. (We shall return to this in Chapter 11.)

These are the main interesting effects of mandatory deposits. Some would add as an unexpected benefit the scavenging of discarded containers as a cheap, additional litter prevention and as a source of income for the homeless. As anyone who gets up early on weekend mornings in mandatory deposit states well knows, there is more container litter than gets measured because the homeless are out scavenging much of it. (Aluminum cans are regularly scavenged even in nondeposit states, simply because there is a strong recycling market for this metal; more on this in Chapter 12.) Are these not *significant* benefits of mandatory deposits? More containers are returned, and they are returned by those with the lowest opportunity cost of time. The problem is that the littered containers represent a "commons" into which too many "herders" will enter. The container commons is overscavenged.[10] I personally would like to think that there are better ways to ameliorate U.S. poverty than mandatory deposits, but scavenging has the advantage of *not* requiring government bureaucracy and *not* requiring the homeless to get themselves organized.

These laws also gave rise to two debates that need a quick look. After much debate, each deposit state has made its decision whether the unredeemed deposits should be kept by the beverage industry or turned over to the state; after somewhat less debate, each state has also decided whether wholesalers should pay a handling fee or not to their retailers to help defray their container re-collection costs. (Of the deposit states, only Oregon does not require wholesalers to pay retailers a handling fee; several states claim the unredeemed deposits, usually earmarked for litter-related activities.) Both debates missed the point. Where there is a mandated handling fee, wholesale prices will just end up higher by the amount of the fee, and the retail markup will be smaller by the same amount; on this account, there will be no effect on retail prices. Either way, retail prices will always go up (at least) enough to cover the added re-collection costs. And where unclaimed deposits are taken from the wholesalers and retailers, and hence are no longer available to help defray the system's re-collection costs, retail prices will be higher—when the government decides to take the unclaimed deposits, it is in effect adding a tax on beverage consumers (McCarthy 1993).

Finally, there was an aspect of mandatory deposits that should have produced debate, but rarely did. What is the optimal number of redemption centers (i.e., the number of places to which empties can be returned and the deposit redeemed)? In almost every deposit-law state until California, any retailer who sold a particular brand of beer or soft drink was required to accept back empty containers of that brand. Implicitly, legislatures were maintaining that the optimal number of redemption centers was equal to the number of stores selling the product. But just because the market decides to have *N* stores selling Brand X does not mean there should be *N* redemption centers for Brand X containers.

California recognized that not every retailer also needed to be a redemption center when it initiated its deposit law.[11] If there is a state-certified redemption center within a half-mile of a store, the store is not required to redeem empty containers. This greatly reduces the number and cost of redemption centers without greatly adding to consumer inconvenience. Although the California return rate is much lower than in other mandatory deposit states, this is due both to the fewer redemp-

Container Redemption Rates of New York City and Michigan

Beverage container redemption rates typically range from 75% to 90% in states where deposits are mandated. There are two interesting exceptions: New York City and Michigan.

In New York City, only half the containers are returned. Apparently, retail outlets there regularly give grief to redeemers (and especially ill-clad scavengers), something that consumer-conscious and competitive stores rarely do elsewhere (Lasdon 1988). It may be that the low auto ownership and congestion in New York City give each retailer more local monopoly power than in other cities. One interesting result of this is the appearance there of a private market for deposit redemption: the private *We Can* redemption centers, which pay out the full deposit and make money on the handling fee wholesalers must pay them; and "two-for-one" middlemen, who stay open 24 hours a day and who give half the redemption value to cash-starved or store-shunned container collectors (Kaufman 1992; Pogrebin 1996).

In Michigan, the redemption rate is almost 100%, and stores in the southern part of the state (bordering depositless Ohio and Indiana) often record redemption rates of more than 100% (Michigan Consultants 1998; 98.5% average during 1990–1996). I would like to think that we Michiganders just are tidier than residents of other states, but there seem to be other, better explanations of the almost perfect redemption rate. First, the Michigan redemption rate is 10 cents, double that of most other deposit states, and one expects some price elasticity to the redemption activity. (Container redemption is a very price-inelastic activity by necessity, because the (legal) redemption rate cannot exceed 100%. If a deposit of 5 cents is doubled, and the redemption rate rises from 85% to 100%, that implies an arc price elasticity of only 0.24.) Two, a 10-cent deposit appears to be sufficiently high to attract illegal imports of out-of-state (and hence non-deposit-bearing) containers. In a returned-container inspection of more than a dozen stores in southeastern Michigan, for example, it was found that a third of the beer containers and a fourth of the soft drink containers were from out of state (Michigan Consultants 1998). Third, the introduction of reverse vending machines (which absorb empty containers through a slot and automatically credit the redemption), has made such illegal redemption possible—even on a large scale, as was once suggested in a Seinfeld episode on TV and has been evidenced by several Michigan arrests. And fourth, since 1990, the state receives 75% of any unclaimed deposits, which means that the storeowner only loses 2.5 cents when a container is redeemed, leaving the store with much less incentive to examine container returns for illegality.

tion centers and to the fact that the deposit is only 2.5 cents per container (Saphire 1994). Several deposit-law states have now adopted this idea, permitting a retailer to refuse to redeem containers for deposit if there is an authorized redemption center within a certain distance.

California's law also solves another of the inefficiencies of mandatory deposits, that redemption centers have to sort out returned containers by brand to recapture their deposits from the brand's wholesaler. From the social viewpoint, this incurs totally unnecessary cost, because all the containers are going to the same recycling places in the end. In California, the state, not the manufacturer, effectively terminates the deposit chain. The redemption center has no need to sort its containers by brand because the state, and the recyclers with which it deals, couldn't care less which brands are being returned. These cost-cutting changes resulted in a collection cost under California's redemption system of 0.2 cents per container, whereas the cost in other deposit-law states is 2.3 cents per container (Ackerman et al. 1995).

Final Thoughts

Handling illegal disposal by standard economic techniques is, to say the least, difficult. Heavy enforcement, high fines, subsidizing legal disposal—none is really satisfactory. Perhaps we need to rely on two cultural changes: We have been learning not to litter during the past half-century, and we are unlikely to reverse that trend because of a few cents a day.[12] In short, we should make solid waste pricing policy initially on the assumption that illegal disposal will not become a problem. Only if that proves a false assumption should we look to a different drawing board.

We do know that mandatory deposits are effective at ending illegal disposal, but they are socially costly ways to collect waste. So far, however, we have been neglecting two important aspects of deposits. One, the re-collected beverage containers go not to the landfill but to recycling. We return to this in Part 2. And two, some products must be kept out of landfills because of their toxicity, and mandatory deposits

Could We Go Back to Refillables?

There is no shortage of literature pointing out how much energy and raw materials could be saved if we were to return to refillable, reusable glass bottles (Saphire 1994; Ackerman 1997). Could we in fact return to these containers? Refillable bottles are still widely used in Europe, but many things are different there—delivery (and retrieval) distances are shorter, the population is denser, and consumers appear to be more environmentally concerned.

It is possible that there are two equilibria for the beverage container mix, 100% refillables and 100% one-ways, but only the 100% one-way equilibrium seems to be stable in the United States. Conceptually, start in an equilibrium like that of the 1950s, with 100% refillables. Then some consumers start to buy the new one-ways and stores start to carry beverages in both kinds of containers, one-ways as well as refillables. Some stores, pressed for space, give up refillables. The number of places to return refillables thereby declines, and hence the inconvenience of returning them increases. More consumers stop using refillables. More stores stop carrying them. And so on. The stable final equilibrium is like that of the 1990s, with 100% one-ways and no places left to buy or return refillables (Schelling 1978).

Could we reverse these dynamics? The best way to answer that question would be to get the prices of energy and raw materials right. The second-best way might be a tax on nonrefillable containers—not a deposit, but a tax. (Such taxes are in fact levied in parts of Europe—e.g., in Belgium, Denmark, Finland, and Germany.) But think how large that tax would have to be to overcome the stability of the current equilibrium. Who is going to start buying refillables when there are not yet places to return them? Of course, there are such places in states with mandatory deposits on beverage containers—every beverage container can be readily returned for deposit. (This is true except for California; that state's move to a redemption center system has effectively conceded that the one-way is here to stay.) In these states, a tax on one-way containers could reverse the trend of the past half-century.

But there is one problem. Almost all the refillable-bottle filling equipment has been retired or sold overseas to bottlers in developing countries—where the old-fashioned voluntary deposit on containers is still very much alive. It would take a huge tax on one-way containers to induce brewers and bottlers to scrap their modern "one-way" equipment. (Anheuser-Busch tried to use the cost-saving refillable bottles in New England—where almost all the states have mandatory deposit laws and where consumers have to return the empties anyway to redeem their deposits—but it finally gave up, because consumers would no longer buy their beer in refillable bottles; Ackerman 1997.)

may be an effective policy for preventing their accidental discharge into the regular MSW stream. We return to this in Part 3.

APPENDIX

A Social Benefit–Cost Analysis of Michigan's Mandatory-Deposit System

Do not litter on the street,
Then our world will be neat ….
So pick up the litter and do what's right,
To make our world pretty and bright.

—Wendy Irizarry, sixth grade student,
Jefferson No. 1 School, Passaic, New Jersey

In December 1978, Michigan's "bottle bill" was launched amid diverse predictions of restoration and ruin (Porter 1978, 1983b). Of the two major expectations, one occurred and one did not in this new regime. Beverage-related litter did fall dramatically, by some 85%, as the rate at which Michiganders returned their empties to redeem their deposits was quite high, nearly 95%. The expected return to "two-way" refillable bottles, however, did not happen. Beverage delivery in cans remained significant, and the bottled beverages stayed largely in one-way glass containers. (In the 1980s, beverages began to be delivered more and more in nonrefillable—and much costlier to recycle—plastic containers.) The prices of beverages, both beer and soft drinks, rose immediately and noticeably.

The social benefit–cost analysis of this mandatory deposit law depends critically on the evaluation of two things. One, how much did Michiganders value (in a willingness-to-pay sense) this reduction in their road, park, and beach litter? And two, how much of the beverage price rise represented a real social cost and how much was oligopolistic opportunism aimed at teaching Michiganders a lesson and deterring other states from copying their example?

The reduced cost of cleaning up public litter and the reduced landfill costs of beverage containers (there was little recycling or incineration at that time) are easy to evaluate. They amounted to about $5 per Michigander per year.[13] The annual willingness to pay of the average Michigander for the reduced "eyesore" of litter stock we will call x for now and think about later.

The higher money and inconvenience cost of beverages induced some consumers to reduce their consumption—beer in deposit-containers declined by 8% and soft drinks by 5% in the year after the introduction of the law. The consumer-surplus loss was about $2 per Michigander a year.

Those who continued to purchase beer and soft drinks paid on average 7.5 cents more per filling as a result of the new law. If all of this represented higher resource costs owing to the need to specially label Michigan containers and collect and transport the containers back from the consumer to the distributor, then this 7.5 cents is a social cost of the law. To the extent, however, that this price increase did not reflect higher cost, then the 7.5 cents was simply a transfer from consumer to

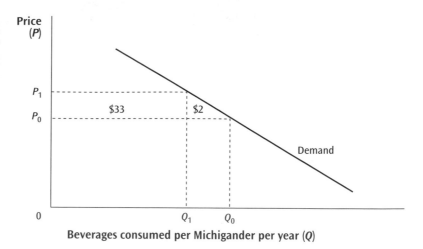

Figure 6-2

Demand for
Mandatory-Deposit
Beverages

Price (P)

P_1

P_0

$33

$2

Demand

0

Q_1

Q_0

Beverages consumed per Michigander per year (Q)

producer and should not be counted as a social cost. The fraction of the 7.5 cents that represented real cost increases we will call m for now and think about later. The social cost of the more complex delivery system necessitated by the mandatory deposit law was about $33m$ per Michigander per year.[14]

The costs estimated in the previous two paragraphs ($2 and $33m$) can be visualized in Figure 6-2. The mandatory deposit law pushed beverage prices up from P_0 to P_1, causing beverage consumption to fall from Q_0 to Q_1. The consumer surplus loss of those who were induced to reduce beverage consumption is the triangle labeled $2, and the increased beverage expenditure of those who continued to buy is the rectangle labeled $33. Of that $33, a fraction $(1 - m)$ is simply transfer from consumers to producers, but a fraction (m) represents real resource costs of meeting the new delivery requirements. So the social cost on this score is $33m$.

Finally, the inconvenience cost to the consumer of collecting empties and returning them to the store needs to be given a monetary value. For some consumers—especially in that pre-curbside-recycling era, this "cost" was negative as they appreciated a chance to recycle their waste. For no consumer was the personal inconvenience cost greater than 10 cents per filling, because those consumers would not in fact return their containers.[15] The fact that other states' 2–5 cent deposits have achieved about the same return rate as Michigan's 10-cent deposit suggests that the marginal inconvenience cost curve is nearly vertical in the 2- to 10-cent deposit range. These considerations lead to an estimate of beverage consumer inconvenience of about $5 per Michigander per year.

This estimated $5 cost can be visualized in Figure 6-3. The area marked δ represents the negative cost (i.e., benefit) of those who would previously have been willing to pay for the opportunity to return empty beverage containers for recycling.[16] And the area marked γ represents the positive cost of those for whom the return was an inconvenience. The marginal cost of net inconvenience is area γ minus area δ. Assume kinked straight lines, as shown in Figure 6-3. And assume, fairly conservatively given other states' experiences, that 80% of the beverage containers would have been returned at a deposit of 2 cents. Then the positive and negative areas to the left of $0.8Q_1$ are each equal to δ and exactly cancel each other. The rest of area γ is about $5 per Michigander per capita of net inconvenience of returning deposit containers.[17]

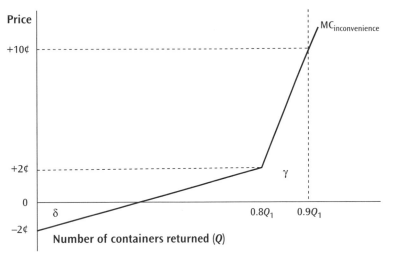

Figure 6-3

The Inconvenience
Cost of Returning
Containers

These social costs and benefits of Michigan's mandatory deposit law add up to $(5 + x - 2 - 33m - 5)$, which is positive or negative as $(x \gtrless 2 + 33m)$. You can now see why informed people still argue about the success of the Michigan mandatory deposit system. If you think $m = 0$, then the average Michigander had only to be willing to pay about \$2 a year for the dramatic reduction in litter in the state's public places. If $m = 0$, the law almost surely passed its benefit–cost test. If, however, the price increases all represented resource cost increases (i.e., $m = 1$), then the WTP had to be greater than \$35 per Michigander per year for the system to pass a benefit–cost test; if $m=1$, the law almost surely did not pass the test.

In short, a social benefit–cost analysis of mandatory deposits did not—and still does not, looking at more recent similar analyses—yield a clear-cut outcome. It is not surprising that both proponents and opponents of bottle bills have long disagreed. Especially when the economics is so inconclusive, emotions and self-interest get free reign. But polls in deposit states consistently show that a majority of the residents support the policy. And perhaps more significant, no state that has adopted mandatory deposits has ever reversed its position (Brugger 2000).

Notes

1. The only country where steep fines for littering *do* seem tolerable is Singapore—often called a "fine" city. First-time littering offenders face a fine of roughly \$500, and for repeat offenders, a fine of \$1,000 *and* a Corrective Work Order, which requires a few hours picking up litter in a park. Moreover those on such a work order must display their status by wearing special brightly colored jackets and sometimes endure local media coverage. Litter is *not* a problem in Singapore (http://www.expatsingapore.com/general/law.htm).

2. A proportional fine is one that takes the same percentage of people's incomes whether they are rich or poor. If everyone earned only wage income and worked 40 hours a week, then a fine of "two hours" would be a 5% tax for everyone, regardless of their wage rates. This simple demonstration ignores differences in work-weeks and in nonwage income, which is the reason for the phrase, "more nearly," in the text.

3. Recall the definitions of "impact" and "incidence." The impact of a fee is on the person who actually pays over the money. The incidence of the fee is on the consumer to the extent

that the consumer's price rises and/or on the producer to the extent that the producer's price falls as a result of the fee.

4. Although we have not yet come to recycling, there is a side benefit to mandatory deposits that must not be overlooked here. Because mandatory deposits require a special, separate re-collection system, what is re-collected is so homogeneous and hence marketable that it practically demands recycling, rather than costly landfill or incinerator disposal. The total *net collection cost* of mandatory deposits has three elements: *plus* the excess of the cost of collecting the item through a special mandatory deposit collection system over the cost of regular MSW collection; *minus* the cost of the landfill to which the mandatory deposit item no longer goes; and *minus* the revenue gained from the sale of the now recyclable item.

5. That the deposit and the redemption values should be the same is not quite accurate (Dobbs 1991; Fullerton and Wolverton 2000). Even if the product is properly returned, there is still a cost to dispose of it (whether by landfilling or recycling), so the redemption value should be less than the initial deposit by the amount of the cost of proper disposal. More and more, however, proper disposal means recycling, and if the cost of recycling is roughly offset by the value of the recovered material, then the *net* cost of proper disposal is zero. In that case, the initial deposit should indeed equal the redemption value (Ackerman et al. 1995; Dinan 1993).

6. These nine are Vermont, Maine, Michigan, Connecticut, Iowa, Massachusetts, Delaware, New York, and California; plus *one* city—Columbia, Missouri (and eight Canadian provinces). Actually, Vermont was the first state to pass a beverage container law. In 1953, the state banned the use of nonrefillable containers because farmers complained that their litter damaged cattle and machinery. Of course, such a ban did nothing to reduce container litter, and the law was allowed to lapse a few years later (Bingham et al. 1989). Exactly which beverage containers are covered by mandatory deposits varies from state to state. All include beer and carbonated soft drinks; some include mineral water; others include wine coolers; Maine includes juices and tea; Delaware exempts aluminum containers from its mandatory deposits (U.S. EPA 2001b). For state details, see http://www.bottlebill.org/USA/States-ALL.htm; for Canadian province deposit system details, see http://www.geocities.com/RainForest/Vines/6156/cdndepos.htm.

7. Some states originally had 2-cent deposits, but all now have 5-cent deposits (on the basic container of 10–16 ounces), except for Michigan (10-cent deposit) and California (2.5-cent deposit). (These are the deposit rates on the smallest containers; in many states, the deposit is larger on larger containers.) In Europe, deposits are comparable and get comparable return rates. It is interesting that in Sweden, a private company is empowered to operate the system of deposits on aluminum cans and can set any deposit it wants, provided it gets at least a 75% return rate; it sets a 6-cent deposit and gets back 85% (Bohm 1994).

8. Under the mandatory deposit law of the province of New Brunswick, Canada, consumers get back the full 10-cent deposit if the container is refillable, but only return half of it if the container is not refillable (i.e., plastic, can, or one-way glass bottle). But this difference had little effect on the trend to one-ways—while the law has been in effect, refillable containers have gone from a few percent of the total to virtually zero percent (Ezeala-Harrison and Ridler 1994).

9. One study concluded that recycling beverage containers under a mandatory deposit system costs $320 a ton, whereas the average cost of all other recycling programs is only $120 a ton (Flynn 1999). Another study found that whereas mandatory deposits increased the volume of recycled containers by about a third, the collection cost of the marginal recycled tonnage was nearly $1,200 a ton (converted to 1997 dollars; Criner et al. 1991).

10. Optimally, if the deposit equals the marginal benefit to society of correct (rather than illegal) disposal, then scavenging should expand to the point where the deposit equals the marginal cost of the effort. But in such a commons, scavenging will expand to the point where the deposit equals the average cost. Because MC is rising, AC < MC, and there is too much scavenging effort.

11. California more or less copied the long-operating British Columbia (BC) system for beverage deposits, except that in BC the redemption centers, the transport, and the recycling are all operated by a private firm whose costs are covered by unredeemed deposits, recycling revenues, and a fee per container levied on beverage manufacturers (Truini 2000a). The BC fee varies with redemption rates, gasoline prices, and revenues from recyclable materials to ensure that the firm's costs are covered.

12. Older readers may remember how serious the U.S. litter problem was a half-century ago, when we were just beginning to get used to the packaging proliferation. (That litter gave rise to the quote at the start of this chapter, a quote that would not be written today.) For those with weaker or nonexistent recollections of 50 years back, visit a country whose gross domestic product per capita is now moving through the $5,000–10,000 level—you will find it "litterly" a roadside mess. It seems to take a generation for states to put out and maintain enough trash barrels and for people to learn to use those trash barrels.

13. The original analysis made all the estimates in 1979 dollars (Porter 1983b). All those figures have been simply converted to 1997 dollars (by means of the GDP deflator) and rounded. Where ranges were given in the original study, only the midpoint of the range is given here. Otherwise, the analysis has not been updated. But the basic findings of this study have only been strengthened by the many others that have followed.

14. Recent studies suggest that this 7.5 cent increase is too great (Criner et al. 1991; and Ackerman et al. 1995, who estimate the social cost to be 3.0 cents per container, when put into 1997 dollars); the Michigan Grocers Association estimates the cost to be 5.5 cents per container (Mathis 2001); a report by Michigan Consultants (1996) estimates the cost to be 5.8 cents per container. I am tempted to set $7.5m = 3.0$ or $7.5m = 5.5$ or $7.5m = 5.8$ and estimate that m is in the 0.4–0.8 range; but I am also tempted to think that this is a measure of the technological improvement in bottle distribution and re-collection in deposit states that has occurred during the past two decades. For example, when Michigan adopted its mandatory deposits, breweries would stop the filling lines to substitute special Michigan labels, which differed only in that they indicated the state and the deposit, whereas today many beverage sellers simply list all the deposit states and their deposits on all their container labels.

15. People whose personal cost of returning containers exceeds the deposit simply throw the container away, as litter or trash. To the extent that these containers are scavenged by low-wage people, it is the lower cost to the *scavenger* of returning containers that counts as a social cost, not the much higher cost to the person who did *not* return the container.

16. Q_1 is the number of containers sold in 1979 (i.e., 390 deposit-bearing containers per capita, about half each of beer and soft drinks). About $0.9Q_1$ of these containers were returned.

17. This rest of γ consists of a rectangular trapezoid of area, $[(0.9Q_1 - 0.8Q_1)(0.10 + 0.02)]/2$. The resulting figure is doubled to put it into 1997 prices (although one could well argue that it should not be so doubled, because the Michigan container deposits have not changed during this period).

7

Exporting and Importing Waste

Under a system of perfectly free commerce, each country naturally devotes its capital and labour to such employments as are most beneficial to each....
By stimulating industry, by rewarding ingenuity, and by using most efficaciously the peculiar powers bestowed by nature, it distributes labour most effectively....
It is this principle which determines that wine shall be made in France and Portugal, that corn shall be grown in America and Poland, and that hardware and other goods shall be manufactured in England.

—David Ricardo, *On the Principles of Political Economy and Taxation*, 1817

One of the few things on which almost all economists fully agree is that free trade is a good thing. It is now more than 200 years since Adam Smith and David Ricardo first formulated the theory of comparative advantage and showed that citizens of a region will improve their lot by specializing in doing what they do best and trading for what they do less well. At first glance, there appears to be no reason for treating waste any differently. Waste, of course, is a "bad," not a "good," and hence a region has to pay to get another region to take its waste "exports" off its hands. But that does not affect the argument. Regions that dispose of waste well can specialize in waste disposal and use the proceeds to import the things they do less well. One purpose of this chapter is to ask why many people who normally think trade is a good thing want to make an exception of waste. (Of course, people who think free trade is a bad thing are perfectly consistent in their opposition to trade in waste products.)

On a more practical level, it is essential that all waste be traded, in the sense that it must be moved from the place where it is generated—usually a place of high population density—to a place where it is less harmful to human health and the environment. Perhaps there was a time when it was sufficient that each household dumped its own slops over the city walls, but no longer. Aesthetics, scale economies, and cheap transport have made more distant disposal feasible and desirable. Transporting waste *is* trading waste. *All waste is traded.*

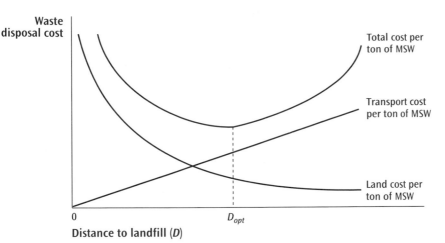

Figure 7-1

Optimal Distance
to a Landfill

Waste disposal cost

Total cost per ton of MSW

Transport cost per ton of MSW

Land cost per ton of MSW

0   D_{opt}

Distance to landfill (D)

How far one transports wastes, to say a landfill, in essence depends on the relative prices of land and transport. In Figure 7-1, land prices fall as one moves outward from densely populated cities, where large volumes of trash accumulate. Transport costs, conversely, rise roughly proportionately to the distance. Figure 7-1 also shows the total cost per ton of landfilling waste, which is the sum of these two (land and transport) costs. The optimal transport distance is the distance at which the total cost per ton is minimized (D_{opt}). (Unlike in most economic diagrams, here the optimal is *not* found where the two curves cross; rather it is found where the marginal decline in the land price—in absolute value—equals the marginal increase in transport cost.) Anything that raises urban land prices or reduces transport costs increases D_{opt}. In fact, in recent U.S. history, urban (and suburban) land prices have been rising relative to rural land prices, and transport costs have been falling. The result, in fact as well as in theory, is that trash gets carried ever further from its birthplace to its resting place. On a planet honeycombed with nations, states, and counties, economic forces urge ever more intercounty, interstate, and international trade in waste.

Nearly 10% of U.S. municipal solid waste enters interstate trade (Repa 1997; Duff 2001c). It is not surprising that New York, Illinois, and New Jersey together generated nearly half of interstate MSW exports in the United States (in 1997). But it is perhaps surprising that nearly half the MSW imports were not into the sparsely populated far western states but into Pennsylvania, Virginia, Indiana, and Michigan (McCarthy 1998). Why states so near the Northeast? Transport costs have been falling—thanks to lighter trucks and super-compaction as well as low fuel costs and low truck taxes—but transport costs are still important to the disposal distance decision (Bader 1999b).

Think for a moment about who gains and who loses when transport costs become low enough that one region can begin to export its waste to another region, one with significantly lower landfill costs. First of all, the tipping fee at its own landfills will fall, because these nearby landfills have to compete with cheaper distant landfills. As landfills fill up, there is no need to replace them if exporting is feasible and cheap. Local landfill owners are hurt, of course, but all the waste-generating

Transport Costs and Waste Trade

How expensive is waste transport? One study carefully estimated the private and external costs of transporting New York City's waste (Konheim and Ketcham 1991). The dollar costs per mile of truck transport (converted to 1997 dollars) were:

Private costs

Truck depreciation	0.22
Insurance (on own truck)	0.07
Highway truck taxes	0.05
Tire, fuel, labor, and maintenance	1.19
Total	1.53

External costs

Congestion and added highway wear	0.34
Uninsured traffic accidents	0.04
Noise, air, and other pollution	0.05
Total	0.43

The total social cost estimate, about $2 a mile, is one way, so if it comes back empty at about the same cost, the cost per one-way mile is doubled. The typical truck carries 22 tons, so the cost per ton-mile ends up at $0.18. (These data have two qualifications: Highway truck taxes were used as a first proxy for road wear; and the study correctly assumed that highway wear was many times larger than highway taxes.)

The study then goes on to calculate how far New York City should send its MSW. The following table shows their estimates of the total MSW disposal dollar cost per ton for varying shipping distances:

Miles from New York	Landfill	Transport	Total
100	72	18	90
200	54	36	90
300	36	54	90
400	34	72	106
500	29	90	119

What these estimates suggest is that MSW transport costs are low enough to allow New York to truck its MSW up to 300 miles away without raising its overall cost; 300 miles means central Virginia, eastern Indiana, or western Pennsylvania. Beyond 300 miles, it is usually less costly to ship waste by rail (according to a personal communication with Robert Wallace of Wastebyrail.com). If plans for rail hauling its trash are not derailed by New Jersey officials, New York City might send its trash by rail as far as South Carolina and Georgia, more than 700 miles away (Geiselman 2001a).

businesses and households in the region benefit. It is easy to prove that the waste generators' gains outweigh the landfill owners losses. In Figure 7-2 are shown the local supply curve of landfill space, S, and the local demand curve for landfill space, D. Price without waste trade would be P_0 and quantity Q_0. When transport costs fall sufficiently that distant landfills offer a lower price (including transport cost), say P_1, then local landfills must meet that price. The total trash generated goes up (to Q_1) but the quantity landfilled locally goes down (to Q_2), with ($Q_1 - Q_2$) becoming exported. Local landfill owners lose rents equal to the area α, but local businesses and households gain consumer surplus equal to the larger area ($\alpha + \beta + \gamma$) (Ley et al. 2000).

Moreover, waste-importing regions also gain from the introduction of waste trade. Your intuition might be telling you that, if one region gains, the other region must lose, so let's also go through that proof. Figure 7-3 shows the local supply curve of landfill space, S, and the local demand curve for landfill space, D. Without waste trade, price and quantity would be P_0 and Q_0. When trade begins (say because transport costs fall), new out-of-state demand for local landfill space appears, pushing local landfill prices up, say to P_1. Total trash goes up to Q_1, with Q_2 being local trash and ($Q_1 - Q_2$) being imported trash. Local trash-generating businesses and households lose consumer surplus equal to area ($\alpha + \beta$), but local landfill owners gain rents equal to the larger area ($\alpha + \beta + \gamma$).

Both regions gain from the waste trade, the exporting region *and* the importing region! So how could anyone be against either exporting or importing waste? There are reasons—some hints now, and then we will explore the source of the opposition to trade in detail. Looking back at Figure 7-2, we see that with a lower trash disposal price, which trade

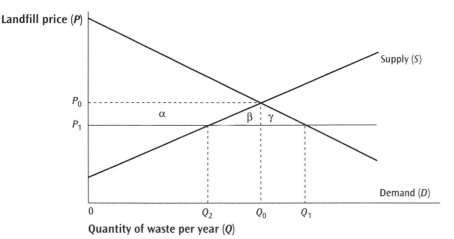

Figure 7-2

Gains and Losses from
Exporting Waste

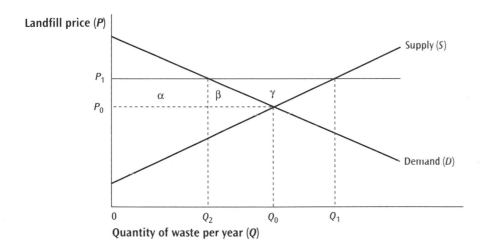

Figure 7-3

Gains and Losses from
Importing Waste

brings about, more trash is generated in the exporting region, and the incentive to recycle is reduced. People there who see these outcomes as backsliding will want to stop exports. And looking back at Figure 7-3, we see that with the new imported waste, more trash is buried in the importing region. People there who see health hazards to trash will want to stop imports.

But attacking waste trade is a very second-best response to these worries. If reducing and recycling are good because their social benefits exceed their private benefits, then subsidizing reduction and recycling is the best way to handle these externalities. If waste burial is bad because its social costs exceed its private costs, then taxing waste is the best way to handle that—taxing waste, not waste imports. Moreover, opposition to trade arises from the fact that, although there are net gains to the population in total, not everyone gains from this trade. Landfill owners in the trash-exporting region will be against trade, and trash generators in the importing region will be against trade. Now, let's look more closely at these arguments.

Restrictions on Importing Waste

If the real resource costs of disposing of waste are higher in region A than in region B (and higher by more than the transport costs between them), then the total disposal costs will be lower if the waste of region A is sent to region B for disposal. If region A pays a price lower than its own disposal cost but higher than region B's disposal cost, both regions will be made better off by the trade. If there are no market failures such as monopolies or externalities in the disposal of waste, there is no reason for opposition to such trade.

In fact, however, with waste there are market failures, and these give rise to resistance to waste imports. The market failure that affects attitudes toward waste trade involves the safeguarding of the waste (whether in landfills or incinerators). Consider Figure 7-4, which shows the marginal social cost and benefit of different degrees of safeguarding. To keep it simple, think of safeguarding as just the expected cancer deaths prevented through groundwater protection in landfills or air pollution prevention in incinerators. The MSC of ever greater safeguarding is ever more costly, because the low-cost safeguards are undertaken first. The MSB of safeguarding is constant, depending on how we value the expected cancer deaths prevented. The optimal safeguarding is S_{opt}.

Suppose, for simplicity, that this is a landfill and that all the waste is imported from other regions. This region will earn a revenue (net of all landfill costs other than safeguarding) of R for accepting this out-of-region waste. Does any $R > 0$ mean that this region should happily accept the waste? Certainly not.

First, consider the worst case, where the safeguarding is totally inadequate (i.e., $S = 0\%$). In this case, the dollar value of the expected cancer deaths will be the sum of the three areas, $(\alpha + \beta + \gamma)$. If we consider the region as a single entity, then the waste should be accepted only if $R > (\alpha + \beta + \gamma)$. But, if R goes to the landfill owner, possibly not even a resident of the region, and the cancer deaths are suffered by residents of the region, then the net benefit to the residents of the region may be as low as *minus* $(\alpha + \beta + \gamma)$.

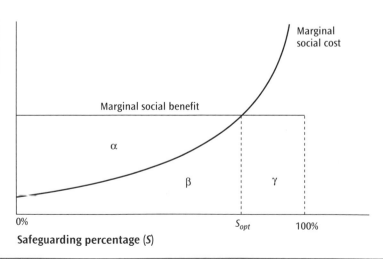

Figure 7-4

Optimal Safeguarding in Waste Disposal

Suppose now that it is possible to constrain the landfill to safeguard optimally (at S_{opt}). Now the profit of the landfill operator is $(R - \beta)$ because the operator must incur the costs of safeguarding. The operator will be anxious to accept the waste if $R > \beta$, but for the region as a whole, the waste should only be accepted if $R > \beta + \gamma$. And again, if the landfill operator is not a local resident, the net benefit to the residents is still negative, *minus* γ. In short, it is conceptually possible that the residents of a region are made worse off by market-driven waste imports into the region.

Where a landfill serves local waste needs *and* imports waste, there are many reasons why the local residents might want to treat the imported waste less favorably than their own local waste. Consider three:

- The municipal landfill may have been deliberately underpriced (i.e., $P < \text{MSC}$) to discourage illegal dumping by locals. In this case, importing waste at this same subsidized price is foolish from the local viewpoint—though it also may discourage illegal disposal, it does so elsewhere and therefore provides no local benefit. Banning waste imports, or at least charging such imports the full MSC, may be good second-best policy (Copeland 1991).
- Even where the municipal landfill prices the local waste correctly at MSC, if the landfill has any monopoly power at all in its waste import market, local welfare is enhanced by charging a higher price for imported waste (i.e., the price where the *marginal revenue* from imported waste equals MSC). Charging out-of-towners or out-of-staters a higher price for services makes sense—from the strict town or state viewpoint—and has a long and unquestioned history in such state services as parks, hunting and fishing licenses, and university education; and extending this to imported waste is logical, albeit parochial (Porter 1982). (This argument raises a logical question: Isn't the landfill even more likely to have monopoly power over local waste? Of course it is, but a properly priced municipal landfill would not exploit its own residents by charging a price above the marginal social cost.)
- If the local landfill is privately owned, its price will be higher and/or it will fill up and close sooner if waste is permitted to be imported from other regions. Even if the local citizenry do not fear external physical damages from imported waste, they may fear monetary damages (Ley et al. 2002).

There are probably even more reasons why people might want to ban—or at least charge a higher price for—waste that is imported from other regions. However, the U.S. Supreme Court has almost always found such bans or price discriminations to be violations of the Interstate Commerce Clause (Article I, Section 8) of the U.S. Constitution (U.S. EPA 1995b). The court's decision is consistent with a national view of welfare—if waste is unwanted locally, then local and imported waste should be equally unwelcome (and hence equally priced).

Local discomfort with imported waste, because of real or wrongly perceived risk, is sufficiently great that there are constant efforts to get around this Commerce Clause deterrent. Some are pernicious; five currently popular proposals are particularly so:

- The U.S. Congress could explicitly permit states to control their own interstate waste trade. The Congress rarely gives this right to states, but it could. The "lower 48" states would then create a jumble of regulations on what kinds of waste and what volumes of waste could pass through or be landfilled in each state. This would dramatically increase the cost of waste disposal, either through big increases in the costs of transporting waste or through the increased landfilling of waste in inappropriate or costly places.

- The states could take advantage of the so-called market participation exception to the Commerce Clause—if the state itself participates in the relevant market, it can discriminate in favor of its own citizens, although the resulting burden on interstate commerce must not be significant (Walls and Marcus 1993). (It is this exception that permits the discriminatory pricing of parks, universities, and the like, which were mentioned above.) Public ownership of landfills might qualify them for this exception. But suppose *both* that private landfills are socially cheaper to operate than public landfills *and* that extensive trade in waste is the best way to handle waste. Then if states take advantage of this exception, there are *two* inefficiencies being perpetrated (Podolsky and Spiegel 1997).

- On grounds of health or security, which the Supreme Court recognizes as legitimate causes of interference with interstate trade, the states could limit the ways in which waste may be transported to landfills, carefully eliminating those ways that are favored by waste traveling long distances. For example, Virginia periodically considers banning barge transport of waste specifically to deter the import of New York City waste into Virginia landfills (Timberg 1999). This could cause New York to choose either socially more costly landfills elsewhere or socially more costly transport methods (e.g., heavily subsidized and air-polluting semis on the heavily congested and accident-prone highways between New York and Virginia). For each ton of waste transported, barges use one-ninth the fuel and emit one-seventh the air pollution of trucks (Timberg 1999). Elizabeth, New Jersey, has tried to reduce imports of trash from New York City by stopping, strictly inspecting, and fining (for such things as cracked taillights) trash-hauling semis (Brown 1999b). As long as *all* semis are equally harshly inspected, this is presumably legal, but it is obviously a socially costly way to deter New York trash (assuming that cracked taillights were optimally penalized before). If it induces New York to circumnavigate New Jersey on its way to Pennsylvania, it clearly adds to overall social costs.

- Specific items could be forbidden in a state's landfills, which means that states that routinely landfill those items must begin to sort them out if they want to use the banning state's landfills. Massachusetts, for example, bans yard waste, batteries, tires, white goods (i.e., large appliances), metal, glass, plastic, and paper from its landfills and incinerators (U.S. EPA 1999b). This of course puts an equal burden on the banning state using its own landfills—except in cases where the banning state is already fully recycling the banned item. For example, states with mandatory deposits on beverage containers could ban the landfill disposal of such containers, because they already have nearly 100% recycling of them. This is not just theory: Iowa has done this, and Michigan is considering it (Truini 2001e). It is a clever way of preventing imports, though from a national viewpoint not an efficient move.

- Onerous landfill regulations will push up landfill tipping fees and drive potential imports away—indeed, they may create *exports* of solid waste. For example, Toronto is rapidly filling up its only landfill and has banned the construction of new landfills, which means that soon all its waste must be incinerated or exported (Brown 1999a). Excessive regulation for extraneous reasons cannot be the best policy choice. All such policies do is raise overall waste disposal costs in an effort to achieve an ill-advised goal.

There is, fortunately, one very sensible way to handle all this. Recall that the basic reason for the NIMBY (i.e., not in my backyard) attitude about waste imports is that residents of the area around the landfill or incinerator will bear uncompensated external costs (even if the operation is optimally safeguarded). The landfill or incinerator should pay a host fee to the neighboring community high enough for these residents to feel adequately compensated for the discomfort of living near potential environmental problems. (In the language of Figure 7-4, the host fee would have to be at least as high as γ, the cost of the expected residual cancer danger despite optimal safeguarding.) This host fee becomes a part of the costs of the disposal operation and becomes incorporated into a single tipping fee for all, both local waste and imported waste. All waste, after all, equally creates the expected environmental damage, so there is no reason for not charging all waste the same price. (An example of such a host fee is offered in the Appendix on page 116.) By the way, notice again in Figure 7-4 that when the landfill operator optimally safeguards, at a cost of β, *and* pays the host fee that compensates neighbors for the residual cancer risk, at a cost of γ, the landfill will only operate if revenue R is greater than $\beta + \gamma$.

Although all waste pays the same fee, not all regions make the same *net* expenditure (per ton of waste) for waste disposal under host fees. For the host community of the landfill or incinerator, the net cost is the waste payment minus the host fee. Distant waste producers pay in cash only, whereas neighbors pay not only in cash but also in potential future health

Bidding to Determine Hosts and Host Fees

Consider five neighboring communities that could each save A dollars a year if they could build their own landfill. Economies of scale dictate that they must get together and plan only one landfill for all of them. The problem is where to site it—because each community would prefer it to be sited somewhere else. Let us assume that each of them values this new landfill sufficiently to be more than willing to provide the home for the landfill a fifth of the time—they just don't want it all the time. But the new landfill, being stubbornly immobile, must be sited permanently in just one of the communities.

Suppose each of them bids an amount that it would have to be paid if it received the landfill. The low-bidding community would be selected for the landfill site and would receive the amount of its bid each year as compensation. Each of the other four communities would be taxed one-fourth of its bid each year to gather the money for the host fee. Notice the basic bidding dilemma each faces: Bid too low, and you are likely to get the site and receive low compensation; bid too high, and you will probably avoid being chosen for the site but will make yourself liable for high tax payments.

Experiments have shown that most people in this situation choose a "maxi-min" strategy—that is, the bid that maximizes the value of the worst possible outcome (Kunreuther et al. 1987). Consider a community that would be willing to accept the landfill for a compensation of B dollars a year (or more). It bids X dollars. If it gets the site, it will net $(X + A - B)$ each year; if it does not get the site, it will net $(A - X/4)$. Its maxi-min strategy requires choosing a bid that equates these two outcomes. The maxi-min bid equals $4B/5$ and guarantees the community a net gain of $(A - B/5)$, whether it gets the site or not—and notice that $(A - B/5)$ is the net gain this community would have gotten if it had hosted the landfill every fifth year, something that is technically impossible but also something we assume it would be willing to do, or else it would simply not participate in this five-community effort to site a joint landfill. Any other choice of X opens up the possibility of a smaller gain, or even a possible loss.

This bidding process gets the landfill into the community with the least dollar distaste for it. Each community pays one-fifth of its dollar distaste, four of them by being taxed that amount and one of them by receiving a host fee of only four-fifths of its dollar distaste. Each of the five is better off with the new joint landfill. The joint landfill authority that handles these taxes and host fees even comes out a bit ahead—monies it can put into, say, education about source reduction.

and environmental damages. The host fee recognizes that and thereby reduces the cash part of the payment. (Ideally, the host fee should be a lump sum or so much a year, and not based on the volume or revenue of solid waste, lest it provide the neighbor community with an incentive to overgenerate trash.)

The correct host fee is elusive. Nevertheless, such reimbursement is being offered more and more. Generous host fees can greatly reduce local opposition to the siting and operation of incinerators and landfills. Because local opposition means, for the facility, the loss of time and the expenditure of legal resources, host fees can even add to profit (Simon 1990). Although a benefit–cost analysis seeks to find out if the benefits of a facility are great enough that everyone's costs *could* be compensated, we can only be *sure* that the social benefits of a solid waste disposal facility outweigh the social costs if everyone *actually is* compensated. The successful negotiation of a host fee with the neighboring community at least guarantees that a majority of the neighbors are in fact better off. (We will return in Chapter 14 to questions of environmental injustice, which poor and/or minority communities suffer when they are systematically burdened by proximity to waste disposal facilities and are uncompensated for the resulting health risks.)

Wisconsin's Host Fee Law for Landfills

Wisconsin has a law requiring that a landfill applicant negotiate with local citizens about social and economic issues raised by the landfill. Host fees are part of this negotiation. Which properties qualify for compensation and for how much are negotiated after an appraiser estimates the property value with and without the landfill. Recall that, in principle, the change in the property value measures the capitalized value of the WTP for the nonlandfill location. Compensation can also be made for impacts on the quality of life through noise, traffic, and ecological destruction.

If the local negotiating group and the applicant are unable to reach agreement, the state's Waste Facility Siting Board serves as arbitrator. The local negotiating group and the landfill applicant each submit a final offer on compensation and operating methods, and the board selects the one it considers the fairer of the two offers—an arbitration technique well known to those who follow baseball salary negotiations. Of the first 55 such negotiations, only 3 failed to reach agreement and were submitted to the board for arbitration (Bacot et al. 1998).

Restrictions on Exporting Waste

It may seem contradictory, but at the same time as some regions are trying to prevent the importing of waste, other regions are trying to prevent its exporting. Why? The answer, in short, is that some communities get themselves into the position of needing waste to fulfill an obligation. These obligations are of two kinds, incinerators and recycling facilities.

In the late 1970s and early 1980s, when it was feared that we "were running out of landfills," especially in the Northeast, many municipalities and groups of municipalities decided to turn to incineration. Future solid waste flows were estimated, and incinerators of the apparently appropriate size were built. The large initial capital costs of the incinerators were usually financed by municipal bond issues, whose amortization and interest costs were to be covered by high tipping fees at the incinerators. Recall that incinerators need to operate continuously at full capacity to run at low cost. But the high tipping fees sent solid waste generators looking for more distant waste disposal opportunities, and the growth of recycling in the 1980s meant that the earlier estimates of the waste flow were too high. For many communities caught in this bind, facing excess capacity at the incinerator and an inability to meet their bond obligations, the answer was sought in a requirement that local solid waste be delivered to the incinerator, called "flow control."

Flow control also became popular in the 1980s, as recycling grew, as a means of keeping recycling costs down and/or expanding recycling services. High tipping fees at the landfill and recycling facility meant high revenue to cover the high recycling costs and to expand recycling services into ever more unprofitable areas. But flow control was then needed to prevent local waste generators from sending their waste to other landfills or recycling facilities.

Needless to say, waste generators were not happy with flow control ordinances. One indication of the added cost of flow control is the extent to which tipping fees at landfills and incinerators come down whenever flow control is ended—in several New Jersey counties, the tipping fees fell \$42–61 per ton when the flow control laws were rescinded (Greenwire 1997b; Smothers 1997).

In any case, flow control is now a thing of the past. Consistent with its decision on controlling inflows of waste, the U.S. Supreme Court found in 1994 that restrictions on waste outflows were also in violation of the Interstate Commerce Clause of the Constitution. Essentially, the court has decided that waste flows are like any other export or import; that states may not restrict them without good health or security reasons; and that, even then, states may not discriminate between their own and other states' wastes.

Let's look at the economics of flow control. Consider first a typical (say municipal) incinerator designed to handle the expected flow (in tons per year) of municipal solid waste to the incinerator (W_{incin}). Its operating costs will be covered by electricity sales, but it has huge initial capital costs (K) and hence high annual interest costs (iK), where i is the interest rate. To cover those interest costs, the tipping fee is set at iK/W_{incin} per ton at the incinerator, and a flow control law prevents local waste generators from taking advantage of the fact that waste disposal costs only P_{landf} (including transport costs) at some out-of-town landfill, and $P_{landf} < iK/W_{incin}$. Assume, realistically, that the incinerator cannot be dismantled and sold for anything like the K it cost—essentially, the incinerator is a sunk cost.

Flow control seems to make sense in this situation. The marginal social cost of using an out-of-town facility is P_{landf} whereas the net MSC of the incinerator is zero because enough electricity is produced to cover operating costs. Each ton of exported waste has a social cost of P_{landf} (not to mention that the exported waste endangers the city's ability to meet its bond obligations). Flow control prevents that wasteful export.

But flow control is *not* the best way to prevent that export. The best way is to lower the incinerator tipping fee to P_{landf} (or slightly below it) and cover any needed bond obligation revenues from the city's general fund. This can be seen by examining Figure 7-5. Start with the flow control restraints, with W_{incin} of waste going to the incinerator at a tipping fee of P_{incin}. Now lower the incinerator tipping fee to P_{landf}; the city's waste grows to W_{landf}, with W_{incin} being accepted at the local incinerator (that's its capacity), and the rest being exported. Consumers are better off by the area ($\alpha + \beta$), with α being the savings on the waste sent to the incinerator and β being the consumer surplus gain on the new waste that is exported. The incinerator is now losing an amount that is equal to the area α, and this must be made up by the city from its general fund budget. But the city's residents gain more on the waste disposal side ($\alpha + \beta$) than they lose as city taxpayers (α). The pricing switch means a net gain of β.[1]

Why did municipalities prefer flow control to marginal cost pricing? There are

Figure 7-5

Deadweight Loss of
Flow Control

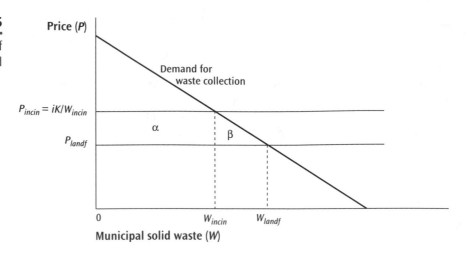

three reasons. First, the deadweight loss of taxation is not zero, and raising monies for the general fund to cover incinerator losses would also involve deadweight loss—plus political cost by making the citizenry explicitly aware of the high cost of their incinerator. Second, Figure 7-5 treats the citizenry of this city as a homogeneous batch of people, each of whom generates the same waste and pays the same taxes as everyone else. In fact, flow control most heavily penalizes businesses, while general fund taxation most heavily penalizes households. That higher business costs get passed on in the form of higher local prices for business services goes unnoticed in the debate between flow control and general fund finance—but the higher property taxation necessitated by marginal cost pricing would *not* go unnoticed. Third, all this debate is occurring at a time when municipalities are committed to encouraging reduction, reuse, and recycling. Lowering the tipping fees at the incinerator discourages all of those things.[2]

Isn't encouraging reduction, reuse, and recycling a good thing? Yes, up to a point. These alternatives to incineration (or landfilling) should be encouraged as long as their net MSC is less than the MSC of the incineration (or landfilling). Once this huge sunk cost of an incinerator is built, and its net MSC goes near or to zero, one no longer wants to encourage the costly alternatives of reduction, reuse, and recycling. Does this mean that we should all build big incinerators and forget reduction, reuse, and recycling? No, it simply means that we should think carefully about building incinerators *before* we build them. Even if we make a mistake in building one, ex post flow control is not the best way to handle the awkward situation. In any case, it is no longer permitted.

International Trade in Waste

International trade in waste raises special questions because there are few international laws, courts, or legislatures to prevent damage in the course of the trade. However, the world does not worry much when one industrial country sends waste to another. And industrial nations send most of their exported waste, even hazardous waste—to be discussed in Chapter 14—to each other. Germany is the world's largest exporter of hazardous waste, and France is the world's largest importer of it (OECD 1997). What causes an "international trade in waste" problem is the exporting of waste—and especially highly toxic waste—from rich industrial countries to poor developing ones.

Economics, as we have seen in the preceding sections of this chapter, provides a strong presumption for free trade, international as well as intranational. Nevertheless, there may be good economic reasons for restricting the exporting of some kinds of waste from rich to poor countries under certain conditions:

- Poor developing countries recognize that they are incapable of distinguishing between waste whose importing would be beneficial and not beneficial. Especially if there is good reason to suspect that recipient poor countries are being unknowingly exploited—say by "sham recycling" that goes instead to a dump—then restrictions are one possible reaction.
- Poor countries recognize that their own governments are incapable of guaranteeing appropriate safeguards to the treatment or disposal of any imported waste. These countries have consistently worried about their ability to prevent "dirty recycling" that endangers the recycling workers' life and health (see the box titled "Every Day One Ship, Every Day One Dead").
- Exported trash may be dumped surreptitiously in oceans. Ocean dumping is particularly pernicious since it uses the seas as a commons, providing benefit (i.e., zero disposal cost) for the dumper at great cost to all other coastal countries. Accordingly, The London Dumping Convention, originally formed in 1972, prohibits, or controls through permits, the ocean dumping of many kinds of wastes. But outright banning of international trade in waste is a very second-best way of controlling ocean dumping.

"Every Day One Ship, Every Day One Dead"

The title of this box—a common saying in Alang, India, which now has the world's largest ship-scrapping yard (Noronha 1999)—captures a vivid tale of waste reborn. When a ship reaches the end of its useful life, after 25 to 30 years, it is scrapped. A ship is made of 95% recyclable, high-quality steel, which is worth about a million dollars. Unfortunately, the other 5% is mostly asbestos, arsenic, lead-based paints, PCBs, and other carcinogenic substances. Until recently, this "shipbreaking" was undertaken in capital-intensive manner in the drydocks of industrial countries. But today, more than half of the world's retired ships are rammed onto a beach in Alang, where they are dismantled by hordes of workers using only simple tools (Burns 1998; Langewiesche 2000).

Why the sudden shift? It is partly due to lower wages in Alang. But it is also the result of the ever higher costs of maintaining environmental, safety, and health standards in the industrial countries relative to the total absence of such standards in India (or more accurately, the total absence of enforcement of such standards). The pluses from India's viewpoint are unskilled jobs (40,000 on the beach at Alang), profit (said to be a half-billion dollars a year), and scrap steel (2.5 million tons a year). The minuses for India are the costs to the environment (toxic substances on the beach and adjoining waters), safety (fires, accidents, and explosions), and health (one worker in four is expected to get cancer from the work in Alang).

Again, what to do is not clear. U.S. ships now cannot leave U.S. waters until all toxic substances are removed, but exemptions are readily available if the removal is difficult, dangerous, or expensive. Yet requiring that we go back to breaking up U.S. ships in the United States does not seem to be the answer. Must every American be buried in the city of his or her birth?

This first of these three situations presumably is the argument that underlies the Basel Convention on the Control of Transboundary Movements of Hazardous Wastes and their Disposal of 1989, under which all exports of hazardous wastes are banned. (For a primer on the Basel Convention, see Krueger 1999.) The problem with the Basel Convention is that waste will not neatly divide itself into hazardous and nonhazardous categories. As a result, the Technical Working Group (TWG) of the Convention has been busy ever since deciding what can be exported from where and to where. So far, the working group has divided the world into six groups (OECD members or not, Basel signatories or not, and Basel implementers or not) and potentially hazardous wastes into three lists (green, amber, and red). Aside from such technical problems in categorizing wastes, there are two other ongoing Basel problems. The first is that what is considered hazardous in one country may not be considered so in another. And the second is that TWG bureaucrats have an asymmetric loss-function, with the result that they are always tempted to put too many items on the banned list. For example, recaptured copper destined for recycling found its way onto the list (Buchholz 1997).

To restate the obvious, not all waste exported from the United States and other industrial countries to the developing world is still waste when it arrives at its destination. Examples abound of rich countries' trash becoming poor countries' recyclables. For instance, in early 1995 Asia experienced shortages of polyester resin just as it was producing increased outputs of polyester textiles for export (Miller 1995b). Asian bidders entered the U.S. market for discarded and collected plastic soda bottles, pushing up the prices here as they bought more than one-fourth of the U.S. supply. Once in Asia, the bottles were sorted and cleaned in a labor-intensive process by small entrepreneurs and then converted into polyester-building chemicals; 25 bottles produce enough for one sweater. In 1988, nearly one-fourth of the waste paper collected by recycling operations in the United States was exported, mostly to Taiwanese, South Korean, and Mexican paper recycling mills, where the cheap labor permits more careful sorting and the resulting homogeneity gives the paper greater value.

But other parts of the rich countries' trash becomes, through exporting, poor countries' trash. Is this bad? Many poor countries produce little of their own trash and contain much excellent landfill area. At the risk of repeating Larry Summers' gaffe, we might say that poor countries are under-landfilled (*Economist* 1992). To prevent the exporting of trash to them is a close kin to protective tariffs that prevent the exporting of labor-intensive finished goods from poor to rich countries.

Of the triad—"reduce, reuse, recycle"—reuse by far gets the least emphasis. The environmental advantages of reuse are clear. Just because the utility of a durable good to its current owner has fallen to zero does not mean that no one else positively values the good. Secondhand markets in general—in particular, international secondhand markets in the present context—provide two possible social benefits:

Bourbon versus Rum

Bourbon whiskey is produced in the United States; rum is made in Barbados. The flavors are very different, for many reasons—including different raw materials (corn vs. sugar cane), different water, and different barrels. Why do the barrels matter?

Whiskey absorbs much of its flavor from the oak of the barrels in which it is aged. Once the whiskey has aged long enough to be bottled and sold, the barrels are waste, usable only as patio flower-stands or firewood. Barbados rum, however, gains its unique flavor by being aged in barrels that have been used to age bourbon. For instance, used Jim Beam bourbon barrels are disassembled, shipped to Barbados, reassembled, and used to age Cockspur rum (Cowdery 1995).

A law in the United States that tried to prevent waste export might end up killing Barbados' biggest export, or at least severely lowering its value among rum connoisseurs. And similar anecdotes could easily be reproduced a thousandfold.

They may add to the useful life of the product, or they may lead to recycling that would not otherwise occur. Both of these postpone the landfill or incinerator disposal of the residual product. Secondhand markets take the first two benefits into account—the transaction only occurs if the price is higher than the current owner's willingness to pay for it and lower than the prospective buyer's willingness to pay for it. But nowhere is the postponement of the cost of disposal taken into account. Again, the fact that disposal is unpriced leads to a market failure. Used goods are disposed of before their socially profitable reuses are exhausted (Aagaard 1995).

Secondhand markets and sales for reuse are well developed in some durable products—houses and automobiles being the most obvious examples. Other potential markets operate inefficiently, informally, or not at all simply because the owner who has tired of a good has the option of costless disposal. Clothing is the clearest example—it is anecdotally claimed that reusable clothing takes up many times the space in landfills as the notorious disposable diapers (Kahlenberg 1992). And much of what clothing reuse does occur in the United States depends upon charitable impulses—Goodwill Industries processes more than 300,000 tons of textiles a year (Todd 1993).

The reuse of durable goods usually involves the gift or sale of the product from someone of higher income to someone of lower income. So it should come as no surprise that much reuse entails transfering used goods from rich to poor countries—200,000 tons of used clothing are exported from the United States alone each year. In developing-world villages, I often see a T-shirt reminder of some 1970s or 1980s Rose Bowl (in which Michigan often played and usually lost). Indeed, textiles there may never become waste—when the T-shirt is worn to rags, it will probably be sold for mattress stuffing or used as a (washable and reusable) diaper and go on providing utility for perhaps another decade.

What does all this diversion into reuse prove? First, trash collection charges might spur socially beneficial secondhand markets. Second, it would be difficult as well as undesirable to ban the exporting of "waste" from rich to poor countries, because there is a fuzzy line between waste that will end up in an unsanitary developing-world dump and "waste" that will keep the sun off a developing-world back.

Does all this mean we should just have free international trade in wastes? Hardly—but it does point out that there is no simple control scheme. So far, the world has leaned toward a system that places no restrictions on trade in ordinary solid waste and completely bans trade in hazardous waste. But as we have seen, not all substances fall neatly and easily into one of these categories. A substance may be waste in one importing country but recyclable in another; or a substance may be waste at one time and recyclable at another.

Final Thoughts

How much trade in waste should there be? Clearly, the optimal amount is greater than zero—it is surely not efficient for each household to bury its own waste on its own property. Equally clearly, one all-planetary landfill would involve too high transport costs and would be far larger than economies of scale could suggest.

Once we have in mind an optimal number of disposal places and optimal distances between them, should it matter whether there are state or national boundaries involved? If the boundaries are proxies for different disposal costs or different tolerances for the external costs of waste handling, then border problems should matter. But capricious sets of international transport rules or state waste taxes can distort the pattern of ultimate disposal away from optimality (Deyle and Bretschneider 1995).

APPENDIX
Host Fees at the Arbor Hills Landfill

Settle one difficulty, and you keep a hundred others away.

—Chinese proverb

The Arbor Hills Center for Resource Management (commonly known as the Arbor Hills Landfill) is operated by Browning Ferris Industries in Salem Township, Michigan—a village of under 4,000 persons, almost all white and with a median annual family income of nearly $60,000. BFI purchased an existing landfill in the late 1980s, and it then negotiated a Host Agreement with Salem Township that outlined its expansion and operating plans and the township's compensation during the next century (Bates 1990; Kilbourn and Mandell 1992; Pyen 1992, 1997; Rzepka 1992; Bui 1995; and personal communication with Kathleen Klein).

Arbor Hills is no small operation—10% of all Michigan trash ends up there, and more comes from Canada. The neighbors suffer the noise, dust, and odors generated by hundreds of trucks a day. And there is always fear of long-run health problems; indeed, the landfill had experienced groundwater contamination leakages under its previous owner, though BFI has built dikes and wells to contain and collect the contaminated water for treatment. To offset all these external costs and fears, a host fee emerged.

Host fees can come in many sizes and shapes, as the BFI agreement with Salem Township illustrates:

- The township receives 2.5% of all landfill revenues and 4% of all compost revenues. Because BFI takes in 4 million cubic yards of waste annually and charges an average of almost $12 per cubic yard, this means annual township revenue of roughly $1.2 million. (BFI charges less than $10 per cubic yard for waste within Washtenaw County, of which the township is a part, and $12.50 for all out-of-county waste; less than a fourth of the Arbor Hills waste comes from Washtenaw County, and compost provides only a small fraction of Arbor Hills revenues.)
- The township receives half of the revenues generated from sales of landfill gases or products produced from those gases, which BFI contractually must collect. This currently adds another $300,000 to the annual host fee.
- BFI must accept free of charge all township waste, without limit to volume, and provide a drop-off facility for recyclables and for compostables. This is worth at least $100,000 to a town of this size. Thus, these first three benefits sum to at least $400 per person a year.
- BFI must prevent its customers from driving their large trucks through the township center (by terminating trucks' access to the landfill if they repeatedly violate this restriction).

- BFI must provide environmental protection to its township neighbors in several ways. In addition to the federal and state restrictions on all landfills, the host agreement obligates BFI to control odor and dust, prevent unsightly growth of vegetation, use "all reasonably available devices" to reduce noise, and "promptly collect" windblown litter (or the township can pick it up and bill BFI for the cost).
- Any damages to township neighbors caused by the landfill must be fully compensated by BFI. For example, if groundwater sources were contaminated by landfill leaching, BFI would have to provide bottled or piped drinking water to all residents whose wells were affected.
- BFI also signed a host agreement with Washtenaw County under which it pays the county more than $1 million a year. The municipalities of the county are guaranteed that Arbor Hills will take their solid waste for the next 20 years, with a discount of at least 15% on the standard tipping fee.

Two additional points: First, the details of the payment are interesting. The county receives 3% of all revenues generated by receipt of out-of-county waste—minus the host-fee monies paid to Salem Township—plus a surcharge of $0.50 per cubic yard if that waste comes from a jurisdiction whose recycling program is inferior to those of Washtenaw County communities. Because the removal of recyclables from solid waste does not affect the potential damage of the waste to the immediate environment, there is no reason for this surcharge beyond do-gooder meddling. Second, one could quarrel with the environmental incentives produced by such a discount. If the regular tipping fee represents the real cost of the disposal, then the discount encourages the county municipalities to recycle too little and to send too much waste to Arbor Hills. Some sort of flat payment would have prevented this distortion.

Going through a host fee agreement like the one in Salem Township shows us that it is possible to make a landfill's neighbors happy with the siting decision and still keep the landfill profitable. And we also can see that there are many ways to do this.

Notes

1. One might think that, since the net MSC of the incinerator is zero, further gain could be achieved by setting the incinerator price at zero. But remember, the capacity of the incinerator is W_{incin}. All waste greater than that amount must be exported, and the MSC of exports is P_{landf}.

2. Flow control has also been used to prevent the export of recyclables so that the municipal recycling facility could charge higher tipping fees and take advantage of greater volume to achieve lower cost per ton of recycling. The inefficiency is similar to the incinerator analysis in that it artificially discourages recycling at the least-cost location and thereby imposes deadweight loss. Moreover, the goal for recycling, about which much more below, is *not* to minimize the cost per ton of recyclable materials.

PART 2

Recycling Solid Waste Products

Market Failure in Recycling

And it all goes into the laundry,
But it never comes out in the wash ….

—Rudyard Kipling, *Stellenbosch*

At long last, at too long last some would say, we come to recycling. Until now, all the policies, prices, taxes, and subsidies we have been talking about have had to do with a world in which recycling was not being undertaken. This has been sort of like the 1960s and 1970s, when less than one-tenth of the MSW in the United States was recycled. By 2000, however, the average recycling rate was nearing one-third, and recycling was not only an option for most households; it was a part of life, even a civic joy or a noxious chore (Franklin Associates 2000; Glenn 1999).

The literature on recycling is vast, but it tends to be written from the extremes. To some writers, recycling is seen as all benefit at no cost, the more the better. Other authors would stop at nothing. The titles reflect this extremism: "Why Do We Recycle?" (Ackerman 1997), "Recycling Is Garbage" (Tierney 1996), "Anti-Recycling Myths" (Denison and Ruston 1996), "Rethinking Recycling" (Black 1995), "Too Good to Throw Away" (Hershkowitz 1997), "Curbside Recycling Comforts the Soul, But Benefits Are Scant" (Bailey 1995a), and "Time to Dump Recycling?" (Hendrickson et al. 1995). The next six chapters will take, I hope, a calmer, more measured approach to recycling. For me, recycling is neither a savior nor an archfiend. To me, two things are very clear: It is not optimal to recycle everything *and* it is not optimal to recycle nothing. The purpose of the next few chapters is to answer several questions: Why do we recycle? When should we start (or more accurately, recommence) recycling? What should we recycle? How, and how much, should we recycle?

We begin with why we recycle. More precisely, we look at why we need to have special government policies to encourage—and sometimes subsidize or require—recycling. Why doesn't "the market" handle recyclables, just as it does broccoli and most other products we produce? Why is waste different in this respect? The short answer is that waste markets experience "market failure," which calls for special public attention.

Market Failure

To get at the essence of recycling, let us consider a farmer who produces two things, broccoli and empty cola bottles. With the broccoli, the farmer must decide whether to take it to market or discard it. This broccoli decision is easy. The farmer checks the market price of broccoli (p), subtracts the cost of transporting the broccoli there (t), and checks to see if the net profit of broccoli sales ($p - t$) is greater or less than the net profit of disposal ($-d$), where d is the disposal cost. For most products (such as this broccoli) and for most producers (such as this farmer), most of us have a lot of faith that the p, t, and d are socially meaningful prices. If p is set in a competitive market, it is set at the intersection of supply and demand and hence represents both the marginal willingness to pay of other consumers for broccoli and the marginal cost of other farmers in producing broccoli. Thus p represents the social value of a marginal unit of broccoli. If t is also set in a competitive market, it reflects the resources needed for the transportation. If the farmer chooses to discard the broccoli, a private hauler would have to be hired from a competitive hauling industry, and this cost, d, would also reflect resource costs—provided the landfill to which the broccoli is headed charges a tipping fee that covers its marginal social cost. All that is straight out of an introductory economics course.

Now look at those empty cola bottles. The farmer must decide whether to discard them, for free pickup by the MSW service, or to recycle them, for free pickup by the municipal recycling service. The price is zero in each case. With discarding, neither the costs of MSW collection nor the costs of the landfilling, neither the private costs nor the external costs, are taken into account in the zero price. And with recycling, the zero price there does not reflect the collection and sorting costs, nor does it consider the social benefit of the recaptured material. The difference between the two markets is immense. One sends signals to the farmer about the social costs of various activities and resources. The other sends no such signals.

Whenever recycling provides a net social benefit in comparison with landfilling or incinerating the waste, a working market should send a signal to the household to that effect. The basic market failure in recycling is that no such signal is sent. Each household continues to choose between the recycling bin and the trash bin on the basis of whim, habit, or conviction—but not on the basis of economic self-interest and accurate signals about social benefits and costs. People will do a lot if they become convinced that it is the right thing to do, but it greatly helps if it is also the self-interested thing to do.

If a product is cheaper to recycle than dispose of, "the market" should tell households so. But "it" does not. As long as municipal solid waste disposal is privately costless—even if recycling is also privately costless—households have no *economic* incentive to recycle. And reliance on noneconomic incentives to recycle will almost certainly achieve a socially incorrect degree of recycling. We shall examine potentially helpful disposal and recycling pricing schemes and other policies in Chapter 10.

Beside this failure to provide households with correct market signals, there are four other kinds of market failure connected with recycling:

■ Most recycling facilities are municipally owned or controlled, and they are not operated for profit. Typically, their goal is to keep down the cost per ton recycled,

Disposing of Old Tires

On average, Americans dispose of more than one old car or truck tire per person each year. Until World War II, worn-out pure rubber tires were melted down and reused. Since the introduction of steel belts and synthetic rubber, however, reuse has become too costly, and old tires have been mostly landfilled. But tires neither compact well nor degrade rapidly in landfills, so most states now either ban tires from their landfills or require them first to be shredded. Continued surface disposal is even less desirable, for it constitutes a fire hazard and a breeding site for rats and insects. The basic problem is that, as long as the *private* cost of these stockpiles is low, there has been little incentive to find alternative means of disposal or to find ways to recycle them.

Socially low-cost, or even socially beneficial, ways of handling old tires exist. Old tires can be retreaded for extended use, but retreading has been declining for decades because real new tire prices have been falling, because retreading steel-belted radial tires is expensive, and because tire disposal has incurred so small a private cost. Tires are eminently burnable, yielding more fuel value per pound than coal. About two-thirds of discarded U.S. tires are now burned. Tires can be used to create rubberized asphalt roads, which cost roughly twice as much initially but provide a more comfortable ride, reduce noise, are temperature-resistant, and last two to four times as long (Serumgard and Blumenthal 1993). (Although—in a present-value sense—it is true that something costing twice as much and lasting twice as long is more expensive, if the road costs twice as much and lasts *three* times as long, it will almost certainly be cheaper in the long run.) So why don't we use more of our stockpile of discarded tires in these beneficial ways? Usually, it is still a costly process from old tire to the beneficial reuse.

The Intermodal Surface Transportation Efficiency Act of 1991 mandated the use of asphalt-rubber pavements and aimed to get rubber into 20% of all federally funded highways. However, the mandate went unenforced and was repealed in 1995. Several states have enacted small taxes, per tire or vehicle, with revenues earmarked for either the disposal of old tire stockpiles or increased research on the use of old tires (U.S. EPA 1993; Jang et al. 1998). But such taxes do nothing directly to stimulate the socially correct disposal of tires (Acohido 1999). Indeed, subsidizing the abatement of old tire stockpiles may, in the short run, simply reduce the market for newly discarded tires and lead to more of them being landfilled or stockpiled.

Some states are creating future tire mines. Having no viable recycling options and resisting the thought of continued burial in regular solid waste landfills, they are permitting tires to be buried in tire monofills (i.e., landfills dedicated to tires alone). It is supposed that, at some near future time, these tires will be dug up and recycled. Unfortunately, dirt-covered tires yield low-quality rubber. This fact, plus the cost of the digging, makes it unlikely that these monofills will soon be mined (Blumenthal 1998).

get the revenues up, and then recycle as much as possible within a constraint on profits—either there is a requirement that they break even or there is a limit to the size of their municipal subsidy. Such recycling facilities will react neither to private profit nor to social profit in their decisions on what, when, where, and how to recycle. The socially correct amount and kind of recycling are unlikely to end up being done.

■ Markets work best when they are stable. Markets subject to dramatic and regular changes do not allocate efficiently. Even libertarians would accept the need for a military draft and other nonprice procurement policies in the case of an all-out war. Although recycling is not the moral equivalent of war, it is subject to sharp changes in supply, demand, and prices—where a steadying, nonmarket hand may be appropriate. We shall look closely at prices and markets in recycling in Chapter 12.

- ■ The periphery of the recycling arena is also littered with incorrect prices and external costs. In a second-best world where they cannot be corrected by direct action, it may be appropriate to attack them through recycling policies. If people consume both of two substitutes, say product X and product Y, and the production of Y generates external costs, then we should tax Y. But if we cannot tax Y for some reason, subsidizing X may be better than doing nothing. We shall explore some of these second-best issues in the next section of this chapter.
- ■ Not only do households receive the wrong price signals; so do manufacturers. We have already discussed the absence of concern by manufacturers for the disposal cost of their products and packages (in Chapter 2). Manufacturers also do not consider the extent to which their products and packages are recyclable. Manufacturers earn no larger profit if they make their products and packages cheaper to dispose of or cheaper to recycle.

To summarize: Recycling is an alternative to disposal. The good reason for recycling is that it is a better alternative than disposal—better in the sense that the net social cost of recycling is lower than the net social cost of disposal, once all the social benefits and costs of each are properly counted. Keep this good reason in mind while we run through some of the bad reasons—or at least not-so-good reasons—for recycling.

Bad Reasons for Recycling

You might think it best to leave bad reasons for recycling alone. But there is a good reason to dispose of these bad reasons: They deflect our attention from the good reason and may lead to distortions in recycling policy.

We regularly hear that the reason for recycling is to conserve resources and thus prevent the planet's "doomsday" when it reaches its "limits to growth" (Meadows et al. 1972; Ruston and Denison 1996). The doomsday argument runs as follows. Our planet is finite and so are its resources. When we have used up these finite resources, our standard of living will fall catastrophically. Recycling means that our resources—and presumably us and/or our living standard—will last longer. This is the recycling corollary of the doomsday hypothesis. The doomsday hypothesis recurs so regularly because it requires so little effort to generate. It needs two pieces of information, the known existing stock of some material (X) and the current annual rate of use of this material (R) . By dividing X by R, we discover that (at the current rate of use) we will run out in X/R years. Because the continuation of our contemporary life is inconceivable without this material, we must anticipate that disaster will descend on us when the material is gone. An example may make the foolishness of such calculations clearer: The world's known reserves in 1973 of aluminum were 23 times the world's annual consumption of aluminum, which means that the world ran out (!) of aluminum in 1996 (Simon 1981).

Much has been written about the dangers of doomsday thinking, and there is no need to rehearse them all here. I want just to look at the recycling corollary. Suppose we really are running out of something. If it is indispensable and irreplaceable, it will become very scarce and valuable in the not-too-distant future. When that happens, it will surely get recycled. The fact that something will be getting extensively recycled in the future is *not* proof that it should be recycled now. Recall the old farm adage: Pick the low-hanging fruit first (which was first adapted to resource use in the 1960s; Herfindahl 1967). Although we are using cheaply extracted virgin raw materials and then throwing them away, the landfills are storing them up for future reuse.

Much of what is in a landfill degrades slowly or not at all. Landfill archeologists are often photographed reading a half-century-old newspaper or nibbling on a mummified carrot. That "landfills are forever" is usually seen as a shortcoming of this means of disposal of waste. But landfills can also be viewed as warehouses of recyclables for future retrieval (Dickinson 1995). Products that now are either technologically impossible or economically costly to recycle may later become feasible and profitable to recycle. At that time, landfills may be mined for their recyclable ore. Indeed, some landfills are already being mined, not so much for their recyclables but rather to extend their lives, to acquire cheap cover soil for ongoing landfills, or to reclaim the land without undergoing a long monitoring and settling period.

Does this mean that we should not consider recycling until we are on the brink of exhaustion of resources or landfill space? No. A fuller (though mathematical) proof is provided in the box titled "When Do You Start Recycling?"; but the basic ideas are easy to come by. Think of recyclable materials as an alternative ore source—sometimes a cheap one, such as aluminum; sometimes not such a cheap one, such as most plastics. Again recall the adage about low-hanging fruit.

When Do You Start Recycling?

Consider the following very simple example: We are running out of landfill space. (Landfill space is a bit out of date as a doomsday resource nearing exhaustion, but it was popular not too many years ago and makes for a simple theory.) You are stuck on a desert island for two years. Each unit of your effort produces one unit of consumption; each unit of your consumption produces one unit of waste. You have one unit of landfill space available, and you can throw your waste in there at no cost at all. Or you can recycle your waste at a cost of one unit of effort per unit of waste recycled. You have one unit of effort available each year. How should you allocate your effort between producing consumption and recycling waste in each year?

We can write all this out algebraically, with C for consumption, L for landfilling, and R for recycling, and with the subscripts 1 or 2 indicating the first or second year. The equations are (Highfill and McAsey 1997):

$$1 = C_1 + R_1 \quad \text{(effort allocation in year one between } C \text{ and } R) \tag{1}$$

$$1 = C_2 + R_2 \quad \text{(ditto for year two)} \tag{2}$$

$$R_1 + L_1 = C_1 \quad \text{(waste allocation of year one's } C \text{ between } R \text{ and } L) \tag{3}$$

$$R_2 + L_2 = C_2 \quad \text{(ditto for year two)} \tag{4}$$

$$1 = L_1 + L_2 \quad \text{(allocation of landfill capacity between year one and two)} \tag{5}$$

Now substitute around in those five equations (and six variables) to eliminate three of the variables (R_1, R_2, and L_2), leaving two equations in three variables:

$$C_1 = (1 + L_1)/2 \tag{6}$$

$$C_2 = (2 - L_1)/2 \tag{7}$$

How you choose L_1 determines your entire consumption path for both years. One obvious candidate is to use the landfill in year one, postponing recycling until it becomes necessary in year two. This means setting $L_1 = 1$ (and hence $L_2 = 0$). This decision would yield you a consumption path of $C_1 = 1$ and $C_2 = \frac{1}{2}$ (C_2 is only $\frac{1}{2}$ because you spend half your time in year two recycling waste). If you have diminishing marginal utility to consumption, this sudden halving of your living standard in year two is a disheartening prospect. But you can avoid it by starting to recycle in year one.

box continues on next page

Indeed, you can equalize your consumption in the two years by using up only half your landfill each year (i.e., $L_1 = L_2 = \frac{1}{2}$, $R_1 = R_2 = \frac{1}{4}$, and $C_1 = C_2 = \frac{3}{4}$). Your two-year consumption total is just as large, and it is leveled out over the two years.

In short, recycling while you still have "cheap" landfill space left may be quite sensible. And notice that in this little model we have not even considered the two main reasons why you might in fact want to start recycling before it becomes physically necessary: If you add learning by doing to recycling, you may be able not only to *equalize* consumption between the two years but even increase the total (because it may no longer take one unit of effort to recycle one unit of consumption in year two if you have been practicing recycling in year one); and what you recycle in year one is available for consumption in year two. But to keep the model simple, we have ignored that "little" bonus from recycling.

Yet it is easy to put these two factors into the model. If recycling takes only half as much effort in year two (because you have "learned" about it in year one) and if you include R_1 as extra consumption in year two, equation (2) becomes

$$1 = C_2 - R_1 + \frac{1}{2}R_2 \tag{8}$$

and, though equation (6) is unchanged (because neither the learning nor the additional consumption occurs until year two), equation (7) becomes

$$C_2 = (4 - 2L_1)/3 \tag{9}$$

Now, to equalize consumption in each year, you would set $L_1 = 5/7$, which would yield $C_1 = C_2 = 6/7$, higher than before.

Applying it to resources, it tells us that as long as virgin ore is cheap, we don't want to recycle. But virgin ore sources become ever more expensive—the remaining fruit hangs ever higher and is harder to reach. We should start recycling when recycling becomes socially cost-effective relative to virgin ore sources.

All well and good—but note what is happening to our standard of living in this process. While ore is cheap, we are using lots of it, living high off the hog. But as cheap ore sources are used up, we find ourselves spending ever more time and resources just to uncover ore, and our living standards decline. When we are reduced to very-high-cost ore and extensive recycling, our living standards will sink very low. If we want an intergenerational smoothing of living standards, we must choose to begin recycling earlier than absolutely necessary.

Another reason often given for encouraging or subsidizing recycling is the need to offset policies that promote the use of virgin resources—such policies as percentage depletion allowances and other tax advantages for virgin materials, public timber sales at prices below social cost, lower costs for transporting virgin steel than for transporting scrap iron, and untaxed external costs in mining (Schall 1993; Denison and Ruston 1996; Hershkowitz 1997).

Whenever public policies favor the use of virgin materials for no good public reason, those policies should be eliminated. About this there is no question. Percentage depletion allowances are nothing more or less than subsidies, and they result in too much mining and too low virgin material prices. If an activity generates external costs, it should be taxed. But the fact that virgin materials are subsidized is not a good reason to subsidize substitute materials, such as recycled materials; and the fact that mining generates external costs is not a good reason to subsidize substitute activities, such as recycling.

However, second-best arguments can be brought into play. Suppose that, for some reason, taxing virgin raw materials is impossible, despite their external costs. Then subsidizing substitute activities such as recycling might be a second-best policy. But such subsidizing is not the *best* policy. Whenever the policy action gets further from the activity it is trying to influence, there is greater opportunity for the infamous slip between the cup and the lip. Here, for example, taxing virgin resources not only encourages a switch to recyclable materials but also encourages a reduction in the use of materials; but subsidizing recycling, though it also encourages a substitution of recyclables for virgin materials, has the side effect of encouraging an increase in the overall use of raw materials in production through a substitution of raw materials for capital and labor.[1]

Not only is the policy roundabout, but the facts are in doubt. Most of the tax advantages of raw materials disappeared in the tax reform of the 1980s (Koplow and Dietly 1994). And even before then, they were probably not big enough to affect producer choices about the sources of their raw materials—for example, the depletion allowance on copper mining was estimated to lower the price of virgin copper by less than 2% (Geiger 1982).

Energy use, pollution, and global warming are also brought out as arguments for especially encouraging recycling. It is widely agreed, even among critics of recycling, that recycling does save energy (Bailey 1995a; Tierney 1996). But what does that prove? Different techniques of producing anything use various amounts of energy, labor, capital, and other inputs. We should not select a technique on the grounds that it uses less of one particular input.

The energy argument must rest on the assumption that energy is underpriced and hence that we use too much of it and incorrectly choose highly energy-intensive processes. That energy is underpriced in the United States is also widely accepted, among economists at least. But here, too, there is a best response: If the social cost of energy use exceeds the private cost, a tax on energy use is called for. It is surely a convoluted, second-best policy either to tax products and activities that are energy intensive or to subsidize ones that are energy efficient. There is also question here about the magnitude of the effects of U.S. energy mispricing; e.g., virgin aluminum uses enormous amounts of energy, relative to recycled aluminum, but the resulting penalty to recycling of artificially low energy prices has been estimated to be only 7% (Koplow 1994).

There is no limit to the number of arguments one can put forward for recycling subsidies by focusing on one particular aspect of recycling versus virgin material use or recycling versus landfilling. For example, consider the fact that more than 80% of the raw material ends up as paper products with newsprint recycling mills, whereas only 25% ends up as paper products with virgin mills (Hershkowitz 1997). So what? Waste minimization is not the sole goal of the economy. If waste disposal is underpriced—and as we have seen, it usually is—then the best policy calls for pricing waste disposal, not for encouraging activities that generate little waste. (In fact, in this example, most of what does not become final product is *not* waste; much is burned to provide energy and heat for the mill.) Another example of the single focus concerns air pollution: Vehicles used to collect recyclables are said to cause less air pollution than those used to collect solid waste (Hershkowitz 1997). Again, reduced air pollution is not the sole goal of society—or, according to this argument, we should scrap all our trash-compactor and recycling trucks and go back to horse-drawn carts. If air pollution is excessive from a social viewpoint, we should tax its generation, not subsidize activities that produce little air pollution. (By the by, it is not always clear that recycling generates less pollution than the use of virgin materials; see Palmer and Walls 1997.) To give specificity to some of these arguments, let's digress to look at paper recycling and tree cutting.

Paper Recycling and the Stock of Trees

We recycle paper to reduce timber harvesting and increase the stock of trees. Right? It may be a little more complicated than that. Consider first a single country that neither imports nor exports wood products. Its demand for paper can be met from recycled paper or virgin pulp. Increased paper re-collection will lower the price of such recyclable paper. This means a larger and cheaper input for the recycled paper industry. Recycled paper becomes cheaper relative to paper made from virgin pulp. As a result, we buy more paper in total (because the average price of all paper is lower), we buy more paper made from recycled paper, and we buy less paper made from virgin pulp. Because less paper from virgin pulp is demanded, fewer trees are harvested to provide the raw material for this industry.

Fewer trees are harvested. But does this mean that there will be a larger *stock* of trees? Not necessarily. Virgin pulp comes from two kinds of forests, wilderness and tree farms. Wilderness forests are made up of old-growth trees that grow back very slowly once harvested. Trees grown on farms are like carrots but with a longer time to maturity—they are planted, nurtured, harvested, and sold by tree farmers (usually large timber corporations). The reduced demand for wilderness harvesting means that more old-growth trees remain. Paper recycling slows the reduction of the wilderness tree stock. But the reduced demand for tree-farm timber means that, in a steady state, fewer tree-farm trees are planted and that the stock of tree-farm trees is smaller.[2] Paper recycling leads to a smaller stock of tree-farm trees (Darby 1973).

If what we want is a larger total stock of trees, then it is not certain that more paper recycling will achieve that. If we want less harvesting of wilderness trees, paper recycling will achieve that, but very indirectly. It is a second-best policy. There are largely unpriced and external benefits to wilderness trees—recreation, watershed protection, species preservation, and reduced global warming. These social values of the wilderness are not captured privately by the forest owners—who, by the way, in the United States are largely the federal government itself, which should be considering social rather than private profit. The best policy urges that these external benefits of the wilderness should be directly confronted by public policy. Direct regulation reducing the wilderness tree harvest, or marketable permits limiting the number that can be harvested, or a Pigovian tax on cutting wilderness trees will all more efficiently achieve the desired goal.

In the United States, however, we do not tax the harvest of wilderness trees; rather, we *subsidize* such harvests by putting too much public timber up for auction, by having the government share the risks to timber firms, and by building logging roads at public expense (Koplow and Dietly 1994). Taxing wilderness harvesting would not only raise virgin timber prices and encourage recycling; it would also discourage paper use and encourage the substitution of tree farms for the wilderness as raw material sources by correctly adjusting prices to account for externalities. Incidentally, if you are interested in the total number of trees, the stock will go up for two reasons: because of a slower reduction of the wilderness tree stock, and because of an increase in the tree-farm tree stock.

Now, consider the effects of trade in timber products. A significant protest group at a recent World Trade Organization meeting in Seattle feared that freer trade in timber would accelerate the rate at which the world's trees are cut down, and as a

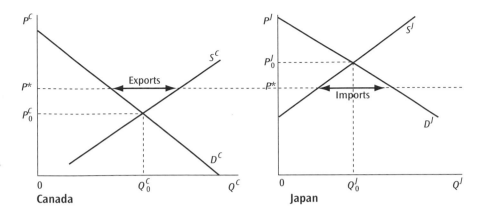

by-product, reduce virgin timber prices and everywhere reduce the incentive to recycle (Carlton 1999). Let's examine the connection between timber trade and paper recycling.

To keep it simple, think of one forest product exporter (Canada) and one forest product importer (Japan). (The country names are not chosen at random; Canada exports about a fifth of world forest product exports, and Japan imports about a seventh of world imports. Each is far and away the global leader in its category; FAO 1996.) Each country is shown in Figure 8-1 with a supply (S) and demand (D) curve for forest products. If they cannot trade, the price in Canada will equate its S^C and D^C and will be P_0^C. The same will be true in Japan for a price of P_0^J. Canada, being the lower-cost producer (with lower demand as well), will have the lower price. With free trade, the price differential is removed by profit-seeking traders; the price becomes P^* in both countries (ignoring transport costs), and Canada exports to Japan.

What happens to the forest product output? It goes up in Canada, and down in Japan. Which change predominates is an empirical question, but most studies conclude that reductions in forest product tariffs—and hence increased trade—would generate only very modest increases in total world production (Barbier 1999; Sedjo and Simpson 1999). In reality, forest product tariffs tend to be already low, so the real-world choice for the future is between very low tariffs and even lower tariffs. No wonder researchers find modest effects for such changes.

What happens to paper recycling? The higher price of forest products in Canada encourages paper recycling there. The lower price in Japan discourages paper recycling there. Again, which effect predominates is an empirical question, but it seems improbable that the net effect on recycling in both countries would be large.

Perhaps the Seattle protesters were thinking only of the United States. We are net importers, but barely so, annually exporting $17 billion of forest products and importing $23 billion. If we think of the United States as being slightly more like Japan in the theory above, then freer trade in timber products would mean for the United States slightly lower forest product prices, slightly lower output from both wilderness forests and tree farms, and slightly discouraged paper recycling.

In short, for forest products, the relationship between recycling, harvesting, trees, and trade is not so simple. To be for recycling or to be against trade in wood products because these *might* increase the total stock of trees or reduce the harvest of wilderness trees is very, very indirect, and even the direction of the impact is unclear.[3] There are many slips between such policy cups and such target lips.

We could go on, but won't. There are many tenuous or bad arguments for recycling. But one is particularly pernicious, and I have been saving a special section for it—that recycling "creates jobs."

Job Creation

Too often, recycling proponents point out that one of the virtues of recycling is its creation of jobs, as in the following:

> Studies suggest that recycling has a strong positive impact on jobs. In North Carolina, it is estimated that for every 100 jobs created by recycling, 13 jobs are lost in solid waste and virgin materials extraction within the state.... And a study of three cities—Baltimore, Richmond, and Washington—found more than 5,100 individuals employed in recycling, compared to about 1,100 employed in solid waste disposal.... (CEQ 1997)

> ... recycling is notable for the jobs it creates. It is estimated that recycling 10,000 tons of material produces 36 jobs, compared to 6 for an equivalent amount of landfilling. (Cohen et al. 1988)

Even the U.S. EPA regularly turns out such statements: "recycling is estimated to create five times as many jobs as landfilling" (U.S. EPA 1994a, 1997a), and "for every 100 recycling jobs created, ... just 10 jobs were lost in the solid waste industry, and three jobs were lost in the timber harvesting industry" (U.S. EPA 1995a).

Three things are wrong with seeing jobs as a benefit. First, the fact that lots of people are needed to carry out recycling programs is basically evidence that recycling is expensive, requiring lots of labor (as well as capital) that could have been used to fulfill other goals of public policy. Second, any jobs created by recycling programs do not reduce unemployment but simply replace jobs elsewhere in the economy. Where these jobs come from we cannot be sure—it depends upon where government spending is decreased or taxation increased when spending on recycling is increased. If new recycling expenditure comes from increased taxation, it depends upon what consumers would have spent their now-taxed income on. Third, even if we were sure that jobs were created, and that the national unemployment rate actually went down as a result of a recycling program, we would still have to be sure that recycling was the best way to achieve this outcome. I still recall vividly the Nixon and McGovern forces in the 1972 presidential election debating whether our involvement in Vietnam increased employment by hiring labor-intensive infantry or decreased employment by investing in capital-intensive helicopters. The sick part of that controversy was that, at times, the two sides seemed to agree that the solution to that issue would go a long way toward settling the question of whether we should be involved at all in Vietnam.

Two more subtle but also weak jobs arguments ought to be disposed of. (The good thing about disposing of weak arguments is that they don't take up landfill space; the bad thing is that they rarely stay disposed but tend to get recycled and reused.) Opponents of recycling often accept that recycling creates jobs, but they then point out that these are not good jobs, being largely low-wage jobs. Perhaps it would be nice if everybody in the United States earned the same wage—and that wage were around $100,000 a year—but it is a fact of capitalist existence that they do not, and the poor do not need the added stigma of being told that their jobs are not worth counting. On the other side, proponents of recycling point out the serendipity of recycling's use of low-skill, low-wage labor:

> [New York] City's recycling and waste management programme will eventually generate between 44,000 and 60,000 jobs in raw materials processing and related industries…. This is significant, since some 700,000 manufacturing jobs have been lost in the City since 1960 leading to … a "dual city" split between an increasingly specialized high-skill economy … and an army of poorly qualified and casually employed people disproportionately composed of ethnic minorities and other disadvantaged groups. (Gandy 1994)

Surely, New York City would not be at a loss to think of other ways to generate low-skill jobs should it put its collective mind to it. The point, in short, is that recycling is not made more or less desirable, from a social viewpoint, by the numbers or kinds of workers it employs. The fact that there are unemployed people *is* a market failure, but it is not one that will be cured by recycling.

Recycling in Cities of the Developing World

Unlike recycling in the United States, which is urged on by non-market activities and policies, recycling in cities of the developing world is a market-driven, for-profit activity. No sooner has solid waste left the household than it begins to be pored over by scavengers, lonely individuals with bags or carts, who seek a modest living by collecting discarded items that have value in recycling or reuse.

Although their numbers and the quantity of materials they remove from the waste stream are usually large—one study estimated that there are 40,000 scavengers in Jakarta recycling nearly one-fifth of the city's solid waste—they are rarely encouraged by the government. Indeed, in Jakarta until recently, they were considered "urban undesirables," grouped with beggars and prostitutes and subject to continual police harassment (Sicular 1992; Porter 1996). In other cities of the developing world, the government recognizes that they not only earn an honest living for themselves but also reduce the volume of waste the city must handle. In Gaborone, Botswana, at least one waste paper recycling firm hires paper scavengers as regular employees and pays them by the hour. This practice seems at first to be expensive and fraught with bad incentives, but the firm has found that its flow of recyclable paper has been augmented and regularized, and hence that it can operate more profitably (Porter 1997).

Why does the profit incentive induce so much recycling in the developing world when government intervention is so often needed to encourage recycling in industrial countries? The answer lies in relative factor prices. Most industrial raw materials are sold on world markets, and thus their prices do not vary much from one country to another. But wages differ greatly among countries, and recycling is labor intensive. Therefore, in poor countries where wages are low, capturing raw materials through recycling tends to be more privately profitable than in rich countries where wages are high.

Final Thoughts

The basic reason why recycling markets do not work is that most households have no incentive to recycle. Given the choice between putting something in the trash or out for recycling, most households face identical marginal costs for the two choices:

zero. Because some trash will be put out in any case, recycling always means extra bins, extra sorting, extra space, and extra time. Households may be educated to recycle or required to recycle, but only a marginal cost on trash collection will give them an economic incentive to recycle. Gung-ho, can-do enthusiasm for recycling may work for a while, but eventually it may wane—indeed, in many cities recycling rates are already declining (Booth 2000).

Notes

1. A similar second-best argument sees external benefits in recycling but an inability for some reason to apply the best policy of a subsidy for recycling. Here too, a tax on virgin materials may be a second-best response, but there is an unintended effect in the wrong direction on overall material use (Fullerton and Kinnaman 1995). In short, second-best taxes or subsidies in the presence of external costs can have many, and even perverse, unintended consequences (Dinan 1993; Palmer and Walls 1997).

2. Reality, of course, is somewhat more complicated. A decline in the rate of return for tree farms would mean several things: Some farms would be returned to agriculture; some farms would continue but would be managed less intensively; and some farms would be switched from pulpwood to wood for furniture and construction, the demand for which would be stimulated by the lower tree prices (Wiseman 1993).

3. An EPA study of paper recycling and tree stocks concluded that an increase in paper recycling would cause an increase in tree stocks in the short run, but that the fall in tree-growing rents would soon reduce the planting of new tree-farm trees. In the long run, the EPA concluded that the rate of paper recycling would have almost no effect on the size of tree stocks (U.S. EPA 1998a).

Economics of Recycling

Do you want to recycle hot dog wrappers? Kitty litter? You could do it,
but at what cost?

—J. Winston Porter, former EPA assistant administrator and
current president of the Waste Policy Center; quoted in Booth (2000)

I f trash disposal is free (or underpriced), people—and "the market"—will recycle
too little. When it is free and recycling is not rewarded, it is just as easy, if not
easier, for people to throw away rather than recycle. When markets fail, there is
an arguable case for public action. And Americans have taken action. Recycling
operations, usually mandated and subsidized, now blanket U.S. urban places. The
question we explore in this chapter is whether the social benefits of these recycling
programs are worth it, relative to their social costs.

The Chief Social Benefits and Costs of Recycling

It would be nice if we could do one definitive benefit–cost analysis of all the recycling in the United States and thereby decide, from an economic viewpoint at least,
that it is a good or bad thing. But, as we shall shortly see, the benefits and costs of
recycling vary greatly in different parts of the country and in different kinds of
cities. So all we can do in a general way is outline the process of doing a benefit–cost
analysis. (An example of such an analysis is offered in the Appendix on page 143.)

Every benefit–cost analysis must start with a list of the relevant benefits and
costs and a plan for measuring them in dollar terms. Recycling is no exception.
Recycling has three major benefits: the recovery and reuse of materials, the reduced
use of a landfill (or incinerator), and the reduced need for solid waste collection.
And recycling has two major costs: collecting recyclable materials and processing
the materials for reuse. There are other benefits and costs, and we will soon come to
them. But let's look first at these five.

Recovering Materials

Recovering and reusing previously discarded materials are major benefits—after all, this is why we recycle. The problem is how to assign a dollar value to the tons of this and the tons of that. The obvious, and the usual, way of valuing recovered recyclables is the revenue gained from their sale. The recovered materials are worth what users of these materials are willing to pay for them.

One problem with using market prices is that they bounce around a lot, which means that the entire benefit–cost analysis may depend critically on what year's prices are used. To some extent, this problem can be avoided by averaging prices over several years. If, however, there are trends in these prices, another problem arises. We do not make decisions to recycle or not every year, so what we are interested in is a measure of the average revenues for a fairly long period of future time. Current prices may not be a good proxy for average future prices.

There is a second problem with using market prices as a measure of the social benefit of recovered and reused materials. Market prices reflect only the private benefit of the materials to their buyers. People often want to include, say, energy savings or reduced mining or logging as additional benefits. As we saw in the last chapter, failures in other markets should be corrected in those markets themselves, but it is conceivable that recycling is the only, albeit second-best, way of getting at these distant market failures. This is a long-time headache of benefit–cost analysis, going back to the early 1970s, when benefit–cost handbooks began to appear (Dasgupta et al. 1972; Little and Mirrlees 1974). Suppose widgets can be produced at location A for $1 each, but political concerns lead to putting the plant at location B, where it inefficiently produces at a cost of $2 each (and presumably is subsidized or otherwise protected). We are doing a benefit–cost analysis of a potential widget-producing plant at location C, where the cost will be $1.50. Does it pass the benefit–cost test because it costs $0.50 less than production at B, or does it fail the test because it costs $0.50 more than production at A? The fluctuations and trends in the prices of recyclable materials and recycled products are important to any economic assessment of recycling, and they are sufficiently interesting to warrant their own chapter (12).

Reduced Waste Disposal

Materials that are recycled do not have to be disposed of, and all the social costs of that disposal are avoided. The *average private* cost of that avoided waste disposal is fairly easy to estimate—to remind yourself how easy, look back at Appendix A to Chapter 4. Unfortunately, what we need is the *marginal social* cost, and marginal social cost is not always easy to estimate.

Consider landfill disposal. As we saw in Chapter 4, the marginal and average costs of landfills are high in the Northeast and ever lower in the South and West—the only problem here is to be sure that the avoided external costs are also counted. With incinerators, however, as we saw in Chapter 5, the revenue from energy generation roughly offsets variable costs, so the avoided marginal private incinerator cost is roughly zero when a (burnable) product is recycled. The benefit of recycling on this score is therefore highest for Northeastern cities that use landfills and lowest wherever incineration is used. There are also current and (expected) future external costs to both landfills and incinerators, and the avoidance of these must be counted as an additional benefit of recycling. Measuring them with any precision is another matter.

Reduced Solid Waste Collection

What is collected for recycling does not need to be collected by the usual municipal solid waste collection system. There is clearly a savings and hence a benefit here—but how much? Again, the average cost of collection is readily estimated, but the marginal cost is certainly lower than the average cost. The MSW trucks must still drive by and stop at each house, even though they pick up less if some is being separately recycled. Even advocates of recycling sometimes admit that there will be little or no savings here due to recycling; see the box titled "Putting Numbers into Benefit–Cost Analyses." The marginal cost saving here is surely much less than the average cost.

The cost of MSW collection varies widely from city to city, depending on such things as household density and overall wage level. Of course, these very factors also affect the cost of collecting recyclables, so what is interesting is how the sum of the costs of the two kinds of collection change with increased recycling.

Recyclables Collection

It is tempting to throw the recyclables collection category of costs together with the just-discussed reduced solid waste collection category of benefits and then think that they cancel each other out. Alas, they usually do not. First of all, there are all the overhead costs of two collection systems where there used to be only one—although they are much less than double because each of the truck fleets is smaller. (At least one case has been documented where the costs were no higher at all; Anderson et al. 1995.) More important, the variable cost of collecting recyclables is much higher than the variable cost of collecting traditional solid waste. Why the difference? Why is a ton not a ton when it comes to collection?

Putting Numbers into Benefit–Cost Analyses

Not everyone is hesitant to assign dollar figures to the benefits and costs of recycling. One careful, balanced study—by someone who favors recycling—estimates the national average dollars per ton of recyclables as follows (Ackerman 1997):

	Benefit (+)	Cost (–)
Recovering materials	+ ???	
Reduced waste disposal	+ 31	
Reduced solid waste collection	+ 0	
Recyclables collection		−123
Processing recyclables		−50

The study (which very conservatively assumes that *no* trash collection cost saving occurs when recycling is introduced) then asked how much must the recycled materials be worth to achieve a break-even benefit–cost for recycling? The answer is $142 per ton of recyclable materials recovered (i.e., $123 + $50 – $31 – $0).

The study concludes that revenues are likely to be less than $142 per ton, but because the average household generates only 440 pounds of recyclables a year, "even at zero revenue, recycling does not impose a crushing burden, costing each household $31 [a year] under average conditions" (Ackerman 1997, 71). Are we supposed to do things that only slightly fail their benefit–cost tests? This $31 per household means that the cost of recycling in the United States exceeds the benefit by $3 *billion* per year—unless the average U.S. household is willing to pay at least $31 per year for the privilege of participating in a recycling program.

A final point: The cost per household is a quite irrelevant number. Given that there is a net cost per ton to recycling, the cost per household will always be lower the less recycling is done, but that does not mean that less recycling is always better. In fairness to Ackerman, we should note that what he may mean here is that, given the uncertainty of the estimates, being close to passing a benefit–cost test may mean that with accurate data it might well pass.

People are often shocked to discover that collecting recyclables costs two to three times as much as collecting trash—usually well over $100 a ton for recyclables versus $50 a ton for trash. They shouldn't be shocked. Consider the following very simplified story. A city sends out two trucks, similar in capacity (in cubic yards), cost, and crew. One truck picks up three-fourths of the tonnage (the trash) and compacts it to a third of its original size; the other truck picks up the other fourth of the tonnage (the recyclables) and doesn't compact it at all. The two trucks fill up at the same time. The two trucks then go to the relevant unloading site (assumed to be the same distance for each truck), empty themselves, and proceed again on the pickup route. Their total costs are identical (ignoring time differences at each stop and differences in gasoline usage owing to the different loaded weights of the two trucks), but the trash collection truck is collecting three times as much material. The difference in the cost per ton basically stems from the fact that you cannot compact the recyclable materials (and still hope to sort them later). As a result, whereas the average cost of collecting recyclables declines as more is collected at each stop, the overall average cost of collecting trash-plus-recyclables rises as ever larger fractions of household trash are recycled—the compaction disadvantage of recycling can never be overcome.[1]

There is empirical evidence of this higher cost of collecting recyclables (Stevens 1994; also see the Appendix). Thus, the cost of collecting recyclable materials is inevitably larger than the savings due to the avoided cost of collecting trash (Denison and Ruston 1990; Francis 1991; CWC 1993; Scarlett 1993; Powers 1995; Ackerman 1997). But as more recyclables are collected, the average cost of collecting them comes down much faster than the average cost of collecting MSW goes up, so the net effect is that total collection costs do not rise very rapidly, according to one careful study (Stevens 1994). Quintupling the recycling rate—from about 5% to about 25%—raises the *total* MSW *and* recyclables collection costs by about a third.

To illustrate: Ecodata surveyed 60 randomly selected U.S. cities in 1993 and found the following average costs in dollars per ton of trash and recycling collected, according to what percentage of the MSW was recycled:

Percentage recycled	MSW cost	Recycling cost	Weighted average cost
0–9	46	312	59
10–19	58	112	66
20 and more	72	102	80

These estimates, which are from Stevens 1994, have been converted to 1997 dollars, and the weighted average has been taken at recycling percentages of 5%, 15%, and 25%, respectively.

Not only are recyclables collection costs high, they also vary a great deal across cities, depending on crew size, truck type, volume collected per household, number of different materials collected, average distances between stops, whether collection is public or private, and so on (Miller 1993; Stevens 1994). To a great extent, the high and varying costs per ton of collecting recyclables is due to the newness of the activity, especially in comparison with trash compaction, where cost-minimizing collection techniques have been evolving for a full century. As cities and recycling collection truck manufacturers learn—from their own experience and from that of other cities and manufacturers—the cost per ton should come down, but never to the level of trash collection, until we find a way to compact recyclables and still sort them.

Processing Recyclables

Recyclables are processed at a materials recovery facility (MRF), where the heterogeneous collection is turned into homogeneous and hence marketable bales of recyclables. The MRF has both capital and operating costs. When the nearest MRF is much further away than the nearest landfill, the MRF also has differential transport costs that must be factored in. In the sparsely populated Rocky Mountain states, this factor alone accounts for the near absence of recycling (Truini 1999).

MRFs are expensive. Inside these unprepossessing buildings, many things are happening: shredding, whereby mechanical force is used to break the collected materials into small, uniform sizes; screening, whereby particles of different sizes are separated; air or water flotation, whereby particles of different densities are separated; magnetizing, to isolate iron materials; and just plain hand-sorting. The first four operations require costly equipment, and the final one is high wage and labor intensive. The good news is that, because MRF technology is still quite young, cost-saving (and especially labor-saving) innovations will probably bring the costs of MRFs down a lot in the near future (Williams 1991).

How expensive are MRFs? Because MRFs are relatively new, and various MRFs do different things in different ways, cost estimates vary greatly. A sampling of studies turned up cost estimates (converted to 1997 dollars) ranging from $30 per ton of recyclables processed to $80 per ton (Francis 1991; Miller 1992; Scarlett 1993; CWC 1993; U.S. EPA 1995a; Ackerman 1997). Almost all such studies, if they look also at MRF revenues, find that the processing cost at the MRF exceeds the revenue gained from the sale of the recovered materials. The MRF, seen in isolation (i.e., ignoring the costs of collecting the recyclables), is not *currently* a commercially viable entity unless it charges tipping fees or receives a municipal subsidy.

There are economies of scale in MRFs. Both the capital cost per ton and the operating cost per ton decline significantly with larger size (Chang 1992). Although small "McMRFs" are available for small cities, a full MRF requires a population of at least 300,000 (Williams 1991). But we should be careful in our use of the word "scale" in this context. Doubling the size and throughput of a MRF, while processing the same materials, indeed much reduces the cost per ton of the operation. But increasing the tonnage processed by handling more different materials does not reduce cost per ton. Indeed, there is, as we shall soon see, a hierarchy of recyclables—some are relatively low cost to process; others are relatively high cost. In short, MRFs display economies of *scale* but diseconomies of *scope*.

Nowhere is this distinction between scale and scope seen more clearly than in the disposal of old personal computers or television sets. Few, if any, municipal recycling programs now accept such items because their daily throughput would be so low that sorting them out from the other recyclable materials and then breaking them down into homogeneous marketable components would add greatly to the cost at the MRF—some estimates say at least $20 per unit (Jung 1999; Salkever 1999). Take computers, for example. Even though they contain such valuable components as gold, silver, copper, and palladium, each computer contains only tiny amounts of each—for example, 200 pounds of circuit boards and costly extraction are needed to yield one ounce of gold, worth a few hundred dollars. One article guessed that all the recyclable materials of a typical personal computer were worth about $6 (Hamilton 2001). In short, adding old computers to the scope of a recycling operation adds much more to the MRF's costs than to its revenues, because scale economies cannot be achieved.[2]

Does this mean that we are doomed to continue throwing old personal computers into landfills? No, for two reasons. First, computers also contain lead, a hazardous substance, and states are beginning to ban their discard in ordinary landfills. Second, economies of scale in computer dismantling may become achievable if the recovery facilities are large enough. This means countywide—perhaps even statewide or regional—specialized computer MRFs. Some private efforts are already appearing, and apparently are already profitable (Paik 1999). Eventually, computers will probably be collected by the regular municipal recycling system—taking advantage of scale economies there—and then shipped immediately to the nearest specialized computer MRF—taking advantage of scale economies there. Profitable TV recycling is more distant, because TVs contain fewer valuable parts and more hazardous parts than computers (Ramstad 2000).

Other Benefits and Costs

Although we have not been able to assign firm dollar figures to most of the benefits and costs discussed in the previous section, it does not look good for recycling to here. The collection of recyclables is two to three times as expensive per ton as the collection of ordinary trash, so the net there is a negative for the benefit–cost analysis. The cost of the MRF is usually larger than the revenues it earns on its materials sales, so there is another net negative. Avoided landfill costs must offset both these net negatives to yield an overall positive benefit–cost result. In most parts of the United States, landfill costs (including their external costs) are not now high enough to do that.

There are two ways to avoid this negative conclusion. The first way is to perceive market failures in other markets—for example, energy and virgin materials—that cannot be directly corrected and hence need the benefit of recycling. If virgin materials are too cheap, from a social viewpoint, because of their neglected external costs, then recycling acts *to some extent* as a tax, in that it pushes down the price that producers of virgin materials receive. In this case, one can attribute some of these indirect benefits in other markets to recycling, and if we attribute enough, it may yield a positive overall net benefit.

The second way is to incorporate households into the analysis. So far, all the benefits and costs have started with the trash already at the curbside. We have ignored both the costs to the household of preparing the recyclable materials for separate collection and the benefits to the household of participating in an activity that may lead to a more sustainable economy. We could readily measure the cost side of this, and some do.[3] It is harder, but not impossible, to put dollar values on recycling's "making people feel good about themselves"—being able to recycle is worth $16.27 a year to the average household, according to one study (Kinnaman 1996). I suspect that these benefits and costs roughly cancel each other, and that omitting them does not bias the benefit–cost results. Not all studies agree—see the box titled "Recycling in Lewisburg, Pennsylvania." In any case, there is evidence that the degree of citizens' environmental concern has significant impact on whether a community recycles or not (Tawil 1996).

In short, recycling probably does not now pay off in a social benefit–cost sense for the *average* municipality in the United States. Although the methodologies vary a lot, many— but definitely not all—empirical studies agree that the bottom line on the *average* city's recycling was negative in the 1990s (Deyle and Schade 1991; Specter 1992; Scarlett 1993; Curlee et al. 1994; Franklin Associates 1994; Kinnaman 1996; Shore 1997). Not only does the net benefit of recycling vary a lot across cities, but there is evidence that cities somehow recognize this in making their decisions whether to adopt curbside recycling. One study examined 80 Massachusetts towns, 31 of which had curbside recycling and 49 did not. It found that the towns that *did recycle* saved on average $70,000 a year because of the decision to recycle, whereas the towns that *did not recycle* saved on average $88,000 a year because of the decision not to recycle (the study's figures have been converted to 1997 dollars; Tawil 1996).

So, the answer to the question of whether recycling passes its benefit–cost test is yes *and* no—it depends on what kind of municipality we are considering. In many places in the United States, recycling already passes the test—it comes arguably close to passing in Ann Arbor, as the Appendix shows. Trash and recycling collection costs, as well as tipping fees, vary greatly across U.S. municipalities, and these costs are a critical ingredient in analyzing the benefits and costs of recycling (Apotheker 1993; Tawil 1996). This suggests yet again the

> ## Recycling in Lewisburg, Pennsylvania
>
> Lewisburg, Pennsylvania, is a town of only 6,000 residents, hardly the place where we would expect to find an economically successful recycling program. Nevertheless, in 1991, in response to a requirement by the state, Lewisburg introduced a minimal curbside recycling program, with a monthly collection of newspaper, aluminum, and glass (drop-off facilities were available for other metals, magazines, and certain plastics). One careful study estimated the annual benefits and costs of Lewisburg's program in 1995 (Kinnaman 2000); in the format we have been using (in thousands of 1997 dollars), they are:
>
> | Category | Benefit (+) or cost (−) |
> | --- | --- |
> | Recovered materials revenues | +9.5 |
> | Reduced waste disposal costs | +16.3 |
> | Reduced solid waste collection costs | +1.6 |
> | Recyclables collection costs | −4.2 |
> | Recyclables storage and processing costs | −11.5 |
> | Administrative and business costs | −13.9 |
> | Net household willingness to pay for recycling | +234.1 |
> | Total | +231.9 |
>
> There are two alternative ways of looking at these estimates: First, the people of Lewisburg love recycling (although they may love it not wisely but too well) and would pay almost anything to do it, so the benefit–cost analysis easily passes. Or second, the average Lewisburg household almost surely wouldn't be willing to pay $100 a year if it had to put its money where its mouth is, so the analysis probably fails.

foolishness of state recycling goals that apply equally to all municipalities in the state, whether large or small, dense or sparse, near to or far from cheap landfills or incinerators or MRFs.

Recycling will be done in more and more cities and towns as time goes on. Recycling is still in its "shakedown" phase. The average national costs of collecting and sorting recyclables will come down over time, as cities learn to operate their recycling programs in a less costly fashion and to use better recycling techniques. Moreover, markets for recyclable materials and for products using recyclable materials are also in their infancy, but they are growing rapidly, partly in response to the growth of recycling and the supply of recyclable materials (we will look at some of these markets in Chapter 12).

In short, over time the benefits of recycling will grow, and its costs will fall. Recycling will probably pass benefit–cost tests with regularity in the not-too-distant future. Furthermore, the time when recycling will pass such tests is nearer because we are already recycling. In a sense, we should count the future net benefits of cost-effective recycling as one of the major benefits of our investments in recycling today.

How Much Should We Recycle?

So far we have been thinking of recycling as an either–or proposition, but in fact each municipality has to decide not only whether to recycle but also how much to recycle. By how much, I do not mean how much of any particular material. Given the economies of scale in recycling, if it is socially profitable to recycle some of a particular material, it is more profitable the greater the participation of the residents in that recycling and hence the more of it that is recycled. The question of how much is really a question of how many different materials should be included in the recycling operation.

"Success" is *not* measured by a municipality's recycling rate. In a report for the U.S. EPA, the Institute for Local Self-Reliance labels 18 U.S. communities "record-setters" and asks not if they are recycling too much but "What can … [we] learn from these record-setters?" (U.S. EPA 1999b, 1). This is the wrong question. Given the diseconomies of scope, more kinds of materials means ever higher marginal cost. As more kinds of materials are collected, the tasks of the curbside collectors and MRF sorters are made ever more complex.

So how many kinds of materials should be collected? In theory, this decision is easy. Assume that all materials (that can be recycled at all) can be acquired either by extracting a virgin material or by acquiring a recyclable material. If the two alternatives are perfect substitutes, society has to make a choice between extracting and disposing of the virgin material each period and extracting it once and then recycling it every period thereafter. (This scenario vastly oversimplifies the choice, but will make the ideas clear; among its other oversimplifications, is its neglect of the fact that no material can be 100% recycled.) The virgin material's marginal social cost to produce is MSC_v, so recycling saves that MSC_v.

Recycling Where Landfills Are Scarce

Because Japan's urban population is dense, landfills near its cities are few and costly. Accordingly, about 75% of Japan's solid waste is incinerated, and its landfills are largely reserved for incinerator ash. How does this reliance on incineration affect Japan's recycling?

The quick answer is that Japan's incineration adversely affects recycling. The national average recycling rate for Japan is less than 5%. Typically, although cities vary a great deal in their procedures, the Japanese separate their trash into two parts: combustibles and noncombustibles. Combustibles include paper and plastic, materials that are usually recycled in other industrial countries. Noncombustibles include metals and glass, also prime candidates for recycling. Noncombustibles are generally sent through a MRF, where many of the recyclables are removed, before the rest is sent to a landfill.

But the quick answer is not a complete answer. All the above percentages refer to municipally collected solid waste, but a great deal of private recycling goes on before the city sees the remainder. Some private recycling is purely market driven. Unsubsidized scavengers work the streets, with a booming loudspeaker announcing their willingness to trade new toilet paper rolls for old newspapers, magazines, cardboard, and telephone directories (Kanabayashi 1982). But much of it is subsidized by the municipality, through payments by weight of collected recyclables—in Hiratsuka City, for example, these subsidies averaged more than $50 per ton of recyclables. A study of Matsudo City estimated that 40% of the city's waste was collected by these subsidized recyclers—40% plus 5% of the other 60% means an overall recycling rate of 43%. Because much of the recycling is privately done, and because preconsumer, industrial recycling data are often lumped with household recycling, the exact Japanese recycling rate is uncertain. Some say 50%, others say only half that (Hershkowitz and Salerni 1987; OTA 1989b; Cohen 1994). In short, the evidence from Japan is unclear on whether incineration reduces recycling.

Recycling also saves the cost of disposing of the virgin material after it has been used, MSC_d. The overall marginal social cost of producing and later discarding virgin material is the sum, $MSC_v + MSC_d$. The marginal social cost of recycling (MSC_r) is rising, indicating that recycling gets ever more expensive the more you do of it—not because you are doing more recycling of the same material but because you are recycling more kinds of materials, and there are diseconomies of scope.

Recycling should be expanded up to the point where $MSC_r = MSC_v + MSC_d$. We should recycle those products where $MSC_r < MSC_v + MSC_d$ and not recycle those where $MSC_r > MSC_v + MSC_d$. There is an optimal amount of recycling, probably greater than zero but less than the maximum amount that could possibly be recycled. Deviations from that optimal amount of recycling *in either direction* are costly. If recycling is too small, resources will be wasted extracting and disposing of too much virgin material; if recycling is too large, resources will be wasted doing too much recycling.

There is no reason for thinking that this optimal amount of recycling will be the same in various cities or at different times. So each municipality that starts to recycle has effectively got to work anew through the decision of how much to recycle. Surveys of recycled materials across cities show great variation in the outcome of this decision process. For example, one survey of more than 600 U.S. cities (that recycled at all) found that almost all of them collected glass bottles and steel, bimetal, and aluminum cans; almost all of them collected old newspaper, but only two-thirds collected corrugated containers, and fewer than one-fifth collected any other type of paper; more than three-fourths of them collected #1 and #2 plastic, but less than one-fifth collected any additional type of plastic; and only one-fifth collected anything other than the materials mentioned above, and these "other" materials ranged widely (e.g., textiles, aluminum foil, auto parts, and aerosol cans; Skumatz et al. 1998).

Some materials are prime candidates for recycling because they are cheap to collect and fetch a good price.[4] Aluminum cans, for example, are easy to sort, especially as the look-alike bimetal cans disappear from the market, and aluminum sells for up to $1,000 a ton (i.e., up to 2 cents per 12-ounce container). Newspapers are compact, also very easily sorted, and sell for up to $100 a ton. Plastics, conversely, are costly to recycle. There are many kinds to distinguish, which complicates the work of households, collectors, and MRF sorters. Plastics are bulky but lightweight, which means that each ton uses up a lot of collection truck space (Lamb and Chertow 1990). And most kinds of recyclable plastic materials rarely earn more than a few dollars a ton. Some materials, such as "mixed residential paper," receive *negative* prices, which means there is a charge to haul them away from the MRF.

Aside from which products are recycled, the sheer number of various products being recycled makes a difference to recycling costs. Recall the diseconomies of scope. Because collectors usually need to sort and do quality control on their routes, they are slowed by this variety. And the more kinds of things that are included in recycling, the more even conscientious households err in sorting—for example, in Ann Arbor, #2 plastic is collected if it is clear, but not if it is yellow. When the recyclables reach the MRF, the more different sorts that are needed, the slower the conveyor belts must move, the more sort bins there must be, and/or the more sorters must be hired.

How important quantitatively is this cost hierarchy? It is not easy to know this, because it is difficult to estimate *marginal* cost. The total costs of operating recy-

Table 9-1

Material	Collection	MRF	Total
Newspaper	72	34	106
Mixed paper		37	
Cardboard containers		43	
Mixed glass		50	
Clear glass	60	73	133
Green glass		88	
Brown glass		112	
Steel cans		68	
Aluminum	581	143	724
Ferrous metals	240	68	308
PET plastic (#1)[a]	1,089	184	1,273
HDPE plastic (#2)[b]		188	

Estimated Costs of Recycling Various Materials (cost in dollars per ton)

Note: Blank cells indicate that no estimates were reported in the sources.

[a]PET = polyethylene terephthalate.

[b]HDPE = high-density polyethylene.

Sources: Miller (1995c); U.S. EPA (1995a). The estimates are averages over a sample of 10 MRFs. The individual MRF estimates varied from 50% lower than the average to 50% higher. The study was conducted by the National Solid Waste Management Association in 1992 and 1993, and (presumably) is reported in dollars of those years.

cling collection trucks and MRFs are clear, but attributing those costs to particular materials is hard. Neither samples of different cities at a given time nor looking at one city over time provides the kind of data on other things held constant that are needed. For what it is worth, the results of one such effort are shown in Table 9-1.

Some of the numbers are curious, and the methodology of the study is vague— presumably the marginal cost of collecting and/or sorting a particular material would depend upon how many (and which) other materials were already being collected. But the basic point is clear: The *marginal* cost of adding another material to the recycling basket is likely to vary greatly across materials.

Despite the general uncertainty about the marginal cost of recycling various materials, one thing stands out clearly: Recycling plastic is expensive. Plastics are made from many different resins, and because they cannot be mixed, they must be sorted and processed separately. Such labor-intensive processing is expensive in such high-wage countries as the United States. Recycled plastic must compete with virgin resins, which are of higher quality, so the usually low price of oil puts a ceiling on the price of recycled plastic. Moreover, in the United States recycled plastic may not be used in food or drug packaging for health reasons, and those packages account for nearly half of all plastic packaging (Leaversuch 1994; Reynolds 1995). There is also anecdotal evidence about plastics recycling. When Philadelphia discontinued the collection of plastics in its curbside program, it cut its collection costs by $400,000 a year (Holusha 1993). When Portland expanded its program to collect all plastic bottles, its incremental recycling costs were estimated to be $706 a ton (Engel and Engleson 1998). Also recall the discussion in Chapter 2 of the German Green Dot and French Eco-Emballage programs. In the much cheaper French system, nearly half of the re-collected plastic is incinerated; and even in the expensive German system, the incineration of plastic may "count" as recycling.

Final Thoughts

We have asked two basic questions in this chapter: Does recycling pass its benefit–cost test? And how much should we recycle? The two questions are interrelated. If we recycle too much or too little, the overall recycling operations are less likely to pass the test.

This may help to explain why private, profit-seeking haulers often recycle presumably profitably, whereas nearby public recycling operations require subsidies. The private recyclers usually handle only a few basic materials, such as metal, glass, and newspapers, but public operations usually handle a larger variety of materials. Of course, private recyclers also may lose money but feel that they gain enough in good public relations to make it worthwhile. But public recycling also might do better on benefit–costs tests if it narrowed—for at least the near future—its scope of operation.

The question that should be asked by public recycling operations everywhere is how to maximize social profit—and that requires estimating the marginal social benefit *and* cost of various possible recycling scopes and choosing that scope that equalizes MSB and MSC. Too often, public-spirited recycling advocates choose the wrong criteria—as, for example, did Naomi Friedman in the study by the Institute for Local Self-Reliance for the EPA: "You want to maximize the amount of material collected per household …. That means collecting as many materials as possible and maximizing participation" (quoted in Glenn 1992).

Recycling "too much" is fine if a majority of the people want to do that, know how much it will cost, and are willing to pay the high price for it. But, of course, most people don't yet know how much it costs. What we should be worried about is that there will be a backlash against even sensible recycling when the costs get unbearably high. One newspaper headline put it tersely: "Recycling: How Long Will a Can-Do Feeling Last?" (Booth 2000).

In fairness—according to the criterion of two wrongs may make a right—urging ever more recycling may be a second-best approach to the basic problem of recycling: that not enough recycling is done by households because the marginal cost of using the trash disposal system is so low (usually zero). Many of the other problems of recycling dealt with in this chapter would be greatly alleviated if more socially correct trash pricing could be adopted. It is to this pricing structure that we next turn.

APPENDIX

A Social Benefit–Cost Analysis of Curbside Recycling in Ann Arbor

It is a good answer that knows when to stop.

—Italian proverb

Ann Arbor's recycling history is typical. Starting from a program in the 1970s where enthusiastic recyclers could bring a few well-sorted recyclables to a volunteer-operated drop-off station, it has grown into a mandatory program with weekly curbside pickup and a multimillion-dollar MRF collecting not only glass, metal, plastic, and paper but even textiles, oil filters, waste oil, batteries, ceramics, and aerosol cans. In a U.S. EPA review of 17 "record-setting" waste reduction communities, Ann Arbor was third in its recycling rate, 30% (U.S. EPA 1999b).

This appendix attempts to assign rough dollar values to the benefits and costs of the Ann Arbor curbside residential recycling effort, as of 1997 (Wu 1998; city budgets and personal communication with Bryan Weinert). The analysis is restricted to residential recycling, omitting not only industrial and commercial efforts but also composting, which in total tonnage is even larger than recycling (see the Appendix to Chapter 13). The basic benefits and costs of recycling are of six kinds:

■ *Revenues from recovering recyclable materials.* Ann Arbor sold its nearly 14,000 tons of recyclables at an average price of $37.65 a ton, which yielded a total revenue to the city and its MRF of $517,000. (This is the revenue from all the MRF's sales of recyclables. By a complex formula, the city receives some of the revenue from the MRF's sales of collected recyclables; in 1997, the city's share was $45,000: $517,000 minus $472,000.) Assuming that the social value of the recyclables is what was paid for them, this means a social benefit of $517,000.

■ *Costs avoided in landfill disposal.* Whatever tonnage of solid waste is recycled is not landfilled, and the resources needed to bury that waste are saved. Ann Arbor no longer operates its own landfill (see Appendix A to Chapter 4) but pays $28 a ton to dispose of it at a nearby private landfill. Because Ann Arbor collects 11,586 tons a year—about two-thirds of a pound per person per day—its residential recycling program directly saves the city $324,000 a year. But the very existence of this program permits the city to add another 1,281 tons of recyclables from other city sources, which means the program permits landfill costs to be avoided to the extent of $360,000 a year. (The drop-off center collects more than 2,000 tons, but nearly half of that is from noncity sources, so the city would not have been responsible for its landfill cost.)

■ *Solid waste collection costs avoided.* Solid waste tonnage diverted to recycling also does not have to be collected as solid waste. Ann Arbor spent $626,000 in 1997 collecting 16,107 tons of residential solid waste, the average cost of such collection is about $40 a ton. If social and private cost were the same, and if the average and marginal cost were the same, this would mean that recycling reduces the total social cost of collecting MSW by $450,000. Social and private cost probably are about the same, but the marginal cost of collecting an extra ton of MSW is almost certainly less than the average cost. It is not easy to know how much less. Assuming marginal cost is half of average cost yields a social benefit on this account of $225,000.

■ *Recyclables collection costs incurred.* The city contracts out the curbside residential collection of recyclable materials at a cost to the city of $1,014,000. Assuming that nonprofit contractor Recycle Ann Arbor breaks even on this activity, it costs about $87 a ton, for a social cost of $1,014,000.

■ *Operating costs of the MRF.* The MRF is also contracted out by the city, and the contractor gives out little information on its finances. But, again assuming that its

Category	Benefit (+) or cost (−)		Table 9-2
Revenues from recyclable materials	+517,000		Benefits and Costs of
Landfill costs avoided	+360,000		Ann Arbor Recycling
MSW collection costs avoided	+225,000		(1997 dollars per year)
Recyclables collection costs incurred	−1,014,000		
Operating costs of the MRF	−739,000		
Transfer costs avoided	+250,000		
Net total	−401,000		

total private revenues equal its total social costs, we can add together what the MRF gets from the recyclable revenues ($472,000) and what the city pays the MRF for its activities ($267,000) to get an estimate of its social cost of operation: $739,000, or just above $63 per ton of recyclables processed. Technically, because the MRF is already built, such a sunk cost should not count in a benefit–cost analysis. What the city pays for the MRF undoubtedly does include both interest and depreciation, so this analysis is counting these sunk costs as if they had not yet been undertaken.

■ *Transfer costs avoided.* In Ann Arbor, there is another benefit to recycling, one not mentioned in the text. Without recycling, all of Ann Arbor's trash would have to be moved by the collection trucks to the private landfill 25 miles away. With recycling, all the trash goes only as far as the nearby MRF. What is not recycled is then transferred and further compacted for its lower-cost trip to the landfill. As a result, recycling saves about 250,000 miles a year of travel for its trash collection trucks. At $1 a mile, this means a benefit of $250,000. Should Ann Arbor have considered a transfer station without recycling? We cannot answer this question, because there is no easy way to disentangle the transfer costs from the recycling costs at the MRF. (The costs of the compacting at the transfer station and of the semi that moves the waste to the landfill are included in the costs of the MRF.)

These six benefit and cost estimates are collected in Table 9-2, which shows that Ann Arbor's recycling program currently has a net social value of about minus $400,000 per year. Does this mean that recycling in Ann Arbor should be abandoned? No—there are four important other considerations:

■ Some of the numbers are only estimates, and two in particular are iffy. First, prices of recyclable materials in 1997 were very low by historical standards. The revenues in the next year, 1998, for example, were more than $1 million. What is important for this benefit–cost analysis is the average price of recyclable materials over the life of the MRF. Second, the estimate of the savings in solid waste collection due to a separate recycling collection—$225,000—*assumed* that the marginal cost of trash collection is half of average cost. In fact, we can only be sure that something between $0 and $450,000 was saved in municipal trash collection. And whatever was saved in 1997, even if we knew it exactly, would only provide a lower-bound estimate of long-run annual savings, as routes, labor, and trucks were adjusted to the new situation and "learned" how to handle the process more cheaply.

- We have not counted either the costs or benefits of recycling to the households that put their home effort into the program. For some, recycling is a pain that they suffer because it is mandatory; for others, it is a joy that offers them a chance to contribute to a more sustainable society. The dollar values of these pains and joys are hard numbers to estimate; but if they net out to an average benefit of more than $4 per Ann Arborite per year, then recycling does pass its social benefit–cost test. Certainly, Ann Arborites would overwhelmingly reject a referendum proposal to simultaneously terminate recycling and reduce property taxes by $4 per person a year.

- Even if recycling now fails its benefit–cost test, there may be learning by doing, and the costs of recycling may be coming down. It is the cost over the life of the MRF that matters, not just the 1997 cost.

- Recycling may fail its benefit–cost test because Ann Arbor is recycling too much. Minor items—such as ceramics, textiles, #3 plastic, batteries, oil filters, and waste oil—raise the sorting costs at the MRF by much more than they raise revenues. But notice the examples here; some—such as batteries, oil filters, and waste oil—are collected not so much to recycle them as to keep them from being carelessly discarded and contaminating groundwater. We should be counting this expected groundwater pollution avoided as a benefit of the recycling program. Still, the inclusion of many minor items in Ann Arbor may tilt its overall benefit–cost result from positive to negative.

One final point, following from the fourth consideration just above. We get a clue about the marginal cost of minor items by looking at a private trash collection and recycling company operating in and around Ann Arbor, Mr. Rubbish. Private companies such as this one may be forced by their customers to offer a recycling opportunity, or they may simply want the positive publicity from offering it; but they do not want to lose much, if any, money on it (Johnson 2000b). Accordingly, Mr. Rubbish requires extensive household separation, charges 35 cents per bag of recyclables, and collects only paper, metal, (#1 and #2) plastic, and alkaline batteries. For a long time, Mr. Rubbish accepted no glass containers, but *clear* glass is now accepted if separated (because collection with the rest of the trash and recyclables tends to break bottles) and unbroken, cleaned, and with cap and ring removed (Woods 1992).

Notes

1. In case this is not obvious, a little algebra proves the point. Suppose each household puts out exactly one cubic yard of waste each week, that it costs $1 for a truck to drive by that household, and that each collection truck has a capacity of C cubic yards. The recyclable truck picks up r cubic yards of recyclables (where r is the fraction recycled) from each household and drives by C/r households before it fills up. It then drives to the recyclables-sorting facility and back, at a cost of Z dollars, and resumes its collections. The total cost of this truck is ($C/r + Z$), where C/r is the variable cost of driving alongside the households and Z is the fixed cost of going to the sorting facility. The average cost per cubic yard of recyclables is ($1/r + Z/C$), which is lower the higher is r.

Meanwhile, the trash collection truck is picking up ($1 - r$) cubic yards at each house and compacting it to ($1 - r$)/3 cubic yards. It can drive by $3C/(1 - r)$ households before filling, at which time it drives to and from the landfill or incinerator (assumed to be the same distance away as the recyclables-sorting facility) at a cost of Z. The trash truck's total cost is $(3C/(1 - r) + Z)$. Because the trash truck collects 3C (where C is precompacted cubic yards) of trash before filling up, the average cost is $[1/(1 - r) + Z/(3C)]$, which is higher the higher is r. The average cost of *both* kinds of collection is the weighted average, $[r(1/r + Z/C)] + (1 - r)[1/(1 - r) + Z/(3C)]$, which can be more simply written as $[2 + (1 + 2r)Z/(3C)]$. This overall average cost rises as r rises. The average cost of collecting recyclables declines as r rises and the average cost of collecting trash rises as r rises, but the average cost of the *total* trash-and-recycling collection rises as r rises.

2. Three anecdotal indications of the cost. First, IBM will promise to reuse or recycle a customer's computer if the customer returns it (at the customer's own expense) *and* pays IBM a fee of $30 (Truini 2000b; http://www.ibm.com/ibm/environment/products/pcrservice.phtml). And two, the Rhode Island Resource Recovery Corp. has begun collecting computer waste for free, if the customer brings it in, but it estimates that even such drop-off collection will end up costing more than $1,500 per ton (Truini 2000c, 2001g). And three, Hewlett-Packard will pick up and recycle *any* manufacturer's computing equipment. The cost depends on quantity—for one old computer and monitor from Ann Arbor, $46 (Truini 2001d; http://warp.external.hp.com/recycle/).

3. Recycling detractors can get some big numbers on this score—up to nearly $3,000 per ton of recyclables collected (Tierney 1996). Conversely, recycling boosters often decry the effort to put dollar values on the cost of civic duty—"life is not a business, and participation in society is not a reimbursable business expense" sounds great but is no excuse for omitting some of the real costs of recycling when assessing the activity (Ackerman 1997, 13). Although I think one should count household costs in principle, I find it hard in fact to get a net cost greater than a few dollars per ton of recyclables for a typical U.S. family.

4. When only one material is recycled, there is no need for a MRF, which can make recycling very cheap. Indeed, a comparison of East Longmeadow, Massachusetts, which did no recycling, with neighboring Longmeadow, which recycled only newspaper (and a little corrugated cardboard), showed that the net (of recyclable revenues) cost per total ton (of both trash and recyclables) in Longmeadow was actually lower than that in East Longmeadow—$77 a ton versus $83 a ton in 1997 prices (Powell 1989).

Policies for Recycling

The truth is never pure and rarely simple.

—Oscar Wilde, *The Importance of Being Earnest*, 1895

There are two ways to encourage a socially optimal degree of recycling—price policies and nonprice policies. We look first, and extensively, at price policies. In Chapter 3, we considered the waste pricing problem when the household faced the very simplest decision: whether to reduce, to reuse, or to legally discard the trash. Then in Chapter 6, we introduced the possibility of illegal disposal. In each case, we looked for the pricing structure that would induce the household to make the socially correct discard decision. Now we expand this choice into the contemporary world, where the household also faces the possibility (for many products) of recycling.

Again, we will move in short steps—the chapter is tough enough reading without big leaps. First, we will review the optimal pricing structures that emerged in Chapters 3 and 6, when households had no opportunity to recycle. Then, we will examine the pricing structure that is needed when households can recycle, first when little illegal disposal occurs and then when illegal disposal is a serious problem.

Review of Pricing without Recycling

Each household must continually make choices between reducing-or-reusing and disposing of its trash (Chapter 3). The decision is made optimally from a social viewpoint when there is a trash collection charge equal to the marginal social cost of collection and landfilling/incineration (whichever is socially less costly). Figure 10-1 (a combination of Figures 3-3 and 3-4) shows the two ways that this MSC price can be achieved—with the cost of collection and landfilling or incineration here called "trash cost." Although practical administrative considerations may dictate one or the other, the two pricing methods are identical in theory. Because any advance disposal fee levied on a manufacturer will be passed on to the consumer through a higher price, the household effectively pays the MSC of disposal, either when the product is purchased or when it is put out for disposal.

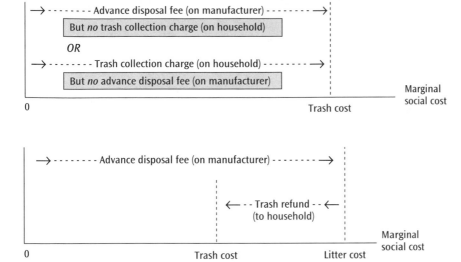

Figure 10-1

Two Trash Pricing Means with No Illegal Dumping and No Recycling

Figure 10-2

Trash Pricing with Illegal Dumping but No Recycling

Once we introduced the possibility of privately costless illegal disposal (in Chapter 6), a trash collection charge on the household becomes counterproductive. Indeed, even if legal trash disposal is also zero-priced, there will be too much illegal disposal because the external costs of litter disposal are greater than the external costs of proper disposal. In theory, there must be a refund to the household for proper disposal equal to the reduced social costs of proper disposal. Figure 10-2 (reproduced exactly from Figure 6-1) shows this optimizing price structure. That's the theory (so far). However, we noted in Chapter 6 the many practical difficulties of giving trash refunds, and we will keep them in mind as we introduce recycling into the household disposal decision.

Pricing with Recycling but No Illegal Disposal

There are *three* policy tools ready for use once recycling becomes a possibility: (a) an advance disposal fee, levied either on the manufacturer at the time of production or on the consumer at the time of purchase; (b) a trash collection charge (or refund) levied on the household at the time of disposal of the package or the remnants of the product itself; and (c) a recycling collection charge (or refund) levied on the household when materials are put out for recycling.

Ideally, these three tools must be chosen in such a way that households buy the right amount of stuff, use (and reuse) that stuff for the right amount of time, and then make the right decision between disposing of the remnants as trash or for recycling. So that the consumer will buy the right amount of stuff, the price ultimately must include not only the cost of producing the product and its package, but also the marginal social cost of disposing of the product and package. So that the consumer will continue to use the product the right amount of time, the benefit to the consumer of reusing the product (e.g., for one more year) must equal the

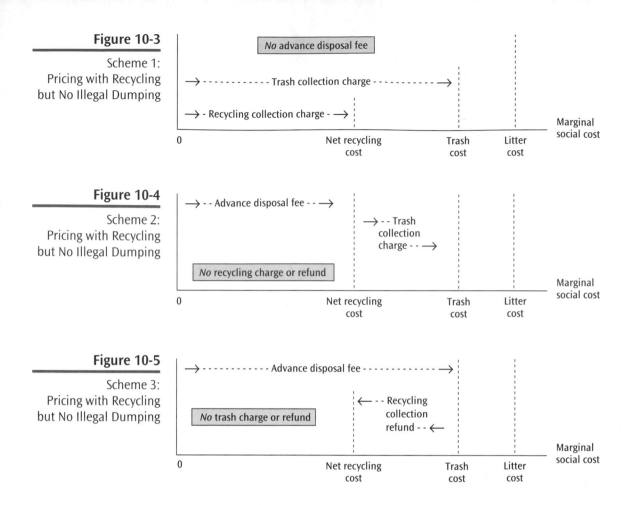

Figure 10-3

Scheme 1:
Pricing with Recycling
but No Illegal Dumping

No advance disposal fee

→ - - - - - - - - - - - Trash collection charge - - - - - - - - - - →

→ - Recycling collection charge - →

0   Net recycling   Trash   Litter   Marginal social cost
    cost   cost   cost

Figure 10-4

Scheme 2:
Pricing with Recycling
but No Illegal Dumping

→ - - Advance disposal fee - - →

→ - - Trash collection charge - - →

No recycling charge or refund

0   Net recycling   Trash   Litter   Marginal social cost
    cost   cost   cost

Figure 10-5

Scheme 3:
Pricing with Recycling
but No Illegal Dumping

→ - - - - - - - - - - Advance disposal fee - - - - - - - - - - →

← - - Recycling collection refund - - ←

No trash charge or refund

0   Net recycling   Trash   Litter   Marginal social cost
    cost   cost   cost

interest charge on the disposal cost (because the reuse postpones that disposal cost for one year). And finally, so that the consumer will choose correctly between trash and recycling, the difference in the price of the two disposal methods must equal the difference in the marginal social costs of the two methods.

All this sounds like a tall order, but it can be done. Indeed, *no* advance disposal fee is needed, simply a trash collection charge that reflects the MSC of trash collection and disposal, and a recycling collection charge that reflects the cost of collection and sorting of recyclables (net of the revenue earned on recyclables; hereafter, we will refer to this as the "net recycling cost."). Call this scheme 1 (Fullerton and Kinnaman 1995). Figure 10-3 shows these necessary fees (with the litter cost drawn in for reference, but it is irrelevant for now because we are assuming no illegal disposal). The figure is drawn on the assumption that the net recycling cost is lower than the trash disposal cost—indeed, if it is not, recycling will probably fail its benefit–cost test (as we saw in Chapter 9; and I say "probably" because some households have a willingness to pay for the pleasure of recycling, which we are now ignoring). In any case, if the net recycling cost were to exceed the trash disposal cost, the only difference would be that the correct recycling collection charge would exceed the correct trash collection charge. (In fact, in many communities where trash and recyclables are collected by a private hauler, there is an additional charge for recyclables.)

Moreover, two other pricing schemes come very close to achieving the optimal household decisions: Schemes 2 and 3 also require the use of only two of the available policy tools;[1] they are shown in the Figures 10-4 and 10-5. These two schemes only "come very close" to optimality because neither offers the correct incentive for the household to reuse—in each of the two schemes, the extra price the household saves by reusing a product is less than the MSC of its disposal, whether as trash or as recycling—indeed, with one of the two schemes, the household actually gains by rejecting reuse and putting the product out for recycling (even though recycling has a positive net social cost).[2]

Look again at scheme 3 (in Figure 10-5). Notice its resemblance to a mandatory deposit system (discussed in Chapter 6). The advance disposal fee, if paid by the consumer at the time of purchase, is akin to the deposit; and the recycling refund is akin to the redemption of the deposit (Palmer and Walls 1997; Palmer et al. 1997). Of course, the deposit system in scheme 3 is administered by the municipality, not the producer of the product, and the collection of the recyclables is operated by the municipality, not the producer of the product (and the attendant wholesalers and retailers). Scheme 3 also makes the point that the whole deposit ought to be redeemed at the end only if the net recycling cost is zero.[3]

Finally, before leaving the theory of these three tax-and-subsidy schemes, we should ask again whether the disposal charges would decline if more of the product were recycled. If extensive recycling of a product affects neither its marginal trash cost nor its marginal net recycling cost, this extensive recycling has no effect on the correct set of taxes or refunds. If, however, more extensive recycling reduces the marginal net recycling cost, as it probably would, then there is a reduction in the recycling collection charge of scheme 1, a reduction in the advance disposal fee (and an increase in the trash collection charge) of scheme 2, or an increase in the recycling refund of Scheme 3. So the answer is, yes, extensive recycling may reduce some charges on a product, but not just because of the recycling rate; rather, because it affects net recycling costs.

Up to the probably unimportant difference in the reuse incentive, the three pricing

Advance Disposal Fee or Recycling Subsidy?

In scheme 3, there are two ingredients, an advance disposal fee and a partial recycling refund of that fee. One study asks whether using one or the other, but not both, of these ingredients is enough to get the desired effects (Palmer et al. 1997). The advance disposal fee encourages source reduction but does nothing to encourage recycling. The recycling subsidy (refund) encourages recycling but does nothing to encourage source reduction. Each, by itself, attacks only one aspect of the problem.

The Palmer study takes as the goal of the policy the reduction of solid waste and asks how big each of the three possible policies must be to achieve a goal of a 10% reduction in U.S. solid waste. (Actually, it looked only at paper, glass, aluminum, steel, and plastic, but these account for roughly half of U.S. MSW.) The three policy tools are an advance disposal fee alone, a recycling subsidy alone, and an advance disposal fee *and* a recycling refund exactly equal to the advance disposal fee. Given their estimated elasticities of responses of households to the various prices involved, the answer is this: An advance disposal fee by itself would have to be $102 a ton—about five cents a pound. A recycling subsidy by itself would have to be $118 a ton—about six cents a pound. But an advance disposal fee that was fully refunded if the product were recycled would need to be only $54 a ton—less than three cents a pound. In short, the two policies are much more effective at reducing solid waste if used together.

schemes are identical, in theory. Are there practical reasons for choosing one of the three over the other two? Let us look at each scheme in turn.

■ *Scheme 1*. This is the ideal scheme, for it is the only one to offer the correct reuse incentive. Moreover, because it does not use an advance disposal charge, the trash collection charge and the recycling collection charge can be tailored to local conditions. It does, however, require two different fees to be levied at the household level, and because there are many more households than firms, the absence of an advance disposal fee means that the more easily collected tax goes unused. The trash collection charge is the highest of the three schemes, being equal to the full marginal cost of trash collection and disposal, which means that this scheme creates the biggest incentive for illegal disposal. Finally, this scheme requires a recycling collection charge—provided the net recycling cost is positive, which it will be for most products for the foreseeable future—which may anger many households, who feel that if recycling is a "good" thing, they ought not to be charged for doing it. Furthermore, in theory, this scheme calls for a different recycling collection charge for each product that is recycled because the fee is supposed to equal the marginal net recycling cost. Because this net recycling cost depends heavily on the price of the collected recyclables, it will differ across products. But it is practically very difficult and costly to charge households different fees for different products. And charging a single sort-of-average fee overencourages households to recycle low-value recyclables and underencourages them to recycle high-value recyclables.

■ *Scheme 2*. This scheme uses the more easily administered advance disposal fee (equal to the net recycling cost) as well as a trash collection charge on the household. The trash collection charge is smaller than in scheme 1, and hence less of an inducement for illegal disposal; but it still provides some such bad incentive. The advance disposal fee must be uniform across regions, though not of course across products, and so must be set at the average national net recycling cost. This means it will be too high in urban areas where recycling collection is cheaper, and it will also be too high in places where prices of recyclable materials are high. There is neither fee nor refund for recyclable materials, which conforms with conventional beliefs and most current practice and makes it easy to administer. But notice that the trash collection charge in this scheme should, in theory, equal the difference between the marginal landfill (or

---

Aseptic Boxes versus Aluminum Cans Revisited

Recall, in Chapter 2, that we considered the unrecyclable aseptic drink box as an alternative to, say, the recyclable aluminum can, and we suggested that there *might* be a role for each in beverage packaging. Let's assume that every household is indifferent about the type of container used and look at scheme 2 described in the main text, with its advance disposal fee and a trash collection charge. The advance disposal fee will differ across products, being equal to the marginal net recycling cost. Because aluminum cans are cheap on this count (owing to the high recycled aluminum price), they will bear a low advance disposal fee. Aseptic boxes, having essentially no recycling potential at this time, will have a high advance disposal fee. Because curbside trash pickup will not distinguish between products, the household will pay the same trash collection charge rate for each. But the aseptic box is lighter and more compact, so its trash collection charge per drink container will end up being lower.

Consumers who can recycle will end up being charged less for the aluminum cans, because recycling the cans will cost them less than trashing the aseptic boxes. Consumers who cannot recycle will end up using aseptic boxes, because their trash disposal will cost less than that of the heavier and bulkier metal cans. Scheme 2 pricing will get aluminum cans used and recycled where there is recycling and will get aseptic boxes used and trashed where there is no recycling.

incineration) cost and the marginal net recycling cost. Because it would be administratively costly to vary the trash collection charge across products, a single sort-of-average trash collection charge would end up being set, underencouraging households to recycle high-value recyclables and overencouraging them to recycle low-value recyclables.[4]

■ *Scheme 3.* The advance disposal fee in this scheme is large, equal to the national average marginal trash collection and disposal cost of the product. Thus, it is too high in some places and too low in others. Of the three schemes, this one most discourages reuse because it actually subsidizes recycling. Indeed, the recycling refund would be most difficult to administer because it should be different for various products, places, and times. Again, the use of a sort-of-average recycling refund would underencourage households to recycle high-value recyclables and overencourage them to recycle low-value recyclables. Moreover, because the advance disposal fee would presumably be collected at a national level and the recycling subsidy paid out at the local level, this scheme would add budgetary burdens to the entire solid waste collection system at the municipal level unless an extensive system of federal grants to local communities were organized.

Without empirical experience to tell us the magnitudes of the various good and bad features of the three schemes, we are left to make a judgment call. I like scheme 2. It seems to me to be closest to an ideal tax-and-subsidy system, it is administratively easy to operate, it conforms with common sense, and it should prove politically acceptable. Moreover, hidden in the preceding paragraphs is another reason for preferring scheme 2: Only it gives the manufacturer a direct incentive to make the product or the packaging more easily recyclable. The manufacturer can reduce the advance disposal fee that must be paid in two obvious ways: by using materials that are less costly to collect and sort for recycling; or by using materials that are more valuable when they end up on the recyclables market. (Actually, there is a third way by which manufacturers can reduce their advance disposal fees: by actively entering and fostering the recyclables markets into which their products and packages ultimately enter; more on this in Chapter 12.)

But if one is pessimistic about the size of the net social benefit to recycling—which means that one thinks the net social cost of recycling is not soon going to be much less than the cost of landfilling or incinerating trash—then scheme 3 will suddenly look very good: It then calls for an advance disposal fee equal to *both* the trash cost of the product and its packaging *and* the net recycling cost of the product and its packaging—the two are thought to be roughly equal. This means that trash and recyclables should both be collected for free. Administratively, this is a serendipitous outcome, for it means that only one tax (or subsidy) is needed—an advance disposal fee—and that tax is the easiest to operate, because there are fewer producers than households. I, however, am optimistic about the net social benefit of recycling in the near future, and scheme 2 recognizes that optimism by setting a trash fee that gives tangible economic incentive to households to recycle. Most, though not all, studies have found that the amount a household recycles significantly depends on the size of its trash collection charge (Miranda et al. 1996; Miranda and LaPalme 1997; Jenkins et al. 2000; Kinnaman and Fullerton 2000).

All the practical complications of the three schemes can easily be summarized. Any advance disposal fee would ideally be equal to some a_{ij}, where i is the product and j the place where it is sold (or more precisely, where it will be disposed of). But in fact, it would be difficult to vary the advance disposal fee by place, so it would end up being a_i, varying across products but not across places. Similarly any trash collection charge (t) or recycling collection charge/refund (r) should be some t_{ij} or r_{ij}, also varying across product and place. But in fact, it is difficult to vary them across products because that would require extensive household sorting or lengthy collector examination; so these would end up being t_j and r_j, varying across places but not across products.[5] Even without considering the possibility of illegal disposal, it is in practice not possible to achieve a best policy—close is as close as we can come (Calcott and Walls 2000).

As I said, I vote for scheme 2—with an advance disposal fee, varying across products and packages, levied nationally; a trash collection charge, varying across places, levied locally; and no recycling collection charge (or refund). What now seems to be appearing in the United States as a practical second-best policy is a variant of scheme 1—with a trash collection charge equal to the marginal trash collection and disposal cost; but with a zero recycling collection charge, despite the fact that there is still a large net cost to most recycling. This overencourages households to recycle, but that may not be a terribly bad idea in the early stages of recycling, when households are being asked to develop new recycling habits.

Pricing with Recycling and Illegal Disposal

If illegal disposal or littering is a realistic alternative to legal disposal either as trash or recyclable materials, then pricing becomes much more problematic. Essentially, once households can undertake costless illegal disposal—and are willing to undertake it—then it is impossible to charge a fee for any kind of legal disposal. There can be no charge for trash, and there can be no charge for recycling (Dinan 1993; Fullerton and Kinnaman 1995; Palmer et al. 1995; Palmer and Walls 1994).

Indeed, if households can litter without cost to themselves, then they must be induced through subsidies to dispose of trash and recyclables by socially more desirable means. An advance disposal fee must be added to the product, equal to the marginal social cost of disposal by *littering*, and refunds must be offered both for trash disposal and for recycling. Each of these refunds must be large enough to reduce the net cost of proper disposal to the marginal social cost of that disposal means. The pricing scheme is shown in Figure 10-6.

Just to glance at this pricing structure is to see its weaknesses. It obviously gives a perverse incentive *not* to reuse. It encourages overloading the trash barrel with nontrash to get the subsidy. And it encourages misplacing trash into the recyclables container to get the larger subsidy. Although huge revenues may be gained at the federal level from advance disposal fees, there are huge new expenditures on refunds—and the administration of these refunds—at the municipal level. The advance disposal fee can vary across products but not across places, and the social cost of illegal disposal will vary across locations. Similarly, the trash and recycling refunds can vary across places but not across products, and the net recycling cost (if not the trash cost) will vary across products.

We come again to the position we took in Chapter 6: It is difficult to use the pricing mechanism to prevent illegal disposal. Adding the possibility of recycling does

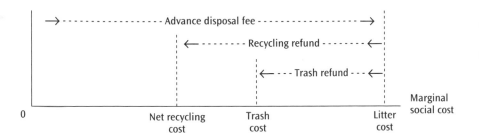

Figure 10-6

Pricing with Recycling
and Illegal Dumping

not diminish this difficulty. Ensuring that there is significant expected punishment for illegal disposal is almost certainly the way to approach this problem.

Although many communities have adopted some sort of trash collection charge, it is still not clear whether illegal disposal is in fact a serious problem. One careful study of the solid waste of Charlottesville, Virginia—after that city introduced an $0.80 price per bag for trash (but no fee for recycling)—showed that the average household put out 1.53 fewer pounds of trash a week and 0.58 more pounds of recyclables (Fullerton and Kinnaman 1996). The difference, nearly 1 pound, represents either source reduction and reuse or illegal disposal. That study guessed that it was 28% illegal disposal. If that were true, and this experience were extrapolated to the entire United States, it would mean about a half-million tons of additional illegally disposed trash a year.

The vast majority of studies of trash collection charges, however, conclude that illegal disposal is not a problem (Deisch 1989; Goldberg 1990; *World Wastes* 1993; Bender et al. 1994). The problem with this "vast majority" is that their results are based on hearsay, often from those who are biased toward trash collection charges. There is also a question of causation. It is very possible that more law-abiding or environmentally minded communities may be the first to embrace trash collection charges. More measurement is needed. But even if illegal disposal is a serious problem, pricing probably cannot solve it.

Other Side Effects of Pricing for Recycling

The possibility of significant illegal dumping is not the only interesting side effect of pricing for recycling. Recycling is bad news for incinerators (Keeler and Renkow 1994). For incinerators already built and operating, if recycling grows more rapidly than the total waste, deliveries to the incinerator will be inadequate to maintain capacity utilization. And for contemplated future incinerators, extensive recycling may make the waste stream left for burning too small for an economical incinerator.

Nor is the impact all in one direction. Not only does recycling hurt the prospects for incineration, incinerators hurt the prospects for recycling. To maintain the capacity utilization of its own incinerator, or to meet the contractual deliveries to a private incinerator, cities may be reluctant to initiate or expand recycling systems. This reluctance also makes sense in an opportunity-cost sense, even if the capacity utilization issue is moot. Every pound recycled saves landfill costs, but every pound recycled saves little if the alternative is an incinerator—the *net* marginal cost of incineration is roughly zero, because operating costs are more or less offset by energy revenues. So for cities with incinerators, recycling offers little or no waste cost savings.

The presence of an incinerator may also affect *what* a city recycles. If metals and glass are recycled, this takes unburnable materials out of the waste stream, which makes incineration more efficient. But if paper and plastic are recycled, this removes energy-intensive materials from the incinerator's fuel supply.

It is not surprising that a trash collection charge—with a smaller or no charge for recyclables—increases both participation in recycling by households and the total amount of recyclables collected. What is not clear is whether a trash collection charge encourages source reduction and reuse. Once a materials recovery facility is in operation, reduction and reuse threaten its capacity utilization—just as recycling threatens that of incinerators. Cities that recycle typically issue large pamphlets explaining how and why we should recycle but tell us much less about how and why we should reduce and reuse.

A little (more) theory tells us why the interactions of trash collection charges, recycling, and source reduction may not be clear. Consider first a city where a trash collection charge is introduced but where no recycling possibility is available. The trash collection charge makes the typical household want to reduce its now costly solid waste. It can only do this by reducing or reusing, so it does some of each. Now, if recycling becomes available, this household will also begin to recycle to reduce its solid waste costs, but it may well decrease its reduction and reuse because recycling has appeared as a personally less onerous means of cutting its solid waste costs. If a trash collection charge precedes recycling, the opportunity for recycling may discourage reduction and reuse. Only if recycling precedes or accompanies a trash collection charge can we be fairly certain that the addition of this charge will increase *all* of recycling, reducing, and reusing.

Setting Recycling Targets

Up to here in this chapter, we have been examining price policies for recycling. Except in those (still few) municipalities with trash collection charges, we have not yet begun to use price policies to optimize our recycling incentives. We have instead relied almost entirely on nonprice policies, to which we now turn.

One favorite way to initiate or expand recycling is to set a target for recycling as a percentage of municipal solid waste. The U.S. EPA has loosely set a target—25% by 1992, announced in a speech in 1988 by the then assistant administrator, Winston Porter; this was later and quietly boosted to 35% by the year 2005 (Porter 1988). Almost every state has also set targets, and they range widely, up to 70% for Rhode Island (Heumann and Egan 1998).

Setting a recycling target does no harm if the state then does nothing to implement it. Many states back their recycling targets with little more than a requirement that cities develop recycling plans or introduce minimal recycling. Others intend to fine municipalities that fail to meet the specified target by the specified date. As of mid-1999, only 69 of California's more than 450 municipalities were meeting the state's year 2000 target—a 50% reduction in landfilled solid waste. Did the state really fine the other 381 towns and cities $10,000 a day starting on 31 December 2000 (*San Francisco Chronicle* 1999)? Of course not. Extensions were granted until the year 2006, giving plenty of time to change the law, grant new extensions, or redefine the denominator—California does not measure waste generation but estimates it from a formula involving the state's income and population (Johnson 2000a).

Even targets without teeth have a way of inducing serious policies to achieve them. Setting a wrong target may well lead to wrong recycling policies. Careful economic analysis is needed to even guess at the answer to the question of how much recycling is desirable. But very little such analysis has gone into setting state targets. Neighbor states with similar living standards and population densities have often set very different targets, strongly suggesting that one—or both—of the targets is wrong. We should expect states with dense populations—and hence more costly landfills and less costly recycling collection—to do more recycling, but there is very

State Recycling Targets and Actual Rates

If there is logic to the setting of state recycling targets, the percentage target should be higher the higher is the density of the state's population and the higher is the state income per capita. That density matters is obvious, but that state income matters is less clear. Higher income per capita means that people generate more recyclable material per capita, and that in turn reduces the average cost of re-collecting it. In fact, there is a slight relationship, as Figure 10-A shows. But when you cover up the two dots to the far right (Rhode Island and New Jersey), almost no correlation remains.

One would also suspect that states' actual recycling achievements should be related to the targets, and they are, as Figure 10-B (on the next page) shows. But the relationship is also weak. And there always lingers the possibility that achievements determine targets rather than the reverse—that is, states that do (or can) recycle cheaply set high recycling targets and states that can only recycle expensively set low (or no) targets.

box continues on next page

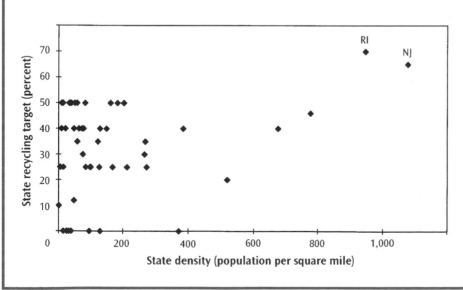

Figure 10-A

Relationship of Recycling Target to Population Density

Figure 10-B

Relationship of
Actual Recycling Rate
to Target Rate

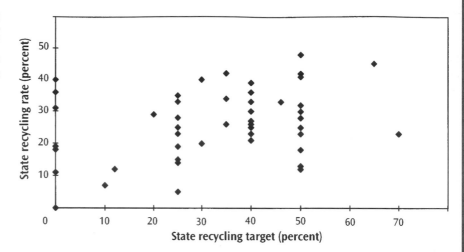

A note for those who have studied statistics: Regression for the 50 states of the state recycling goal on the natural log of population per square mile and the natural log of state disposable personal income per capita yields an R^2 of 0.07 (U.S. Bureau of the Census, various years; Heumann and Egan 1998). The R^2 in Figure 10-B is 0.11. Some states' goals involve the recycling rate, others' the waste reduction rate. They are not the same thing, being related by this equation (Wenger et al. 1997):

$$\delta_t = \frac{[(1-\rho_0)Q_0]-[(1-\rho_t)Q_t]}{Q_0} = \left\{(1-\rho_0)-[(1-\rho_t)(1+g)]\right\}^t$$

where ρ_t is the recycling rate in time t, δ_t is the waste reduction rate in t (relative to time 0), Q_t is the volume of municipal solid waste generated in t, and g is the annual growth rate of total municipal solid waste. Although conceptually distinct, the two variables are nearly the same if ρ_0 is near zero and either g or t is small. To further complicate comparisons, four states count source reduction and incineration as recycling, seven others count source reduction but not incineration, and three others count incineration but not source reduction; indeed, half the states do not count composting as recycling (Rabasca 1995).

little correlation between states' recycling goals and population densities (see the box titled "State Recycling Targets and Actual Rates"). When the state then urges its chosen target on each of its cities and towns, it is also failing to recognize that the optimal degree of recycling will vary across municipalities of different density and location.

Where states have taken serious steps to increase recycling, these steps have been of two basic kinds, supply side and demand side. Supply-side policies stimulate greater collection and processing of recyclable materials; and demand-side policies stimulate demand for those recyclable materials or the products made from recycled materials.

Supply-Side Policies

The market failure that affects recycling is basically a supply-side failure: Households supply too little of their waste for recycling because those households are able to use the municipal solid waste disposal service for free. The best response to this is also on the supply side: charging for solid waste service so that households have an incentive to seek other waste disposal techniques—reduction, reuse, and recycling.

Without pricing waste disposal and gaining the stimulus it provides to recycling, other supply-side policies are almost inevitably going to be at best partially effective. Let's look at a few of the more popular supply-side approaches:

■ About half the states offer either tax credits or direct subsidies to encourage municipal recycling operations (Sparks 1998). There seems to be no limit on the imagination—tax-exempt bond financing for recycling facilities, property tax exemptions on land and buildings involved in recycling, subsidized land sales for recycling facilities, grants or low-interest loans to finance the purchase of recycling equipment, and so on (O'Leary and Walsh 1995). Notice, however, what they all have in common—they affect things connected with recycling, but they do not directly affect recycling. Each relies on indirect linkages to recycling, so each is prone to cup–lip slippage. In each case, some input price is distorted, so each stimulates a distortion in the input choice. Recycling operations become too capital intensive or too land intensive, and facilities become located where the tax advantages are best, not where costs are lowest.

■ Products may be taxed if delivered to landfills, or in the extreme case, completely banned from landfills (see the box below titled "Rhode Island Bans Recyclables from Landfills") (Rabasca 1995). Of course the hope is that these products will, as a result, be recycled. But the incentive offered by the tax or ban is not an incentive to recycle, but rather an incentive to *not landfill*. Illegal dumping, exporting, and incineration are also stimulated. Because landfill arrivals are often weighed, taxing overall landfill weight is not a big administrative problem, but taxing or banning *individual* items in those deliveries requires an expensive inspection system. States that have tried it have generally repealed or simply not enforced the ban or tax.

■ The construction of new landfills and incinerators may be banned, as in Massachusetts (Johnson and McMullen 2000). This has the advantage over an immediate ban in that it gives waste handlers time to adjust to the new situation, whatever time is left in the existing landfills or whatever life is left to the existing incinerators. But eventually, this too becomes a ban on landfilling and incineration, and it then becomes a stimulus not only to recycle but also to export or to dump illegally. In Massachusetts, for example, within the past five years, exports of solid waste have risen from practically zero to about 20% (Johnson and McMullen 2000; MDEP 1997). It is not surprising that—as its landfill space becomes tighter and its incinerators work at full capacity—Massachusetts is considering lifting its ban on new landfills and incinerators (Daley 2000). In the—seemingly but not really—strange-bedfellows department, not only environmentalists but also owners of existing landfills and incinerators oppose lifting this ban.

■ Households may be required to recycle. Mandated recycling, without any financial incentive to recycle, may "work" because people are basically law-abiding, but it works against the self-interest of most households. (The word "most" in this and the next sentence is necessary because, for some households, the psychic benefits of recycling are sufficiently great that no price is needed—nor is mandated recycling.) Mandating recycling means that most people recycle because they are told to, not because they are better off if they recycle and hence want to recycle.

■ Products that are not recyclable may be forbidden to be sold. In 1989, Minneapolis banned the sale of products that were sold in packaging that could not be locally recycled—which, it is not surprising, proved impossible to enforce. Massachusetts and Oregon nearly passed referendums requiring that all packaging be at least 50% recycled. Laws like these ignore, or discount to zero, the fact that nonrecyclable packaging may serve an important other purpose. For example, shrink-wrap and other plastics containing medical supplies or foodstuffs usually cannot (now) be recycled, but they may be necessary for the sterility or security of the product.

To return for a minute to the German Green Dot program (discussed in Chapter 2), we see now that it implements the ultimate supply-side policy. In principle, every producer is required to physically take back every product and package it produces. The idea, of course, is that once they are forced to take it back, they will recycle it. Indeed, as long as recycling it is cheaper (net) than landfilling it, the producer will recycle it. But that is no guarantee that the correct volume of recycling will take place. Recall that the Green Dot collection process is duplicative and expensive. For many products and packages, Green Dot–type programs force uneconomic recycling because the recycling decision does not take into account the excess collection costs.

Rhode Island Bans Recyclables from Landfills

In 1989, Rhode Island prohibited landfills from accepting commercial solid waste with more than 20% designated recyclables in them. The forbidden items included cardboard, paper, wood, yard waste, and laser cartridges. Loads were inspected as they entered, and if they failed, the haulers were refused entry and were liable for fines. Because Rhode Island has essentially only one landfill, it was expected that businesses would be forced to comply.

Alas, the greedy incinerators of Massachusetts were not far away. It did not help that—with declining waste inflow and increasing expenditures on recycling—the Rhode Island landfill doubled its tipping fee, to nearly $50 per ton. Its revenues plummeted. After a year of this, the tipping fee was lowered, and the landfill stopped inspecting the incoming trucks (Raymond 1992).

Does this mean that bans are inherently unworkable? No, but they are heavy weapons, essentially infinite taxes on the banned items, and those who use them must anticipate that loopholes will be sought, found, and used. A ban sounds like a cheap policy, but lots of enforcement effort is needed to make it effective.

There are almost no Green Dot–type programs in the United States at this time, but there is no shortage of effort to introduce producer take-back responsibility. In California, for example, the Silicon Valley Toxics Commission is collecting signatures for a petition to require electronics manufacturers to physically take back their discarded products (http://www.svtc.org/cleancc/e_platform.htm). If this sort of take-back operation were the only way to keep old—and toxic—electronic equipment out of landfills, then it might be socially sensible. But it is not the only way. The municipal collection of electronic equipment (discussed in the previous chapter) would surely be cheaper than the United Parcel Service costs of returning equipment, one by one, to the manufacturer. And the optimal number of U.S. electronic recycling centers is surely smaller than the number of computer, monitor, and printer manufacturers.

Demand-Side Policies

The higher the price of recyclables—and hence the greater the value of recyclables to their purchasers—the greater is the optimal amount of recycling. Demand-side policies seek to artificially raise the price of recyclables to stimulate recycling. In the perfect world of competitive equilibrium, artificially stimulating demand would overstimulate recycling. But in a world where the supply of recyclables is artificially held back, increasing demand may be a sensible second-best approach.

In any case, demand-side policies abound. States, counties, and cities regularly mandate minimum recycled-input percentages for products such as the use of old newspapers in making new newsprint, and governments at all levels establish procurement preferences for recycled products such as office paper, re-refined oil, and retreaded tires.[6] Some simply mandate a minimum percentage for recycled product purchases—often 100%—whereas others give a 5–10% price preference to recycled products. Although the percentage price preference is desirable, because it limits the potential for inefficiency, all such procurement policies are second best—the market failure essentially is in the supply of recyclable materials, not the demand.

It is worthwhile taking a long paragraph to see what's wrong with mandated recycled-content percentages. Consider a product made from virgin material, recycled material, and labor (Palmer and Walls 1997). Because households are not charged for their solid waste, they do not recycle the socially optimal amount. As a result, the firm gets too little recycled material and uses too much virgin material. Mandating a higher recycled-content percentage can offset this, forcing the firm to offer a higher price for recycled material. This higher price might induce cities to do more recycling and could, in theory, establish the socially optimal ratio between recycled and virgin inputs. (However, the higher price of recyclable material will not change household behavior if it continues to be charged zero for both its trash and its recyclables. Scheme 2, conversely, would encourage greater recycling by increasing the trash collection fee when the market value of recyclable materials rises.) But it has also pushed up the costs of the firm and made labor relatively cheaper than material inputs. The firm's total output may fall, and its labor use may rise relative to its material inputs. All kinds of indirect, unwanted, and unpredictable secondary effects may take place, calling in turn for a corrective output and/or labor tax (or subsidy)—cup–lip slippage again.

Aside from the second-best quality of demand-side policies, they have other problems. For one, those whose purchasing decisions are forced to change will resist the mandates and may find ingenious ways to circumnavigate them—governments themselves regularly fail in their efforts to get their own purchasing agents to buy more expensive recycled products. But a second problem is that it is all so ad hoc. For those recycled products for which there is a means of stimulating demand, demand is stimulated, but for others no stimulation is available. There is no guarantee that the right products—from the viewpoint of net social benefit—will be the ones to get the stimuli. If the federal government completely stopped buying paper and spent the equivalent amount of money on beverages in refillable bottles, it could lead the entire United States back to refillable bottles within a decade. Alas, the federal government seems to prefer paper.

If demand-side policies must be used, they should be applied through marketable permits. Consider newsprint. The government mandates, say, an increase to 40% recycled content. Newspapers comply. We have no explicit picture whether that was an easy target and perhaps should be raised still further, or whether that was an extremely costly move that has much raised the price of newspapers for little recycling gain. If, however, the 40% target were marketable, newspapers in big cities near recycled newsprint mills could raise the recycled content beyond 40% and sell the surplus to newspapers near virgin pulp newsprint mills (Bingham and Chandran 1990). The 40% target would be reached *on average*, but at lower total social cost. Moreover, from the price of the marketed percentage, we would get a clear idea about the marginal cost of achieving the target.

Final Thoughts

This has been a short, but hard, chapter. Hard to read (and hard to write).[7] But it is easy to summarize: Too few policy tools are chasing too many policy goals. If illegal dumping is serious and unstoppable by other means, some sort of deposit system may be appropriate. Of course, it does not matter whether the original deposit is paid by the manufacturer as an advance disposal fee or by the consumer at the time of purchase (except that the latter is a more visible reminder that a deposit may later be redeemed). What is critical is that a deposit is redeemed at the end for correct disposal. But deposit systems are expensive means of litter control, as we saw in Chapter 6; and if illegal dumping can be controlled by other means, trash charges can better attack other goals.

Scheme 2 (see Figure 10-4) is the one of the three that I prefer, although any of three systems (or an infinite number of combinations of them) will work in theory. With scheme 2, the advance disposal fee on the manufacturer of a product equals the marginal net recycling cost of the product and package (i.e., collection and processing cost net of revenue earned on the recyclable materials), and the trash collection charge on the household equals the excess of the marginal landfill or incinerator disposal cost over the marginal net recycling cost (with all costs being social, not just private costs). Scheme 2 gives the correct incentive to reduce and recycle and gives some (albeit inadequate) incentive to reuse. It gives manufacturers an incentive to use fewer and more readily recyclable materials. It is explicable in one-syllable words, is administratively manageable, and does not require either a fee or a subsidy on recycling.

The fact remains that price policies are impossible to get exactly right. It is therefore tempting to forget them and rely entirely on nonprice policies. Indeed, nonprice policies may have helpful cheerleading effects as recycling is just getting started. But all nonprice policies suffer one great defect—they all try to get people to do something that it is not in their economic interest to do. When states mandate recycling targets, it is only the fear of the fine that induces cities to react. Low-interest loans for recycling facilities may be snapped up, but only because cities love low-interest loans, not because it makes them love recycling. And requiring purchasing officers to buy recycled materials will be ferociously resisted unless the recycled materials are priced competitively.

Notes

1. There are of course also an infinite number of optimal prices when all three tools are used. There are only two constraints on the optimal prices: the sum of the advance disposal fee and the trash collection charge must equal the MSC of trash collection and disposal; and the sum of the advance disposal fee and the recycling collection charge must equal the net MSC of collecting, sorting, and selling recyclable materials. We will consider only the three schemes pictured in Figures 10-3, 10-4, and 10-5, because two fees (or refunds) are easier than three to administer (and easier to examine pedagogically).

2. For the sake of completeness, we should look at how the prices change if the net recycling cost exceeds the trash disposal cost. In scheme 2, the advance disposal fee becomes larger than the trash cost, and a trash refund is appropriate. In scheme 3, the advance disposal fee remains equal to the trash disposal cost, but a recycling charge, rather than a refund, is called for.

3. This consideration provides the conceptual rationale for California's partial (rather than full) redemption of its deposits on beverage containers (see Chapter 6).

4. If the marginal landfill cost and the marginal net recycling cost are close to each other—that is, when recycling shows only a small social profit—then the trash collection charge under scheme 2 should also be small. Indeed, it is perhaps so small as to not make its introduction and administration worthwhile. If so, scheme 2 reduces to a single kind of tax, the advance disposal fee.

5. In this respect—and in this respect only—a mandatory deposit and refund scheme has a big advantage over a trash collection charge plus a zero recycling charge. Mandatory deposits can easily be set at different levels for different products—higher, when the social costs of improper disposal are higher. Littered batteries, for example, are much more costly than littered Popsicle sticks. But the problem, recall, is that mandatory deposits require a special waste disposal system that is duplicative and costly.

6. In an effort to ensure greater consumer recycling, most of these recycled-content mandates specify a *postconsumer* recycled content. But there is no difference in the value of recycling postindustrial, preconsumer waste and postconsumer waste, and hence these mandates should not differentiate between the two types of waste. Indeed, if strong enough postconsumer mandates were to be enforced, they would actually drive preconsumer waste into the landfills!

7. Moreover, this chapter has only scratched the theoretical surface of waste and recycling pricing possibilities. For those who want more (and much more challenging) theory, try Dinan 1993; Fullerton and Wu 1998; Choe and Fraser 1999; Fullerton and Wolverton 2000; Calcott and Walls 2000; and Walls and Palmer 2001.

11

Logistics of Recycling

The difference between theory and practice is bigger in practice than in theory.
—Jay R. Ashworth

So far in Part 2, we have been talking about the big picture—why to recycle, when to recycle, how much to recycle, and how to price recycling. Here we turn to the little picture, pointing out the many small decisions needed to determine *how* to recycle. But before turning to these multifarious details, let's look at why so many decisions must be made. If a new city were, somehow, to spring up overnight, and you were asked to establish a trash collection system, you could call in a dozen consultants, and they would all give you about the same advice. We have been collecting garbage in the "modern way" for nearly a century now, a best way has emerged, and a consensus on it has been reached.

Call in a dozen consultants on recycling, and you will get reports recommending vastly different methods. On recycling in the modern sense—as opposed to low-wage, free-market scavenging—we have not reached anything like a consensus on how to do it. Probably in the next few decades, possibly in the next few years, this consensus will emerge. In the meantime, both private and external recycling costs will continue to fall, for two reasons. First, the costs of the best technique will fall. Second, inferior techniques will become recognized as such and will be discarded, to be replaced by better or cheaper techniques. Let us now look at the many choices that a decision to recycle entails.

Buy-Back and Drop-Off Centers

Historically, the rebirth of recycling in the 1970s began with buy-back and drop-off centers. Buy-back centers are privately operated enterprises that pay consumers to bring in and sell certain items. The range of items purchased is usually small, the distances are usually great, and the prices are usually low. Drop-off centers—sometimes private, sometimes public—are usually run by volunteer labor and with a municipal subsidy. They usually collect more items than buy-back centers, but they do not pay for the materials, and they insist on extensive household presorting. Neither kind of center ever achieves much recycling, because rarely do more than 10% to 20% of the area's households participate (Stevens 1994).

Large U.S. cities have outgrown drop-off centers—by 1997, 83 of the 100 largest cities provided curbside pickup of recyclables at least once a week (Ezzet 1997). But for small towns, drop-off centers remain the only affordable recycling method. The private costs to the town government are low, and the costs to others are ignored. These ignored costs can be high, but some are offset by unnoticed social benefits. When workers volunteer to staff the centers and when households voluntarily deliver recyclable materials, they are revealing the fact that their enjoyment of the activity is worth the costs to them.

Nevertheless, the external costs of these activities are not necessarily offset by personal benefits, especially litter and auto-related pollution. Drop-off centers are often unattended and the drop-off bins often become overfull or overblown between collections. Unattended bins attract trash as well as recyclables, especially where there is a trash collection charge, and this increases the sorting costs and reduces the value of the recovered materials. The auto-related problems arise from the external costs of driving. Recyclers pay for their own gasoline, use their own time, and undergo risk to their own lives; but there are also external costs—largely air pollution and accidents (to other than the drivers themselves)—of at least 3 cents a mile (Porter 1999).

How do we get that 3-cent figure—which is used in the calculation of the next paragraph? Consider only accidents that cause the death of someone other than the driver; there are about 20,000 such deaths a year. Valuing each life (V_L) at $3 million (see Appendix B to Chapter 1), and dividing by the 2 trillion miles Americans drive each year, yields 3 cents a mile as the external cost from highway death alone.

Think about what these external costs amount to. To get some rough figures, suppose someone in the average dropping-off household drives an extra five miles to go by a drop-off center once every two weeks and to deliver 20 pounds of recyclables—that's an optimistic 10% of the 200 pounds of solid waste generated by the typical American household per fortnight. This means at least $15 per ton of recyclables just on highway-death external costs, and the collection cost is far from complete at this point because the materials still have to be taken from the drop-off bins to a processing facility. More and nearer drop-off locations cut these automotive externalities, but only at the expense of more residential area noise and litter and of greater municipal subsidies.

Commingling and Co-Collection

The next stage in recycling is often commingling or co-collection. Commingling means just what it sounds like: The recyclable materials are collected along with the municipal solid waste, and the waste is picked over and sorted later for recoverable materials. Although commingling requires no effort by the household, it greatly diminishes the volume and value of the recyclables. Mixing paper with wet garbage reduces its value, and compacting glass yields many unrecoverable shards. Also, the cost at the sorting facility is high per ton of recoverable items because all the municipal trash has to go through the sorting process.

Co-collection is an effort to escape these problems by requiring the household to place the recyclable materials at the curb in a separate (clearly marked) bag. They are then picked up in the regular trash collection, and the recyclable bags are removed at a materials recovery facility before the collection vehicle makes the trip to the landfill. Although the glass is usually kept separate and not compacted, the compaction process damages the ease of recovery and the value of the recyclables—some co-collecting cities report that a fourth of the incoming recyclables are contaminated and thus unmarketable (O'Leary and Walsh 1995).

The advantages of these kinds of collection are obvious. Participation rates are much higher than with drop-off centers, as would be expected, because much less is demanded of the household's time and car. And little new equipment or labor is needed in the municipal waste operation, because only the glass is handled any differently than before. The disadvantages are equally obvious. Although participation rates are high, the costs of recovery are also high, and the volume of recovered materials low—especially where municipalities or private waste haulers require the participating households to pay for the bags in which they set out the recyclable materials (Steuteville 1993). Still, where population densities are low and distances to landfills and materials recovery facilities great—and where a landfill and MRF are near each other—commingling or co-collection may be financially necessary, and even socially appropriate (Miller 1995a).

Curbside Pickup

Moving to curbside pickup of recyclables using separate vehicles increases the participation rate of households and the amount households recycle, but it also greatly increases the total collection costs of trash and recyclables (Jenkins et al. 2000; Kinnaman and Fullerton 2000). After all, a separate curbside pickup means that every house is driven by twice as often. But what is collected is much more cheaply sorted at the MRF, and much more of it is marketable. In large, dense cities, especially where disposal costs are high, moving to curbside pickup may be a compelling final stage in the collection of recyclable materials.

Curbside pickup, however, is not a binary, either–or decision. Quite aside from the decision about what materials to include in the recycling program—and more can be reasonably included once separate, specialized trucks are doing the collection—many ancillary decisions must be made about the collection process.

One such decision is the frequency of the pickup. Usually, curbside pickup is made weekly, sometimes for trash and recyclables on the same day, which is convenient for household memories but can lead to vehicle overlap and congestion. Conceptually, the argument for biweekly pickup (i.e., once every two weeks) is that it saves a lot of money, and the argument against it is that households will recycle less because of forgetfulness, the inconvenience of longer storage, or the doubled weight of the recyclables container. (On the forgetfulness front with biweekly pickup, look at http://www.libertynet.org/~recycle/calendar.html to see how Philadelphia reminds its residents which week it is.) In fact, none of these arguments is incontrovertible. A few studies show that recycling rates are about the same whatever the time gap between pickups—for one sizable sample, the rate was 13.1% for cities with weekly pickup and 12.8% with biweekly pickup (Stevens 1994).

Pickup costs do not fall by anything near a half when pickup goes from weekly to biweekly. Administrative costs fall little. The trucks fill up faster because they are collecting nearly twice as much at each stop. The driving time while picking up is halved, but the total time at household stops and the total time driving to the MRF are little reduced. When Philadelphia went from weekly to biweekly pickup (and dropped plastic), its collection costs only fell from $160 to $125 per ton (Egan 1998).

How Much Household Sorting?

The next decision is how much sorting to demand from the participating households. Some require nothing more than the separation of recyclables from trash, placed in a single separate bag or container. Others require quite extensive separation. For example, at one time in Berlin, New Jersey, all recyclables were put out in a single container, which was provided by the town. Meanwhile, in Woodbury, New Jersey, only 20 miles away, residents had to sort their recyclable materials into *nine* different containers, the containers were not provided, and stiff penalties were imposed if a household did not sort correctly (Salimando 1989).[1]

Greater source separation imposes a greater burden on participating households. Greater separation requires more time, more sorts, more containers, and more storage space. These are real costs of a recycling program, but because they are all imputed costs, they are not easily put into dollar terms. Although measuring this cost may be difficult, the fact of household reaction to this burden is clear. The higher the household cost, the lower the amount of recyclable materials collected. Although systematic large-sample evidence has not yet been examined, anecdotes suggest that the number of household sorts makes a big difference in the amount collected—in Los Angeles, for example, collection of recyclables in a single container, instead of having the household separate paper from other materials, increased the flow of recyclables by nearly 150% (Bader 1999a).

Commingling all recyclables in one container maximizes the amount collected, and hence maximally reduces solid waste collection costs and landfill costs. So, why ask the household to do any separation? Commingled collection increases recyclable volumes and hence recyclable collection costs; commingled recyclables cost more to separate at the MRF; more

> ## Recycling Logistics in Europe
>
> Whereas in the United States recyclables are usually collected together on a single day in a single truck, Western Europeans often have different ways of collecting different recyclables. Paper is usually collected in compactor trucks, but glass is more likely to be collected in drop-off containers, called igloos, which are publicly placed in regularly spaced groups, with different igloos for different kinds of glass and plastic.
>
> Collection of recyclable materials is sometimes done by the municipality itself, but often it is contracted out to a profit-seeking firm or an environmental organization. When contracted out, the collection is financed by a municipal payment, either per month or per ton collected. Such payments are based on the avoided cost of disposal for the materials collected, and the collector gets to keep most or all of the revenues from the sale of recyclable materials.
>
> In Southern and Eastern Europe—and also in France and the United Kingdom—recycling containers and recyclable collection programs are less common. Urban centers may have some igloos and some residential recycling collection, but these are not widespread. Only in cities with a strong environmental ethos or progressive mayors does one typically find curbside collection, and then of few materials. Where the beverage or packaging industry supports buy-back programs, waste scavengers collect and sell materials, primarily metals and paper. But their activities are sporadic. When market prices are low, the scavengers disappear, only to return when prices are higher.

recyclables arrive at the MRF too contaminated to recycle—for example, wet paper and broken glass; and the average revenue of commingled-collected recyclable materials is lower than the average revenue of those that have been source-separated (Apotheker 1991).

In one study of voluntary recycling programs, for example, the measured household reaction was dramatic. When the recyclables collection was weekly, the collectors went to the garage for the recyclable materials, and households put out all their recyclables in one container, 319 pounds of recyclables were collected per capita each year (Judge and Becker 1993). But when the recyclables collection was biweekly, the households had to bring their recyclables to the curbside, and households had to sort their glass, plastics, newspapers, and metal into four separate containers, only 11 pounds of recyclables were collected per capita each year.

Here again, we see that recycling experiences rising costs. In voluntary programs, to get households to recycle more, the municipality must make it easier for the households to do the recycling. But making it easier for the households means making it more costly later on in the recycling process—more trucks and workers for weekly or backyard pickup or truckside sorting, or more capital and labor at the MRF when recyclables arrive there still completely commingled. The total costs of the collection process go up, but more is being collected, so it is not immediately clear whether the cost *per ton* goes up or down. In fact, in the study of the preceding paragraph, the average collection cost (with truckside sorting if the recyclables were completely commingled) fell from $713 per ton of recyclables when collection was biweekly, curbside, and sorted to only $59 per ton when collection was weekly, backyard, and commingled (Judge and Becker 1993). Why the huge drop? Because the increased costs of collection per household are much more than offset by the increased collection of recyclables per household— the collection cost per household, when going from the least to the most conven-

---

Seattle's Two Different Recycling Systems

For many years, Seattle was one of the few big cities to use two different recycling systems (Kamberg 1990). The city was divided into northern and southern areas, and each had its own contractor to collect and process its recyclable materials. In the north, households put out their materials weekly in three different bins, for newspapers, other paper, and hard materials (glass, metals, and plastic). The contractor received a fixed fee per ton. In the south, all the recyclables were collected monthly commingled in a single container. The contractor received a fee per ton that was based on the past prices of recyclables, and in each year the contractor and the city evenly shared any deviation from that price.

Each system had advantages and disadvantages. The north collected more often and hence collected more material. Its curbside separation also yielded cleaner materials. However, the collection was much more expensive than that in the south. Not only was it four times as frequent, but also the household separation into three bins meant that the trucks filled up sooner (because each truck must drive to a MRF or transfer station as soon as one of its three bins is filled). In the south, commingling meant lower collection and transport cost, but it also meant more broken glass, which meant more wet newspapers and more difficult glass sorting.

Recycling participation rates have always been high in both parts of Seattle, not least of all because households are charged by the can for their regular trash. The recycling participation rate is slightly higher in the north because, it is said, of the higher income and educational levels there. But it may also be relevant that there are more renters in the south, and in apartment buildings, landlords rather than the renters themselves usually pay the trash fees. These trash fees are, of course, passed on to the renters, but the point is that renters do not pay any *marginal* cost for their trash, and hence they save nothing by recycling.

In any case, all this has now changed. Throughout Seattle, curbside recycling pickup is biweekly, the materials are collected commingled, and the number of recycled items has greatly expanded (Acohido 2000a, 2000b). The changes will save $2 million annually, about $4 per capita.

---

ient type of collection, less than doubled, but the recyclables collected increased nearly 30 times (Skumatz 1996).

Does this suggest that we should make it as convenient as possible for the household to recycle? Hardly. What ultimately determines the optimal amount of recycling is not the average cost per ton recycled but the *marginal* cost per ton of recyclables (net of trash collection cost reductions). And the marginal cost will at some point begin to rise. Going from weekly to daily pickup, for example, would increase cost per household by roughly sevenfold but would probably not increase the volume of recyclables collected by much.

Asking more effort from households risks getting smaller participation, fewer recyclable materials, and less efficient sorting at the household level. Multiple sorting by the household also adds time to the job of the collection vehicle—as a rule of thumb, it is said that each additional sort adds five seconds to each stop. Moreover, multiple sorts by the household—or at the collection truck by the workers—means that the truck must go to the MRF when its first compartment fills. Compartment sizes can be matched to *average* collection volumes, but the actual collection volumes will vary a lot from week to week. In short, more household sorting means more cost throughout the collection process, even if it does save costs at the MRF.

This makes it sound like the one-container sort is a clear winner. After all, the MRF has specialists and scale economies unmatched by households. Indeed, anecdotal evidence is constantly appearing that cities that formerly required households to do multiple sorts are turning to the one-container sort, where households do no sorting beyond separating recyclables from nonrecyclables. The economics of recycling is still driven by collection costs and landfill costs, and the one-container sort greatly reduces the marginal collection cost of recyclables and greatly increases household participation in recycling (Truini 2001a).

One Truck or Two?

Trash and recyclables can be either collected together in one truck, with both a compaction compartment for trash and a bin or bins for recyclables, or collected by two separate trucks, with a compactor truck collecting the trash and a second truck collecting the recyclables. The conventional wisdom is that one truck is less costly than two. Let's think about why. Consider a neat little town with graph-paper-gridded streets and evenly spaced houses, each of which generates the same amount of trash and recyclables each week. If one truck is sent down each street, it will stop for N seconds at each house; if two trucks are sent, each will stop for $N/2$ seconds, half as much because each truck picks up trash or recyclables, not both—no cost difference there. The one truck fills up twice as fast and therefore has to drive to and from the MRF and landfill (assumed next to each other) twice as often as each of the two trucks—no cost difference there. But the two-truck system requires that two trucks drive over the entire grid of the town each week, whereas the one truck system requires only one such drive-over. One truck is cheaper.

But not so fast. Reality is not as simple as that "neat little town" of the previous paragraph. First, when a truck and crew is specialized for one task, either trash or recyclables collection, the trucks will be cheaper and the crews may be able to cut the pickup time to less than half. Second, the mixture of trash and recyclables varies from week to week, and the one truck, even if it is proportioned correctly on average, will find its trash compartment fills before its recyclables, or the reverse, in any given week. This means trips to the MRF and/or landfill when the truck is not yet completely full, and hence this means more trips in total than with the two-truck system. Third, not every household will put out recyclables every week, and this means that the recyclables collector can drive right by some houses in some weeks, speeding its operation and cutting costs—it is the same kind of cost cutting that biweekly trash or recyclables collection would achieve.

It would be nice to conclude with empirical evidence about whether one truck or two wins, but not much comparable data have surfaced yet. One study, of Madison, Wisconsin, concluded that the two-truck system was cheaper—$9.52 per person each year versus $9.70 for the one-truck system—but Madison is hardly a typical U.S. city (Anderson et al. 1995). There, the introduction of recycling reportedly lowered the *total* costs of collection!

Multifamily Housing

Curbside recycling always starts with single-family dwellings and only later, if at all, moves into the parts of the city with multifamily dwellings. On the surface, this seems surprising, because recyclable material is more densely, and hence more cheaply, available from apartment houses. But studies of who participates in voluntary recycling remove the surprise (Hong et al. 1993). People are more likely to recycle where the residence is owned rather than rented, the number of people in the residence is greater, and the wage of the oldest female is lower. None of these correlations is unexpected. More people means children, for whom recycling is an easy chore. Lower wage means lower opportunity cost of time. And owning rather than renting means the personal benefits as well as the costs of recycling accrue entirely to the owner. Recycling is also higher where there is a trash collection charge, and although apartment house pickup is often priced, the cost of marginal trash is either buried in the rent or shared with the other occupants of the building. Finally, space may be a greater problem in small and often garageless housing.

The result is that multifamily housing is a poor source for recyclables. Most but not all studies show the recycling rate of multifamily housing is lower than that of single-family housing. Because of this, the cost per ton of collecting recyclables is much higher—one study of 40 cities that

---

A Different View of Multifamily Recycling

Mark Shantzis, president of Hi-Rise Recycling Systems, takes a different view of the high-rise apartment house (Bader 1997). He sees it as the cheapest way to collect recyclables. There is no need to waste time and gas driving a truck from family to family when you collect from hundreds of families in a single multistory building. Inherently, multistory buildings should be the cheapest source of recyclables.

And he has a system for getting apartment dwellers to participate more fully in recycling programs. His firm's system consists of the existing trash chute, a compactor, and a turntable that positions several containers for various recyclables and for unrecyclable trash. Residents take their trash and recyclables to the chute on their floor and push a button on a panel, which repositions the turntable to the correct container, and then dump the appropriate material into the appropriate container. There is no need for residents to store their recyclables, package them, and carry them perhaps a distance to a container. Meanwhile, below in the basement, the compactor works continually on each container and automatically calls maintenance when one is full.

offered or mandated recycling to multifamily housing found that the cost for collecting recyclables averaged $177 per ton in multifamily housing, versus only $127 in single-family housing (Stevens 1998). And the difference is not due to the configuration of the housing, for the cost per ton of picking up the trash was, as one would guess, slightly lower for multifamily housing, $63 versus $69. Everybody's business is nobody's business. Cost-effective recycling in multifamily housing must find a way to make individuals responsible for the group's actions—the classic free rider problem again.

Voluntary or Mandatory Recycling?

The advantage of voluntary recycling is that the people who participate do so willingly, which means that the personal benefit they get from recycling *always* exceeds any personal cost they undergo. With voluntary recycling, participants can be asked to do a fair amount of sorting at home, and they can be counted on to do it well. If recycling is mandated, enforcement is needed, less sorting can be demanded, and what sorting is done is less well done.

Why, one might wonder, would any city force recycling on those who do not want to do it? The answer lies in volume. Mandatory recycling achieves larger volume and at very little extra cost—the average cost per ton of collecting recyclables comes down a lot, according to one study (put in 1997 dollars), from $164 with 25% of the households participating to $121 with 75% participating (Miller 1995c). One sample of U.S. cities that recycle showed that cities that mandated recycling gathered 100 pounds more recyclables per household per year than cities in which recycling was voluntary—550 versus 450 pounds on average (Miranda and LaPalme 1997). Because the marginal cost of collecting and sorting these 100 pounds is each less than the average cost—and the additional materials will earn revenues—the typical mandatory program makes the recycling economics look better.

It may look better, but maybe not be better. To begin with, the 100 pounds may be, partially at least, illusory. It may mean simply that in cities with great enthusiasm for recycling, the enthusiastic majority is willing and able to force the reluctant minority into recycling, whereas cities without this enthusiasm stick with a voluntary program. There are two pieces of evidence that the difference in the cities is more important than the difference in the policies. First, in the study of the preceding paragraph, it was found that in the cities with voluntary recycling programs, the average household generated nearly 400 pounds more total trash and recyclables than in the cities with mandatory programs, even though there is nothing in the mandating process that induces source reduction and reuse (Miranda and LaPalme 1997). Second, another study found that communities that enforced recycling laws with fines ended up with no more recycling than other towns without such enforcement (Duggal et al. 1991).

Moreover, any increased total recycling due to mandating the recycling reduces the average cost of collecting recyclables, but it increases the total cost of collecting all trash and recyclables. The loss of welfare to reluctant recyclers must also be recognized as a cost. The benefits in saved landfill costs and recyclables revenues may or may not be worth these costs.

Enforcing Mandatory Recycling

In most cities with mandatory recycling, nothing more than a reminder is given those who fail in some aspect of their recycling. But some have tried to enforce the mandate sternly. It is neither easy nor cheap.

Most commonly, the trash collectors visually examine the trash for evidence of recyclables—and recycling collectors also examine recyclables for evidence of trash, especially in cities with trash collection charges. This slows their work up considerably.

Elsewhere, bizarre efforts have been made. In Rockland, Massachusetts, for example, the trash collectors will not stop at your house unless you have put out a recyclables bucket—even an empty bucket will do (Gust 2000). Essentially, the city has mandated putting out a recycling bucket every week, not putting out recyclables. The system wastes not only the time of nonrecycling households but also the time of the recyclables collectors, who have to stop at every bucket, even though many may be empty for weeks. People who generate few recyclables should save their own time, and the recyclable-collector's time, by only putting the bucket out when it is full, But the Rockland system precludes this. Empty buckets are more prone to blow away, forcing a $6 replacement cost on the household, so a household might start to buy recyclable materials just to weigh down its bucket every week—in short, a bizarre system.

Finally, mandating recycling is a clumsy nonmarket tool for encouraging recycling. In principle, if one does not recycle all one's recyclables, week after week, one is liable to fines (though few cities in fact have gone beyond "courtesy" warnings to serious nonrecyclers). Mandating is made necessary by the absence of any market incentive to recycle in cities without a trash collection charge. Such a charge fine-tunes the recycling decision. If trash costs x cents per pound, then each pound put out as recyclables earns the household x cents in opportunity cost. Only people for whom recycling is so onerous as not to be worth x cents per pound would fail to recycle, and if x cents is the cost to society of this failure, why should we not permit them to not recycle? With the pricing of trash, voluntary recycling is both cost-effective (because each household will recycle up to the point where its own marginal cost of recycling equals the trash collection charge) and correct in volume (if the trash fee is set at the correct level—see Chapter 10—so each household's marginal cost of recycling equals the marginal social benefit of recycling). Mandating recycling is unlikely to be cost-effective or optimal in volume. Notwithstanding, roughly half of cities that recycle in the United States mandate that recycling.

Operations of Materials Recovery Facilities

Early recycling operations relied almost entirely on hand sorting. Often the only capital equipment beyond (or rather below) a roof was a forklift and a baler. As volumes grew, conveyor belts were added, but most of the sorting and baling was still labor intensive. Now, the state-of-the-art MRF changes almost daily, and always in ways that substitute capital for labor. (For the ultimate capital-intensive scenario, try this one: Could MRFs some day use bar codes to sort out recyclables—entirely automatically?)

Where commingled recyclables are collected, paper products can be diverted by means of a trommel, a drum-shaped rotating screen. Magnets can be used to separate "tin" cans and other ferrous materials. Mechanical plastic sorters are becoming available, but they are still very expensive. Somewhat less expensive is the separation of aluminum by means of eddy-current induction (Bader 1999a). For separating glass by color—done because the revenues of mixed glass are usually negative—the machinery is still a ways off.

Mechanization poses two kinds of problems. First, how big does a MRF have to be to make large and expensive pieces of capital equipment better than hand sorting? The minimum size of cost-effective MRFs will undoubtedly grow with mechanical sorting inventions, and it may well be that as a result, recycling will become less, not more, socially profitable in sparsely populated areas. Second, given the rapid expected pace of invention in MRF capital, any investment now bears the risk of being surpassed by something cheaper or better in the near future. Choosing when to invest remains more of an art than a science.

Still, for the foreseeable future, MRF sorting will be largely a piece-by-piece, hand-to-bin operation—very labor intensive. Although the labor need not be skilled, the pace of the conveyor belts, the concentration required, and the occasional danger are sufficiently high that the workers usually earn well above the minimum wage. Moreover, a sizable fraction of the so-called recyclable material that households put out is misplaced—in some MRFs as much as a third—and simply goes expensively by a roundabout route to the landfill (Stewart 2000b). Sorting heterogeneous recyclables into homogeneous marketable bales is not cheap.

Are Mandatory Deposits and Recycling Compatible?

When mandatory deposits on beverage containers were enacted in many states in the 1970s, there were no recycling programs at all in most places and only drop-off centers where there was recycling. Indeed, when the programs were introduced, mandatory deposits were not viewed so much as a recycling stimulant as a means of making bottlers and brewers either wash and reuse the containers or pay for the landfilling of the one-ways. Although little return to reusable containers occurred, recycling markets for the re-collected aluminum and glass appeared almost immediately. Mandatory deposits serendipitously demonstrated the viability of today's massive recycling programs.

But mandatory deposits, as we saw in Chapter 6, are an expensive way to collect recyclables. The question today is whether such deposits are sensible in an era when much lower cost municipal recycling programs have become widespread. Even recycling enthusiasts who were in the vanguard of those proposing mandatory deposit laws 20 years ago are now asking this question, because such deposits seriously hurt the financial viability of municipal recycling programs (Hawes 1991; Ackerman et al. 1995). Adding more beverage containers to the curbside pickup would not raise collection costs much, relative to the revenue gains that would result from sales of high-priced aluminum. In fact, where there is recycling but no mandatory deposits, beverage container scrap provides nearly half of all MRF revenue (GAO 1990).

If mandatory deposits are added to an existing recycling scheme, both the total collection costs and total recyclable tonnage collected increase. The relevant question is whether the marginal benefit (of recyclables revenue and landfill cost avoided) exceeds the marginal cost (of collection and processing) when mandatory deposits are added. There are estimates of this marginal cost—the cost of collecting the extra recyclables by adding mandatory deposits to a system that already recycles has been estimated to be (converted to 1997 dollars) $290 a ton in Vermont and $790 in New York (Franklin Associates 1988).[2] The marginal benefit of the added tonnage is surely less, because it would include much more glass and plastic than aluminum. However, only about half of all Americans receive curbside recycling collection. And even for those who do, the collection rate for beverage containers *is* much higher when there are deposits on them than when only curbside recycling is an option. Curbside recycling collects less than half of all beverage containers, whereas deposits achieve much higher return rates (Ackerman 1997). This result is hardly surprising. Deposits are typically five cents per container, whereas the money a household saves by recycling a container is at most a small fraction of one cent—or zero if there is no trash collection charge.

If the goal were to maximize beverage container recycling, then deposits work much better than other recycling procedures. A 5-cent deposit works; a 50-cent deposit would work even better. But that is not a sensible goal. If there was a trash collection charge, and that charge properly reflected the fact that landfilling containers is more socially costly than recycling them, then the optimal number of containers would be recycled. (Of course, that optimality depends also on the removal of all distortions in the energy and virgin raw materials markets.)

One very good argument remains for adding or retaining mandatory deposits after recycling has become widespread. Mandatory deposits are an antilitter policy, which recycling is not. Neither the availability of curbside recycling nor a trash collection charge does anything to prevent beverage container litter (unless people are paid to recycle). Quite the opposite is true for a trash collection charge, which encourages both recycling and illegal dumping—that is, litter. If beverage container litter is considered to be a serious problem, then mandatory deposits are a much more effective policy than either the availability of recycling or a trash collection charge, and the coexistence of all three may make sense.[3]

However, there is a lot of evidence that litter, and especially beverage container litter, is nowhere near as serious a problem today as it was two decades ago, and total roadside litter is about the same in deposit states and nondeposit states [according to studies surveyed by Michigan Consultants (1996)]. How can this be? Beverage containers have become a small part of our litter problem, being replaced by paper—gum wrappers, fast food packaging, computer printouts, and the like. For recyclable items, the availability of curbside recycling has reduced littering as well as legal trash disposal. Because beverage containers account for only a small percentage of roadside litter, even if a mandatory deposit law reduced such litter to zero, it would only reduce *total* litter by a small percentage.[4]

At the bottom, there are two problems with the coexistence of mandatory deposits and curbside recycling. First, mandatory deposits have been shown to be an expensive antilitter program (in Chapter 6), and an even more expensive one once recycling has been introduced. And if the growth of recycling has led to the reduction of the beverage container litter problem, then mandatory deposits have become an even more expensive antilitter program, because the programs' costs to producers and consumers of beverages are not reduced by the existence of recycling. Second, mandatory deposits not only reduce litter and landfill costs; they also reduce MRF revenues. If this loss of recycling revenue delays or undermines the operation of socially profitable recycling programs, the final cost of any litter reduction will be even higher (Ackerman and Schatzki 1989, 1991). This is not just speculation; one empirical study found that where mandatory deposits were in effect, communities were 18% less likely to start curbside recycling collection (Kinnaman and Fullerton 2000).

If retailers really hate mandatory deposits as much as they (frequently) say they do, why don't they collectively offer the mandatory-deposit states a deal: replace mandatory deposits with a small tax (one or two cents per beverage container), whose revenue is earmarked to subsidize recycling and antilitter programs? In Michigan, for example, such a one-cent tax would generate $50 million a year, probably enough to make every existing recycling program in the state *privately* profitable—if the cities could resist pressure to blow the money away through excessive recycling.

Final Thoughts

How recycling will shape up over the next few decades will depend on empirical information, which is even now being gathered from the thousands of different recycling systems that are under way across the United States. The data, not speculation that we might undertake here, will determine which scheme ultimately will be generally adopted—though that scheme will probably vary from large to small urban areas and will certainly be different in rural areas.

Still, it is hard to resist one speculation. What now makes recycling inherently more costly than ordinary trash collection alone are the twin facts that recycling requires a second truck and that the recycling truck does not compact its contents. Better collection techniques will help. Better recyclables markets, and hence better prices for recyclables (which we turn to in Chapter 12), will help. But the biggest help of all would be getting rid of the second truck and the compaction disadvantage.

It is not impossible to conceive that recycling will ultimately achieve that. Already, there is co-collection and commingling of recyclables, as we discussed above. Ironically, these have been generally discarded as inferior in large cities, but they may reemerge as the ultimately low-cost collection techniques, once MRF equipment has progressed to the point where it can efficiently sort the resulting input (Miller 1995a). Compacting recyclables is harder to picture now, but it may be just around the corner.

Think what it would means if trash and recyclables collection were equally costly. It would be efficient to recycle materials if only the cost of MRF sorting were less than the *sum* of the landfill costs avoided *and* the revenues earned on the recyclable material. Current recycling targets, which are mostly unreasonable today, might become quaintly old-fashioned.

Notes

1. Nine household sorts is probably not the record. Guinness doesn't keep this statistic, but I have been told that one collection firm in Portland, Oregon, once required 10 sorts—newspapers, magazines, corrugated cardboard, mixed waste paper, aluminum, tin cans, plastic milk jugs, and green, amber, and white glass—and I have read about a town (unnamed) with 12 sorts (Stewart 2000b).

2. It is sometimes argued that adding mandatory deposits to an existing recycling program is a good thing because it reduces the overall cost per ton of the collection (Franklin Associates 1991). But this absurd result is reached by counting the additional tonnage of recyclables col-

lected and not counting the additional costs of the mandatory deposit collection system. By this kind of reasoning, putting mandatory deposits on everything would reduce all municipal solid waste costs—and hence cost per ton by this measure—to zero.

3. Where there is both recycling and mandatory deposits, it is important to remember that these are two separate policies aimed at two separate goals. The amount of the deposit should be a function of the marginal social cost of illegal disposal—there is no good reason for the deposit to be lower for more widely recycled materials or lower for potentially refillable containers, as is often suggested (Cohen et al. 1988).

4. Another reason why the roadside stock of litter in deposit states may seem nearly as high as in non-deposit states may be that deposit states pick up their litter less frequently in response to the lower litter rate. This would also be a social benefit of deposits, but it is a benefit that would not show up in a lower observed litter stock.

Markets for Recycling

Many people love recycling. It seems to meet some deep need to atone for modern materialism by saving some of the materials from the rubbish bin. Unfortunately, people do not feel quite the same craving to buy products made of recycled materials.

— Cairncross (1993)

Recyclable materials not only need to be collected, sorted, and processed. There are two further steps to recycling. The recyclable materials processed at the MRF must be sold to someone who will convert them into something useful, and these useful things must find customers who will buy and use them. These markets for recyclable materials and markets for recycled products are the concern of this chapter.

Why aren't these markets just like the market for broccoli, with no cause for public concern? After all, there *are* markets for all the recyclables being collected in the sense that these recycled materials are all being sold to *someone*. Well, almost all. Occasionally we hear of recycled materials that are sold at a negative price, which means that someone has to be paid to haul them away. Even more shocking, we sometimes hear of materials collected for recycling that end up in the landfill. Indeed, if they would "earn" a large enough negative price from being recycled, there is a good case for sending them to a landfill.

So, it is not really the absence of markets for recyclable materials but the discouragingly low prices there. As if low prices were not bad enough for recycling operations, the prices in markets for recyclable materials also fluctuate a lot, making budget planning difficult.

Markets for Recyclable Materials

Prices of recyclable materials tend to be low and variable because the markets are *secondary* ones. In almost every case, a recyclable material is a close substitute to, and hence is competing with, a virgin raw material that is cheap to produce. Sometimes the virgin material is cheap because it is abundant. The 1970s' doomsday

Are Recyclables and Virgin Materials Close Substitutes?

The closeness of recyclables and virgin materials as substitutes varies with particular materials, so no general answer can be given to the title question of this box. But we can answer it for a particular material. Consider glass (Williams 1991; Truini 2001c). The average glass container consists of 30% used glass (called cullet), but that percentage could go as high as 70% before technical problems set in. (In aluminum production, the percentage of used beverage containers in the input mix is even less constrained than the percentage of cullet in glass production.) Glass manufacturers prefer cullet because it melts at lower temperatures, saving fuel costs. But the percentage of cullet must be constant or the furnace temperature must be changed, which causes wear and tear on the furnace. So basically the instability of the cullet supply limits its use.

The quality of cullet also matters. Some 90% of municipal solid waste glass consists of beer and soft drink bottles. These come in three colors: green, amber, and clear. The glass has to be hand-sorted by color—broken glass is prohibitively costly to sort, and mixed-color glass is of no value. Because the raw material for glass is cheap and plentiful, the advantages of cullet are offset by its sorting costs and the low price of its substitute. Virgin-glass raw materials cost about $50 a ton, which means that the price of cullet can never rise much above $50 a ton.

Recycled and virgin inputs are highly substitutable in glass and aluminum—but not so with paper in the short run. Paper mills are designed to use either virgin (wood) pulp or recycled (paper) pulp. Only in the long run are the two input sources substitutes, for in the long run the kind of mills being built will respond to the relative prices of the two sources of pulp input.

prophesies about the planet's running out of basic raw materials have been followed by three decades of almost steadily falling real prices for most basic raw materials. Even cartel efforts, such as the Organization of the Petroleum Exporting Countries, have been unable to achieve and sustain higher prices.

The market cheapness of virgin materials, however, is not entirely due to natural abundance. For many virgin materials, the *social* cost is well above the *private* cost, for two reasons. First, harvesting and extracting virgin materials usually generates external costs. If extractors were forced to pay those external costs—either by being forced to generate less external cost or by compensating the victims or by paying Pigovian taxes—the market prices of virgin materials would be higher, and recyclable substitutes could more easily compete with them. Second, government provides many hidden subsidies to producers of virgin materials, through the provision of either free services or special tax advantages.

A few examples may help to make the point. Almost half of the U.S. virgin aluminum output is produced with hydroelectric power by the Bonneville Power Administration (BPA). Hydropower is cheap power. In a private market, if a producer had some advantage over its competitors, and could produce cheaply, it would still charge the market price for the product. But the BPA prices at no profit. Furthermore, the BPA gets low-cost federal government loans. Because primary aluminum production is highly energy intensive, this amounts to a subsidy of 5–12% of the primary aluminum price (Koplow 1994).[1]

Virgin paper production, even if from tree farms, imposes external costs on the surrounding land, water, wildlife, and recreation. It also receives hidden subsidies. One study estimated the extent of various subsidies on the cost of production: tax policies favoring timber harvesting (0.59%), below-cost federal timber sales (0.32%), below-social-cost energy provision (1.31%), and water subsidies (0.02%) (Koplow and Dietly 1994).[2] The total differential subsidy is nowhere near as large as for aluminum, but here too it favors virgin production.

Federal and state legislatures have made a habit of ignoring the external costs of virgin material production. And federal subsidies to virgin raw materials go back a long way. Percentage depletion deductions for minerals began in 1913. A *cost* depletion allowance makes perfectly good sense, for it allows an extractor to recover the initial capital cost of opening the mine or well before paying taxes on its net revenues. But a *percentage* depletion allowance is something else. This permits the mine to deduct from taxable income a percentage of its gross income, regardless of how much initial investment was involved.

Below-cost federal land timber sales began in 1891. These sales have been defended on a variety of grounds, none of which is sufficiently meritorious to detail here. Below-cost mining leases date even further back, to 1872. The Tax Reform Act of 1986 eliminated many of these, but many remain. All of these subsidies affect recyclable materials markets, albeit in many cases only slightly. If the United States is a major world producer of the product (i.e., of both virgin and recycled materials), then a subsidy to virgin production pushes the supply curve of virgin output down and to the right, lowers the price of the virgin materials, and increases the amount of virgin output. The lower price of virgin materials lowers the price paid for the closely substitutable recyclable materials and discourages their collection and recycling.

In short, the price of virgin materials is often below the social cost of their production. This not only encourages excessive virgin output, it discourages what could be socially profitable recycling. Currently, some effort is made to counter this advantage to virgin materials, by undertaking unprofitable recycling and by mandating that (some) purchasers must buy products of recycled materials even where they are more costly. These may not be bad ideas, they are just less than the best idea. The best idea is to directly remove the artificial stimuli to virgin outputs, by setting Pigovian taxes on their external costs and removing their subsidies and tax preferences.

Prices of recyclable materials also fluctuate greatly. This also derives from the fact that recyclables markets are secondary ones. Most raw material output is from virgin stocks, and ensuring a steady flow of these virgin materials is an essential concern of manufacturers. Indeed, most basic materials producers are vertically integrated back to the new material source for just this reason. Aluminum producers own bauxite mines, paper producers own forests, and so on. Recyclables are used as a marginal supplement as needed (Ackerman 1997). Thus, when demand for a material rises, demand for recyclables rises manyfold; but when demand for a material falls, demand for recyclables greatly diminishes and may even disappear.

Figures 12-1 and 12-2 show these fluctuations in the prices of recyclable materials. In Figure 12-1, the prices during the past half-century of old cardboard containers (OCC) are

The Price Spike of Old Newspapers in the Mid-1990s

The price of old newspapers (ONP) was about $20 a ton when it began to climb in mid-1994. The newsprint price, in response to newsprint shortages, less than doubled, but the ONP price reached a peak of $200 a ton in June 1995. It then fell—to $100 by October 1995, to $50 by November 1995, to never above $40 in 1996, and to $20 by July 1997. There are many theories about what caused this spike—mostly involving lagged and excessive capacity responses to changes in paper and pulp prices, unexpected shifts in paper demand, and irrational speculation—but none of them do very well in explaining the extremely high prices of 1995 (Ackerman and Gallagher 2001).

High prices cause recycling problems, too. For a while in 1995, some city sanitation departments had to assign personnel to crack down on scavengers who were going ahead of the city recycling trucks to steal valuable recyclable materials—ONP was not the only spiking price (e.g., see also Figures 12-1 and 12-3). Los Angeles estimated that it was losing two-thirds of its usual curbside newspaper collection, which meant a loss of more than $2 million a year in recycling revenues (*Newsweek* 1995).

Figure 12-1

Price Indices for Old
and New Cardboard
Containers

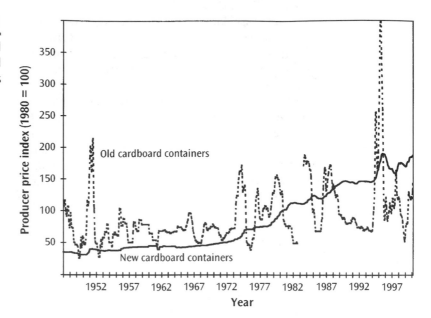

Figure 12-2

Price Indices for
Old Newspapers and
New Newsprint

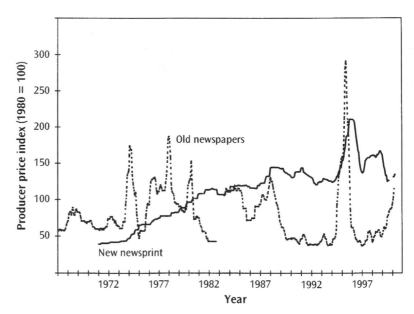

shown along with the prices of new cardboard containers.[3] The greater fluctuations of the OCC price are obvious. Less obvious is the declining price of OCC relative to the price of new containers. You can see it by noticing that the OCC price index is above the new container price for most of the first quarter-century, but below for most of the next quarter-century (with the two indices equal by definition, at 100, in 1980).

Figure 12-2 tells a similar story. There, the prices for the past third of a century of old newspapers (ONP) are shown along with the prices of new newsprint.[4] ONP

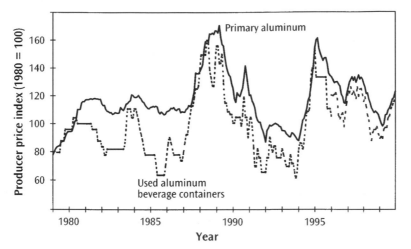

Figure 12-3

Price Indices for
Aluminum Used
Beverage Containers
and Primary Aluminum

prices display greater fluctuations, and the ONP price falls on average during the period relative to the price of new newsprint.

Does this mean that we are doomed forever to see fluctuating and declining relative prices for recyclable materials? No. Aluminum used beverage containers (UBCs) have already broken out of this bind. In Figure 12-3, the prices for the past 20 years of aluminum UBCs are shown along with the prices of primary aluminum.[5] The UBC price has tracked quite closely with the primary aluminum price, especially in the 1990s. The UBC market has clearly developed into a well-functioning market. Other recyclables markets will presumably also become more stable in time.

Looking back on the 1980s and 1990s, we can see specific but temporary reasons for the fluctuating and low prices of recyclables. Recycling was booming all over the United States, which brought ever-increasing supplies of recyclables to the marketplace. When supply outgrows demand, lower prices result. And prices fall the most when supply—or demand—is price inelastic. Municipal recycling has not been very motivated by price, so it is no surprise that studies show the supply of recyclables to be very price inelastic (Edwards and Pearce 1978; Edgren and Moreland 1990). But recycling supplies will surely grow less rapidly in the future as there are fewer municipalities left to begin recycling and many already recycling municipalities begin to reduce the items they recycle in order to save money. Price fluctuation in the past two decades has also been a phenomenon of this startup of recycling. When the numbers are small, there is an "integer problem." Twenty years ago, there were few recycled paper mills in the United States (Alexander 1994). A little recycling fulfilled their needs. But as recycling grew, with a fixed number of mills, the ONP and OCC prices fell, and the mills became ever more profitable. Eventually, these prices fell low enough that new mills were established, and the boost in demand ratcheted ONP and OCC prices upward. But ever increased recycling again soon saturated the market, and so on.

Can governments intervene to speed the development of efficient recyclables markets? Yes, but at some risk. For example, a direct subsidy to recyclers should give them incentive to establish and encourage the markets where they buy these subsi-

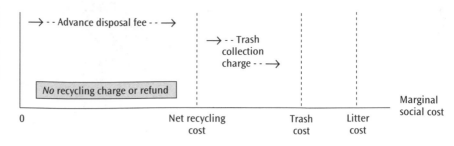

Figure 12-4

Scheme 2:
Pricing with Recycling
but No Illegal Dumping

dized materials. The demand for recyclables has also been found to be price inelastic, so subsidies in such markets transfer a lot of income for very little increase in quantity (Kinkley and Lahiri 1984; Edgren and Moreland 1990). The risk lies in the method of subsidy. Often, governments provide the subsidy through sharing the initial capital costs or offering low-interest loans, but these give only indirect incentive to buy and recycle more material. But there are also risks in subsidies based on the volume of recycling, which in principle is the right subsidy if there are social advantages to the faster, earlier growth of recyclables markets. Such "infant-industry" subsidies can prove hard to remove when the market has grown up.

Concern for the development of markets for recyclables adds another argument for scheme 2 (in Figure 10-4). Recall that scheme, reproduced here as Figure 12-4. Scheme 2 was the only one to base the advanced disposal fee on the net recycling cost of the product and its package. This gives each manufacturer an incentive to enter, or at least encourage, the markets for the recyclables generated by that manufacturer's products and packages. Although it would be costly to have each manufacturer buying back its own discarded products and packages—the producer "take-back" responsibility we discussed in Chapter 2—it might be helpful to "urge" manufacturers to concern themselves with their recyclables markets. Scheme 2 does that. If the manufacturer

Door-to-Door Markets for Recyclables

The prices paid for recyclables depend largely on the price and availability of the substitutable virgin materials. In Japan, few forests are left. Virgin wood products are almost all imported, from Canada or the United States, a long distance to transport a low value-for-weight product. Accordingly, wood and paper products are expensive in Japan. This encourages paper recycling.

It encourages recycling so much that the recyclable paper market does not just depend upon the civic-mindedness of households for recycling. *Chirigami kokan*—toilet paper exchangers—cruise the streets of residential areas with loudspeakers blaring, offering to exchange new toilet paper for old recyclable paper of all kinds; 6,000 such exchangers work in Tokyo alone. They sell their recyclable paper to junkyards, which compress it for recycling. The yards also provide them with the toilet paper, also of course from recycled paper, for the next day's exchanges (Kanabayashi 1982).

can alter the package or product so as to reduce the cost of collection and sorting or to raise the price of the recyclable residual, the manufacturer's advance disposal fee will be lowered.

Before leaving markets for recyclables, we must notice *international* markets as well. Although recyclables are usually of low value for weight, and hence resist lengthy transport, developing countries often provide sizable demand for U.S. recyclables. Consider paper. Many poor countries are simultaneously experiencing rapidly expanding literacy and rapidly shrinking forest areas. Their low wages

The Sorry State of Plastics Recycling

Proponents and opponents of plastic packaging agree that less than 10% of all postconsumer plastic is recycled in the United States and that the percentage is declining in recent years (Beck 2000; Denison 1997). This is at first glance surprising, considering how much effort we now put into recycling. And it is even more surprising when one notes that almost all of the preconsumer plastic waste is recycled. Noting the difference between preconsumer and postconsumer plastic removes the surprise.

Preconsumer plastic means factory waste. Because factories use only a few kinds of plastic in any particular process, sorting waste into homogeneous piles is easy, and such homogeneous waste, especially in large quantities, is readily recyclable and hence valuable. Postconsumer plastic waste emerges from households in small and very heterogeneous dribbles. It is expensive to collect and sort and reprocess. Moreover, much postconsumer polystyrene—#6 plastic—has been used to serve or contain food or drugs, and this plastic cannot be reused for these same purposes because the recycling process cannot guarantee that any contamination has been removed. How expensive is seen in the price difference. In 1999, virgin polyethylene terephthalate (PET) resin was selling for about $3,000 a ton, recyclers were only getting about $140 a ton for PET bottles, and the resulting recycled PET resin was still selling for 50% more than comparable virgin resin (Toloken 1999; Warren 1999).

Although most plastic is made from fossil fuels, it emerges in many different chemical forms and from many different industrial processes. There are literally thousands of different plastics—not just the six that the numbers stamped inside the (very misleading recycling) arrows suggest. In fact, more than 99% of the recycled plastic containers are #1 (PET, such as Pepsi bottles) and #2 (high-density polyethylene, such as milk bottles). Finally, many plastic containers are emptied away from home, with the result that recycling collection rates are low, except in states with mandatory deposits on beverage containers. Not only is the U.S. recycling rate for plastic containers low, it is also *falling rapidly*, from more than one-third in 1995 to barely one-fifth in 1999 (NAPCOR 2000).

Adding to these recycling difficulties are two more setbacks for recycling. First is the now-appearing plastic beer bottle. It has a special chemical coating to prevent oxygen from getting to the beer, which adds to its recycling costs, and its color, caps, and labels all demand special sorting. And second, Gerber Products is switching after 60 years from glass baby food containers to plastic. Not #1 or #2 plastic, but an almost never recycled #7 plastic. The #7 classification is reserved for multilayer containers, and those made with other than the #1–6 resins (Truini 2001b). Would advance disposal fees have influenced either of these two business decisions?

Not only are plastic containers largely unrecyclable, but they also generally replace glass containers, so much so that many plants producing glass bottles shut down in the 1990s. Each such closing reduces the demand for glass cullet, and cullet prices have plummeted—to the point where many recycling operations are considering no longer collecting and recycling glass (Truini 2001f).

make the sorting and processing of old paper more economical there, and their high cost of capital favors less capital-intensive recycled paper mills (Cohen et al. 1988).

There are success stories. Back in the 1960s, Lady Bird Johnson, then U.S. first lady, went on a major crusade to stop the eyesore of discarded motor vehicles littering the countryside. For decades, the price of new raw materials had been declining and the wage rates of potential recyclers rising, so there was no profit in recycling autos once their mobile life had ended. Some European countries (and New Mexico) initiated mandatory deposits on new automobiles, which were refunded when an auto hulk was properly turned over to an authorized junk dealer (Bohm 1981).

Today, however, only 5% of defunct automobiles are abandoned, and of those that are properly disposed of, about 80% of the vehicle (by weight) is recovered for recycling or reuse (Keoleian et al. 1996; Bigness 1995). This dramatic change in junked car disposal during the past three decades has come about from a variety of sources, such as the development and proliferation of dismantlers, shredders, and steel mini-mills that have increased the efficiency of the recycling and reduced the transport distances involved. Nevertheless, 100% of 5% of the junked cars and 20% of the other 95% still impose external costs in the form of street and highway cleanup costs, "eyesore" damage, and the (at least partly) unpaid costs of landfill space usage. One study guessed that this implicit subsidy to the disposal of automotive waste products amounts to $4 billion a year (Lee 1995).

Cars are recycled not only for recyclable materials but also for reusable parts. Currently, there are more than 5,000 mom-and-pop auto parts junkyards in the United States (Hoffman 2000). As "lemon" laws and end-of-lease recoveries generate ever larger numbers of cars still in basically good condition, rebuilding "wrecks" and reusing parts becomes ever more important than recycling what's left. As with many other traditionally prosaic small-business industries, the junk-car industry has been consolidated by the Internet. Two companies now handle half the business, and auto manufacturers themselves are becoming interested—Ford has begun to buy up auto salvage yards (Greenwire 1999a; Bailey 2001).

Not Everyone Wants Recycling

Continental Aluminum Incorporated buys up discarded doors, windows, and road signs, extracts aluminum from them, and sells the resulting ingots to Ford and General Motors for use in new cars and auto parts. It does this without subsidy at any stage. You would think the firm deserved applause from all.

Well, almost all—melting aluminum can produce chlorine, hydrogen chloride, and hydrogen fluoride, which can pose health hazards if not properly controlled. Continental was cited so often for its fire and pollution violations at its original plant in downtown Detroit that it built a new plant 40 miles west, in sparsely populated Lyon Township (Pearce 2000a, 2000b, 2000c, 2001). Now, its new neighbors are complaining of smoke, odors, and toxic fumes and are demanding compensation *both* for the physical and psychological harm *and* for the resulting reduced property values. (Assuming that there is merit to the case, only one of the two kinds of damage should be compensated. The decline in property values presumably is a present-value dollar measure of the flow of material and health damages, not a separate source of damage.) Ironically, the move to Lyon Township was financed by the Michigan Economic Development Corporation, a state entity that makes tax-exempt financing available to small firms that contribute to a greater public purpose and that have the financial commitment of the local community. Recycling operations can cause external costs—just as virgin materials processing does—and they may need to be taxed, to be regulated, or to provide host fees to their neighbors to internalize these externalities.

Markets for Recycled Products

"Recycled products" are products that use recycled materials, that is, materials that have been previously used. Consumers have long been trained to think of previously used stuff as inferior. This training goes back centuries, to the times when clothing made of recycled rags was called "shoddy," and the very word "recycled" became a synonym for second rate. This instinctive dislike of recycled products is not unique to Western culture. In Japan, when a forward-looking magazine printed on recycled paper, it received complaints that turning pages made an irritating sound, and the magazine reverted to new paper (see http://www.wtco.osakawtc.or.jp/market/item/printing.html).

Indeed, recycled products are almost always at least somewhat second rate in comparison with products made from virgin materials. Getting recyclable materials perfectly clean and homogeneous would cost so much that the resulting product would be too expensive to market, and stopping at somewhat clean and homogeneous leaves the finished products with structural or visual imperfections. To find customers, many recycled products must be priced below the comparable products made of virgin materials.

This is not hard to do. In theory at least, products made from recycled raw materials should be cheaper to produce. Transport costs are lower, because the recycled material is "mined" near both the factory and the customers. Recycled materials can generally be processed with less energy, and they usually generate less air and water pollution. But with newly established recycled-materials manufacturers still learning and using technology that will rapidly improve, working in factories that are still too small, any ultimate cost advantages are temporarily more than offset. Even if recycled goods leave the factory with a competitive price, they may be sold at retail on such a small scale as to require higher markups there. Low demand begets high price, which in turn perpetuates low demand. Until recycled goods are "mainstreamed"—which means getting good positions on retail shelves and bearing low markups—they will carry unnecessarily high prices.

The federal government and almost all the states are making an effort to accelerate this mainstreaming by imposing buy-recycled requirements on themselves. Unfortunately, so far this has proved easier said than done. According to one survey (by Raymond Communications) of the 47 states that require purchase of recycled copier paper and plastic products, only 30 of the state bureaucracies actually buy recycled paper and only four actually buy recycled plastic products (Leroux 1999). The reason for the discrepancy? It costs little to issue an executive order or pass a law, but it requires serious budget to enforce it, even when the violators are the government itself.

Some U.S. laws actually discourage the purchase of products with recyclable content. Here are a few examples. The procurement guidelines of half the states restrict the purchase of office equipment with reprocessed parts (Fishbein et al. 2000). Many states require auto insurers to provide only new parts when crashed cars are being repaired (see http://www.insure.com/auto/aftermarket/aftermarketlaw.html). In many states, rules for waste operations also apply to industries whose input is—or would have been if not recycled—waste. In Tennessee, for example, any "solid waste processor" must get approval of the local government before it can operate. Thus, if a firm produces, say, chipboard from new wood, it does not need local approval, but if it uses recycled wood, it does (Truini 2000d).

Final Thoughts

I argued, I hope successfully, in Chapter 10 that supply-side and demand-side policies to increase recycling were second-best policies. But there is no question that they have provided spurs to both supply and demand for recyclable materials. And the resulting increased familiarity of most Americans with recycled products has begun to change our attitudes toward them. Products no longer cover up their recycled content, but rather advertise it in an effort to boost sales.

In the end, however, it is—somewhat ironically—not high prices for recyclables and recycled products that will make recycling a success, but rather low prices. Low prices for recyclables make the products they produce less costly; and less costly recycled products will gradually begin to be generally preferred over higher cost virgin-input alternatives. Extensive recycling, if that is what is to eventuate, requires that recycled products be not only good but cheap.

Notes

1. Virgin aluminum production uses 300,000 kilowatt-hours of energy per ton, whereas recycled aluminum uses only 15,000 (Williams 1991). When other-than-energy subsidies are considered, the price advantage of primary aluminum amounts to 23% of its price (Koplow and Dietly 1994).

2. The original study gives high and low estimates of the subsidy; the percentages given in the text are averages of the two estimates. The production of recycled paper uses much less energy and water than does that of virgin paper, which is why the subsidies there have differential impacts on the costs of the two kinds of paper.

3. A principal use of OCC is the production of "new" cardboard. The OCC price is the Bureau of Labor Statistics (BLS) producer price index (PPI), WPU09120311 from 1947 until 1986 and WPU091203 from 1967 until 2000 (the series are identical for the overlapping years). No price index was recorded for several of the months of the period 1981–1983. The new paper boxes and containers price is the BLS PPI, WPU091503. For each series, the base year is 1980 = 100.

4. A principal use of ONP is the production of "new" newsprint. The ONP price is the BLS PPI, WPU091201. No price index was recorded for several of the months of the period 1981–1983. The newsprint price is the BLS PPI, WPU091302. The base year for both indices is 1980.

5. Aluminum UBCs are used to make new aluminum. The UBC price combines two series compiled by Connecticut Metal Industries and *Resource Recycling* (http://www.recyclemetal. com and http://www.recyclemetal.com/historic.htm), converted into an index (the actual prices, not weighted by quantities, averaged $955 a ton for the entire period). When more than one price quote was given for any month, the quote nearest the middle of the month was used. The primary aluminum price is the BLS PPI, PCU3334#. The base year for both indices is 1980.

Yard Waste and Composting

Compost happens.

—Anonymous

You may already be asking, why do we need a separate chapter for yard waste? Until 1980, the answer would have been, we don't. But three things happened in the 1980s to change our attitudes about yard waste, and hence our handling of it. Landfills began to close, tipping fees began to rise, and landfill operation began to become more costly (the three phenomena are of course related). Municipal solid waste authorities thought immediately of yard waste, mostly grass clippings, which accounts for more than a fifth of U.S. MSW—more than half in Los Angeles (Miller, 1995b). Keeping yard waste out of landfills would make a significant difference in a municipality's cost of waste disposal.

Nothing fancy, expensive, or newfangled is necessary to keep yard waste out of landfills. It is called "composting" and is one of the oldest reuse techniques known. (Anything organic can be composted, so counting paper, wood, and food waste, about two-thirds of the MSW stream is compostable. Currently, however, only yard waste is being composted to any extent in the United States.) Composting means organic waste piles where fungi and bacteria feed. In this feeding process, a pile of yard waste gradually decomposes into humus (decayed organic matter), carbon dioxide, and water. The humus, less than a third the volume of the original yard waste, then has a variety of uses. It can be used as topsoil, mulch, and soil nutrient—even as landfill cover. Mixed into soil as an amendment ("mulch"), it can modify soil temperatures, improve moisture retention, and reduce erosion and weeds.

Compost happens, but *useful* compost doesn't just happen. The process requires care, and hence cost. The microorganisms need oxygen, so the yard waste needs to be shredded initially, then turned regularly ("aerated"), and its temperature, moisture, and carbon-to-nitrogen ratio controlled. Otherwise, the decomposition becomes anaerobic (i.e., without oxygen), which slows the pace and increases the odor (Hlavka 1993). The whole process takes from several weeks to several months, depending on the size and care of the operation. The final humus also needs more screening and grinding to prepare it for use.

The benefits from composting are that it avoids landfill costs and produces a useful product. The costs are the additional MSW collection cost when yet a third stream of waste is separately collected and the cost of the composting operation itself. Municipalities more and more see the benefits as outweighing the costs, and composting is growing rapidly in the United States. In 1988, only 0.5 million tons of MSW were composted; by 1995, 11.1 million tons were composted (U.S. EPA 1995a). Let us take a closer look at the economics of a separate yard waste collection, the composting process, and the use or sale of the humus product.

Collecting and Processing Yard Waste

Composting usually begins very informally. Environmentally concerned households may buy lawn mowers that produce finely shredded clippings that can be returned directly to the lawn—called "grasscycling." Leaving the grass clippings on the lawn decreases the amount of time and fertilizer needed to maintain the lawn. Those with sufficient space and time may undertake "backyard composting," which converts yard waste into useful mulch, but it takes time and space, and it produces odors that many households (and their neighbors) will not tolerate.

Cities became involved with composting in the 1980s, but the early compost programs were voluntary and drop-off, and the participation rate was low. Although cheap to the city, drop-off composting is socially costly when one recognizes its heavy demand on household time and gasoline. These early drop-off programs did, however, demonstrate that economies of scale were possible in moving from backyard composting to large, centralized, citywide composting. These programs used machinery to turn and aerate the compost—piled in long narrow "windrows"—and this reduced the need for labor and land (per unit of output), reduced the odor problem (by moving the compost away from the densely populated areas), and greatly reduced the time needed to convert yard waste into marketable mulch (Hlavka 1993).

The final yard waste collection process is curbside pickup. As with recyclables, there are many ways to gather yard waste, and the curbside programs are still too young to have found the cheapest way (Snow 1998). Most cities gather leaves once or twice in the fall by having households rake them into the street, where they are collected by vacuum trucks. (Seemingly slick, vacuum truck collection is not without side problems: The leaves blow around, they clog storm drains, they can be set afire by vehicle catalytic converters, and they can become contaminated with road litter that reduces the marketability of the final mulch.) Other yard waste is set out in paper or plastic bags or in specially marked bins and collected by special yard waste trucks, weekly or biweekly. Bins are becoming preferred because they do away with the bags that must be removed lest they prevent (plastic) or retard (paper) the necessary compost process.

Three final issues must be resolved when there is curbside pickup of yard waste. The first question is, how much to pick up? All organic materials can be composted, but as with recycling, there are diseconomies of scope despite the economies of scale (Kashmanian and Spencer 1993). Leaves are the least costly to handle, but complexity and hence costs rise as one moves into grass, brush, Christmas trees, food scraps, and finally into nonrecyclable paper. Obviously, the denser the population—other things equal—the more yard waste can be sensibly collected and composted. But other things are never equal. The denser the population, the smaller are the yards, and the less yard waste is generated.

The second question is, should backyard composting be encouraged once curbside pickup is available citywide? Given the economies of scale, the answer is almost always no, but because backyard composting saves some (though not much) city budget cost, most cities continue to encourage it (U.S. EPA 1994c). This encouragement is usually of the carrot type, where the city subsidizes the sale of backyard composting bins and/or offers free technical advice (Roulac 1995). But stick-type encouragement is also possible, by the addition of a fee for yard waste put out for collection. (More shortly on yard waste collection charges.)

And the third question is, how to keep hazardous wastes out of the compost stream? If toxic substances (heavy metals) or contaminants (plastic and glass) creep in, even in small quantities, the resulting mulch is much reduced in value (Kasmanian and Spencer 1993). This means households must take care that hazardous wastes never get into yard waste at the start of the process. One state (New York) is so concerned about this problem that it will not permit compost operations to start unless the area has a household hazardous waste (HHW) collection program in place and approved. But HHW collection programs are expensive, and few households participate—more on HHW in the next chapter (Duxbury 1992).

Backyard composting is a little more than stacking the yard waste and waiting—the compost pile needs to be turned about once a year, and the final stabilized mulch emerges in about three years. A municipal compost operation stacks the raw material higher, wets the compost, adds nitrogen, aerates with blowers, and turns frequently with specialized windrow-turning equipment. And the process is complete within a few months.

Composting Food Waste

A half-century ago, food waste was picked up regularly at curbside by farmers seeking the free feed for their pigs. Because households usually then had to carry their own waste to the dump, or pay privately for someone else to do it, the farmers' services were welcomed. But the process came to an end because of health worries—the health of humans who might eat diseased pork and the health of pigs who might eat contaminated feed.

Food waste can also be composted, but it is more costly than other organic composting because of the need to guard against rodents, insects, and odor. As a rule, for the small amounts available from house to house (in industrial countries), food composting is considered too costly.

There are exceptions to that rule. The New York State Department of Corrections found that it generated an average of one pound of food waste per prisoner a day (Fishbein and Gelb 1992). With 60,000 prisoners, that's 30 tons a day. Hauling and tipping fees ran (in 1991) $115 a ton—more than a million dollars a year. The compost equipment cost only $20,000 (also in 1991), and prison labor is "free." Because composting food waste alone is not the best practice, the prisons arranged to get yard waste as a bulking agent from the roadside grass mowing of the Department of Transportation. The resulting mulch is used on prison farms and in horticultural training programs, or given to community service programs and prison workers.

Can urban food composting become widespread? San Francisco is considering a food waste collection and composting program (Wiley 2000). The plan is that residents would put out three carts, a green cart for compost-bound food scraps and yard waste, a blue cart for MRF-bound recyclable materials, and a black cart for landfill-bound everything else. The principal perceived advantage seems to be that adding food scraps to the city's recycling program will get the percentage recycling rate up to 50%, preventing the threatened imposition of a state fine. The disadvantage is that the monthly household cost of the waste program will go up from $11.68 to $18.09. Assuming that the waste collection charge increase is entirely due to the added net cost of composting food scraps, that means that adding food scraps costs more than $300 a ton (net of the compost value)–very costly.

What does the collection and processing of compost cost? Studies abound, and they vary greatly, from under $10 per ton of yard waste up to $140 per ton, with a rough average around $70 per ton (Blumberg and Gottlieb 1989; CalRecovery 1989; Kashmanian and Spencer 1993; U.S. EPA 1994c; Segall et al. 1994; Miller 1995b, 1998; Caldwell and Rynk 2000). Yard waste collection and composting in Ann Arbor (see the Appendix on page 194) comes in at the low end of the range, at a cost of about $25 a ton.

The compost cost estimates vary greatly for three reasons: the usual collection variances due to different wage rates and city size and density; the newness of composting operations so that different programs are at different stages of learning curves; and the fact that many cities and studies count only cash costs, which means that land already owned by the city may not be counted as a cost and that any costs that are subsidized by the state or by environmental organizations will be omitted. No study (that I have seen) attempts to value the costs associated with the odor and fungi that compost can produce. So the true social cost of composting is certainly higher than the estimates mentioned above. But remember, these costs are offset, at least partly, by the landfill tipping fees that composting avoids and the value of the mulch that composting generates.

Selling Mulch

Composting began as a way to save on landfill space and tipping fees, not to provide a marketable product. As a result, early mulch was generally of low quality and often polluted or contaminated as well. Markets for such low-quality products were weak, and most of the mulch in the late 1980s and early 1990s was stockpiled, used as landfill cover, or landfilled (Kashmanian and Spencer 1993). Composting only to landfill the end product may seem hopelessly inefficient, but recall that the volume of mulch is only a fraction of the original yard waste volume. Using mulch as landfill cover saves about $10 per cubic yard for otherwise purchased cover.

An EPA study in the mid-1990s showed that even then, many facilities were still selling mulch for less than one dollar per cubic yard, and some were paying brokers to take it away (U.S. EPA 1995a). Conversely, where facilities paid attention to mulch quality and actively sought markets, some were earning up to $30 per cubic yard, principally from landscapers who used it as a soil supplement (Glenn 1990; Taylor and Kashmanian 1989). High-quality mulch also found a use in the city's own parks.

Nearly half the states have procurement policies favoring compost products. In California, for example, 40% of what state agencies spend on topsoil and organic materials must be spent on compost. Such policies, as we have seen with recycled products, may be helpful as "infant-industry" stimuli to the expansion of mulch markets, but they serve no good purpose once such markets are well established. In the long run, if mulch cannot compete with other soil supplements without subsidy, it means that cities are charging too low fees for yard waste collection.

Benefit–Cost Analysis of Composting Yard Waste

We are now in a position to see the complete picture of the benefits and costs of composting. The principal benefit is landfill cost avoided—recall that yard waste

Category of benefit (+) or cost (−)	Montgomery County, MD	Wellesley, MA	Seattle, WA[a]	Woodbury, MN	Omaha, NE
Landfill cost avoided	+62	+69	+41	+41	+8
MSW collection cost avoided	+37	0	+47	+44	+20
Mulch revenue	+6	0	+3	+4	+11
Yard waste collection cost	−113	−15	−60	−58	−52
Compost processing cost	−25	−15	−29	−20	−5
Overall benefit–cost	−33	+39	+2	+11	−18

Table 13-1

Benefit–Cost Analysis of Composting in Five Communities, 1987–1988 (1997 dollars per ton of yard waste collected)

Note: 0 means less than $0.50; totals may not add due to rounding.

[a]The cost of the Seattle yard waste collection is assumed to be equal to the average for the other four communities.

Source: Taylor and Kashmanian 1989.

accounts for more than a fifth of U.S. MSW. Another benefit is the revenue from the sale of mulch, small still for most cities, but perhaps more important as mulch markets develop. A final benefit of composting may be that people *want* to compost and would be willing to pay to be able to do it. Evidence of this willingness to pay is the number of households that—in the absence of citywide yard waste composting—operate their own backyard compost piles, despite the costs involved. The principal cost arises from the fact that three, rather than two, trucks are now passing each house, so that the savings in trash collection costs is more than offset by the new yard waste collection costs. There is also the cost of the composting process itself.

Because cities differ in many critical ways, it is not possible to do a single benefit–cost analysis of composting for all of the United States. Clearly, composting is uneconomical in sparsely populated areas where landfilling is cheap. At some point, as population density and landfill prices rise, composting becomes sensible. See the Appendix for a benefit–cost analysis of one such composting operation (in Ann Arbor), one that clearly passes its benefit–cost test.

Taylor and Kashmanian (1989) studied several composting operations in the late 1980s. Although their data often omitted land costs (when the government already owned the land) and usually counted equipment expenditures rather than properly annualized capital costs, the results are instructive. First of all, the sampled communities collected only (a weighted average of) 12% of their total generated yard waste, so that as participation rates have since then improved, the benefit–cost assessments should also have changed. Table 13-1 shows the benefit–cost results for the five communities for which there are sufficient data for a crude test. (I have assumed that the marginal cost of trash collection in these communities is half the average cost, and I have converted the figures to 1997 dollars.)

Although Table 13-1 raises a lot of questions, three of these benefit–cost analyses do end up positive, even though all were at that time very new programs. More mature programs would almost certainly show even more favorable results. More recently, the U.S. EPA studied nine communities' compost and landfill operations and found, despite a wide range of costs, that collecting and composting yard waste cost an average of $53 per ton *less* than would collecting and landfilling yard waste—though not one of these communities earned revenue from their compost

(U.S. EPA 1995a). Even more recently, a study of seven municipal composting operations estimated their average cost at $53 per ton (Renkow and Rubin 1998). In all but one of the communities in this study, the resulting compost earned almost no revenue, and the avoided landfill costs were not netted out. In all these studies, it is implicitly assumed that the transport distances to the landfill and to the compost operation are the same, although compost operations today are often much closer (as in Ann Arbor; see the Appendix).

Composting seems to pay, though the final proof of that must await more experience and study of more mature composting programs. Still, municipal composting will never pay off in sparsely populated areas, and states ought not to be encouraging composting everywhere. Nevertheless, about half the states have banned the landfilling of yard waste, and others offer startup subsidies to local communities for new composting programs. Such bans and subsidies are probably unnecessary in large, densely populated cities, and they may perversely encourage uneconomical composting programs in small, sparsely populated towns.[1]

Pricing Yard Waste Collection

The benefit–cost analysis outlined above gives us our clue as to how yard waste collection should, in principle, be priced. If composting passes its benefit–cost test, this means that composting yard waste is socially preferred to landfilling it. This, in turn, means that the price (e.g., per bag) of yard waste should be lower than the price (for the same-size bag) if the yard waste were to be set out as ordinary trash and sent to the landfill.

Composting passes its benefit–cost test if the sum of its two benefits—the value of the resulting mulch and the avoided cost of collecting and landfilling the yard waste—is greater than the cost of collecting and composting the yard waste. The size of this net benefit tells us how much cheaper it should be to put yard waste out for compost rather than put it out as ordinary trash for landfilling.

To see this more clearly, consider the numbers for Woodbury, Minnesota, in Table 13-1. (These numbers are averages, and we should be using marginals; but they will do for an illustration.) Per ton of yard waste collected, the value of the mulch is $4 and the avoided cost of collecting and landfilling the yard waste is $85, whereas the cost of collecting and composting it is $78; the net benefit of composting rather than landfilling is $11 (i.e., $4 + $85 − $78 = $11). Yard waste put out as compost should be charged $11 a ton less than yard waste put out as trash. Because few people think in terms of tons of yard waste, let's convert that to pounds. Compost should be charged about a half cent per pound less than ordinary trash. In fact, this charge would have to be further converted to a price per bag or per can. Where there is no trash collection charge and composting passes its benefit–cost test, people should be paid for putting their yard waste out for composting!

With the correct yard waste collection charge, households get the proper incentive—that is, the incentive to choose the socially efficient generation and disposal of yard waste. Households have three choices:

1. They can reduce their yard waste (perhaps by grasscycling) to avoid the yard waste charge if the cost to them of doing this (including any forgone pleasure from the less neatly manicured yard) is less than the charge.

2. They can begin backyard composting, again if the cost to themselves in time, lost space, and odor is less than the charge.
3. They can put their yard waste out for the special yard waste collection, paying the yard waste collection charge.

Note that, as long as the yard waste collection charge is less than the trash collection charge, banning the commingling of yard waste and trash would be redundant for all but a few households. Because the personal cost of separating yard waste is always small—much less than separating recyclables since yard waste is generated at a different time and in a different place than most trash and recyclables—almost all households will choose to take advantage of the lower yard waste charge. Why do half the states, as well as many cities and towns in other states, bother to ban the disposal of yard waste in landfills (Kashmanian 1993; Skumatz et al. 1997)? (Many also ban the disposal of yard waste in municipal incinerators, but even more redundantly; yard waste is poor fuel.) Because in most communities, neither trash nor compost is priced, so households have no incentive to put yard waste into a cheaper compost pickup. Bans force people to do the socially correct thing. Proper waste pricing would make people *want* to do the socially correct thing.

Administrative convenience, political acceptance, and fear of illegal dumping of yard

> ## Composting in Western Europe
>
> Most of Western Europe, unlike North America, makes the collection of compostable materials for transport to centralized composting facilities an important aspect of the waste management and source separation program (Bonomo and Higginson 1988; Segall 1992; UNEP 1996). Centralized composting has long been practiced, with some facilities dating back to the 1960s. Typically, households are required to separate noncompostable waste from compostable waste—all compostable waste, including food waste (Scheinberg and Smoler 1990). Often, to keep costs down, collection of the two separate bins is biweekly—bio-waste one week and the rest the next week.

waste may suggest that the yard waste charge be set at zero. Speaking of illegality, the difference between the yard waste collection cost and the trash collection cost provides an incentive for households to illegally put noncompostables in the compostable bag (or can). But this is not hard for the collectors to spot, and all they need to do is just not pick up yard waste that is contaminated. They do exactly that in Japan, and it solves this potential problem (Hershkowitz and Salerni 1987).

A Different View of Compost

Throughout this chapter, we have been comparing the composting of yard waste with disposing of yard waste in a landfill, but implicitly that landfill has been one from which landfill gas was *not* collected. Consider how the comparison changes, once the possibility of generating energy from the landfill is introduced. It changes a lot because the decomposition of yard waste in a landfill generates a lot of gas.

Redo the conceptual benefit–cost analysis of composting with the generation and capture of landfill methane considered. The value of the electricity, net of the costs of its generation, becomes an opportunity cost of compost. Currently, electricity is fairly cheap, just as is mulch, but if energy shortages appear, landfilling yard waste—as well as food waste—may become a socially preferred activity to composting. Moreover, if the government begins to mandate not only the collection of methane from landfills but its use to produce electricity, then the landfill of yard waste gains even more. With the MSW collection system *and* the landfill gas-to-electricity system *already* in place, the cost of the electricity that could be produced from yard waste would be very low.

Final Thoughts

From an economic viewpoint, composting is just another way of recycling. The social benefit–cost analysis is conceptually identical. But the quantitative magnitudes of the various benefit and cost categories are quite different. The separate collection and "recycling" of yard waste principally adds to total collection costs and avoids landfill costs. The cost of operating the compost pile and the revenue from the resulting mulch are relatively small. Though considered minor by most people, in comparison with "regular" recycling, the benefit–cost analyses of early composting programs suggest that they are already much more successful, in a net benefit sense, than most recycling programs.

APPENDIX

A Social Benefit–Cost Analysis of Composting in Ann Arbor

It is not the horse that draws the cart, but the oats.

—Russian proverb

For years, the City of Ann Arbor has collected leaves, yard waste, and Christmas trees, along with its other solid waste, taking it all to its own landfill. As it became clear that the city landfill would soon have to be closed, the looming high cost of transport to Salem Township and the tipping fees at the Arbor Hills landfill there strongly urged the introduction of composting. Gradually, during the 1990s, city composting has become a full-scale operation (Jendryka 1992). This appendix attempts to put rough dollar values on the benefits and costs of the current (1997) Ann Arbor composting program (data from city budgets and personal communication with Ray Ayer and Bryan Weinert).

The basic benefits and costs of composting are of six kinds:

■ *Landfill costs avoided.* Whenever yard waste, leaves, and Christmas trees are composted, the resources needed to landfill that waste are saved. Ann Arbor pays about $28.00 a ton to dispose of its waste at Arbor Hills. Because Ann Arbor collects more than 14,000 tons a year of compostable material—almost 300 pounds per person a year—its compost program directly saves the city $401,000 a year in avoided landfill charges.

■ *Solid waste collection costs avoided.* When waste is collected as compost, it does not have to be collected by the usual MSW collection system. Because Ann Arbor spends $626,000 a year collecting 16,107 tons of residential solid waste, the average cost of such collection is about $40 a ton. If yard waste and solid waste cost the same per ton to collect, if social and private cost were the same, and if the average and marginal cost were the same, this would mean that removing yard waste from the MSW flow would reduce the total social cost of collecting MSW by $140,000. (More than 3,500 tons of the compostables are collected by the MSW collection system; the rest, the leaves and Christmas trees, are collected by Street Maintenance in exactly the same way as they would be collected if there were no composting. The only difference here is that the leaves are taken to the compost facility rather than to a landfill.) Social and private cost probably are about the same, but the marginal cost of collecting an extra ton of MSW is almost certainly less than the average cost. It is not easy to know how much less. Assuming marginal cost is half of average cost yields a social benefit on this account of $70,000.

■ *Transfer costs avoided.* Without composting, all of Ann Arbor's yard waste would have to be moved by the collection trucks to the private landfill 25 miles away. (In Michigan, landfilling or incinerating yard waste is now illegal, so Ann Arbor's alternative to compost would mean an even more expensive transport of yard waste out of state. The question we are effectively asking here is, would Ann Arbor composting make sense if there were no such law?) With composting, all the yard waste goes only as far as the nearby compost facility. As a result, composting saves about 100,000 miles a year of travel for its MSW collection and Street Maintenance trucks. (The 14,000 tons of compostables would require about 2,000 round-trip truck trips of 50 miles each.) At $1 a mile, this means a benefit of $100,000.

■ *Value of the mulch.* In the process of composting, the 14,325 tons of inbound yard waste shrink to about a fourth of that tonnage of marketable outbound mulch. The city sells much of this mulch for $12 a cubic yard and earns $40,000. Assuming this mulch was worth to purchasers at least what they paid for it means this $40,000 is not only a city revenue but also a social benefit.

■ *Yard waste collection costs incurred.* The city spends $71,000 a year on yard waste collection costs. (Recall, this is only for the 3,500 tons of yard waste; the leaves and Christmas trees are still collected in the same way.) Adding in depreciation and interest on the vehicles used brings the cost up to $78,000.

■ *Costs of the composting operation.* The operating and maintenance cost of the compost operation is $136,000. The initial land and development costs were about $510,000, which at an interest rate of 4% yields an annualized cost of $20,000. The operation uses about $770,000 worth of equipment. Assuming an equipment life of 10 years and an interest rate of 4% yields an annualized equipment cost of $108,000. The sum of these three costs is $264,000. (Because there have been no complaints about compost odors from the neighbors, who are not that close, we ignore any such external costs.) Adding and subtracting these six benefit and cost estimates:

Landfill costs avoided	+401,000
Solid waste collection costs avoided	+ 70,000
Transfer costs avoided	+100,000
Value of the mulch	+ 40,000
Yard waste collection costs incurred	−78,000
Costs of the composting operation	−264,000
Net total	+269,000

Composting is a clear winner, with a *net* benefit of about $20 per ton of yard waste collected and composted. Landfilling yard waste is expensive business, especially when the landfill is some distance away.

A final point: The yard waste collectors are nonunion and are paid significantly less in cash and benefits than the solid waste collectors. To what extent is the positive outcome for composting due to the mere substitution of low-paid for high-paid workers? If the yard waste collectors were paid the same wages and benefits as the solid waste collectors, it would have added about $20,000 to the yard waste collection costs—and hence reduced the net benefit of composting by only 7%, from $269,000 to $249,000.

Note

1. Such perverse incentives are not new to economists—Adam Smith in the *The Wealth of Nations* (Book IV, Chapter V): "The bounty to the white-herring fishery is a tonnage bounty; and is proportioned to the burden [cargo capacity] of the ship, not to her diligence or success in the fishery; and it has, I am afraid, been too common for vessels to fit out for the sole purpose of catching, not the fish, but the bounty."

PART 3

Special Waste Categories

Hazardous Waste

The nice thing about standards is that there are so many of them to choose from.
—Andres S. Tannenbaum

We turn now to hazardous waste, stuff that is a serious threat to human health and the environment. Not that we haven't been dealing with hazardous waste up until now. The reason for the regulations on landfills and incinerators, discussed earlier, is that a small percentage of their waste is hazardous to health and hence needs to be prevented from reaching humans through the air, water, or soil. But that was about a "small percentage." In this chapter, we deal with high-concentration hazardous waste.

Hazardous waste—almost completely uncontrolled until two decades ago—is now defined and regulated by two federal statutes, Subtitle C of the Resource Conservation and Recovery Act of 1976 (RCRA) and the Hazardous and Solid Waste Amendments (HSWA) of 1980 and 1984.[1] RCRA defines hazardous waste as

> a solid waste, or combination of solid wastes, which because of its quantity, concentration, or physical, chemical, or infectious characteristics may cause, or significantly contribute to, an increase in mortality or an increase in serious irreversible, or incapacitating reversible, illness … or pose a substantial present or potential hazard to human health or the environment when improperly treated, stored, transported, or disposed of …. [RCRA, Section 6903(5); http://www4.law.cornell.edu/uscode/42/6903.html]

RCRA and HSWA not only defined hazardous waste but also developed standards for those who generated, transported, stored, treated, or disposed of hazardous waste. Most important, a tracking system was created for the location and movement of hazardous waste.

Hazardous waste is no small part of the volume of U.S. waste. According to the U.S. EPA Biennial Survey of Hazardous Waste, more than 50 million tons are created annually (U.S. EPA 2001a). The data say that Americans generate much more hazardous waste than do Western Europeans, but that is probably due to the stricter definition of hazardous waste in the United States. There appears to be little downward trend in the volume of U.S. hazardous waste, despite the increasing controls and concomitant costs, but that is probably due to the steady increase in the number of substances considered hazardous. Indeed, all these numbers are at best suggestive since much hazardous waste is treated on-site and avoids RCRA scrutiny.

Unlike municipal solid waste, to which nearly everyone contributes significantly, most of the hazardous waste is produced by a very few firms. Of the more than 20,000 so-called large-quantity generators (LQGs) of hazardous waste in the United States, the 50 largest ones produced four-fifths of the hazardous waste. (Roughly, an LQG is anyone who generates more than one ton per month of hazardous waste.) Most of the hazardous waste is produced by the chemical and petroleum industries—the 17 largest hazardous waste generators are *all* from those two industries (U.S. EPA 2001a).

Despite the seemingly hard data, we really do not know with any certainty how much *hazardous* waste is generated in the United States. Nor do we really know how much gets out in the public where it shouldn't be, how it gets out, or how much damage it does. There are very few and mostly vague guesses about those issues. One Public Interest Research Group report added up government data from various sources to estimate that there were 60,000 chemical plant accidents and spills *per year* during the period 1987–1996, killing more than 250 people per year (Baumann et al. 1999).

It is more promising for us to look at what we label hazardous waste, how we then handle it, how we price its treatment and disposal, and how we penalize those who mishandle it. In the course of this examination, the question we keep asking is, are we setting up the right incentives to create only the optimal quantities of hazardous waste and then to handle it optimally?[2]

How Does Waste Become Hazardous Waste?

How does a waste get the title "hazardous"? Subtitle C of RCRA specifies four criteria, and if a material meets any one of the criteria, it becomes hazardous waste. First, when tested, is the material shown to be ignitable, corrosive, reactive, or toxic? Second, is the material known to contain anything that is designated as toxic in the United States? Third, has the material been shown to be acutely toxic in studies of humans or animals? ("Acutely toxic" means that very small doses are lethal to human beings or, lacking human evidence, to laboratory mammals.) And fourth, if an entire mixture consists in part of a hazardous waste, then the entire mixture becomes a hazardous waste. In short, there are many tests, and the long list of materials designated as hazardous is quite diverse. A few examples: mercury, lead, and cadmium; wastewater treatment sludges, spent cyanide solutions, and plating bath residues; oven residue from the production of chrome oxide green pigments, still bottoms from the distillation of benzyl chloride, and emission control sludge from secondary lead smelting; acetic acid, nitrogen dioxide, and tetraethyl lead; DDT, benzene, and formaldehyde (U.S. EPA 2001a, Appendix D).

On the surface, this series of tests seems sensible—where there are risks to health and the environment, it seems better to err on the side of caution. But there are many problems with the approach of Subtitle C.

The first problem is that each material tested becomes labeled either a hazardous waste or a nonhazardous waste—a binary, zero–one, either–or decision. In this decision, there is no concern for the degree of toxicity. Other things remaining equal, material that is more toxic should be handled more carefully. Moreover, in the either–or decision process, there is no concern for the extent of the benefits that using the material may be conferring on society, there is no concern for the cost (or toxicity) of substitutes for the material, and there is no concern for the size and proximity of the potentially affected population. Declaring a material to be a hazardous waste subjects it to expensive handling and disposal procedures that will almost inevitably reduce its generation. Reducing the generation of hazardous wastes is, of course, a desirable side benefit of making the generators pay the true social cost of generating dangerous material. But how much its generation should be reduced depends partly on its benefit (or the cost of less toxic substitutes), neither of which criteria enter the current either–or decision process.[3] Indeed, the very language of RCRA precludes any balancing of costs and risks.

A second problem with the either–or decision is that it does not consider the magnitude of the toxicity. Recall the adage, "The dose makes the poison." Many substances provide beneficent treatments of diseases in small doses, but in large quantities are deadly. Declaring a substance to be hazardous ignores this distinction. It raises its disposal costs and hence reduces its use—both its beneficial and its deadly uses. The failure to consider the degree of a material's toxicity is essentially an implicit acceptance of the belief that there are no thresholds with toxics—if a million drops are fatal, then one drop is also fatal.

Assuming no threshold adds a third problem to the either–or decision. Almost

Why Mining Waste Is Not Hazardous Waste

Hazardous waste regulations are strict and expensive. Mining generates nearly half the total waste of the United States. These two facts, plus intense lobbying pressure from the mining industry, explain the Bevill Amendment to RCRA in 1980, which excluded mining waste from regulation as hazardous waste until the EPA could prepare a full report on their proper treatment. That report emerged five years later and found that almost all mining waste could continue to be treated as ordinary solid waste.

The U.S. EPA gave several reasons for this conclusion: There was so much mining waste that treating it as hazardous waste would be financially too onerous; mining waste was rarely transported or treated but remained on site; most mines were distant from densely populated regions and were in arid locations; and mining wastes, though more dangerous than most solid waste, were far less dangerous than most other hazardous wastes (Tilton 1994).

(Two other major categories of potentially hazardous waste are also excluded from treatment as hazardous waste under Subtitle C of RCRA: agricultural waste and household hazardous waste. The reasons they are excluded are different from mining. Often in agriculture, it is hard to know which farmer's fertilizer or pesticide is causing which pollution, as we discussed in Chapter 2. And the monitoring and enforcement necessary to prevent household disposal of toxic materials as ordinary solid waste are considered too costly. Household hazardous waste will be discussed later in this chapter.)

Are the EPA's reasons good ones? As for the first, the ability of an industry to pay for pollution abatement should not be a determinant of its regulation. If the external costs of producing a product are such that, if internalized, consumers would be unwilling to pay for it, then the product should not be produced. We should then find substitutes for the minerals or import them at higher prices from places where either their extraction generates less external cost or their mines' neighbors are more tolerant of the risks. As for the second reason, whether a waste is transported or treated is at most a minor determinant of the size of its external cost. The third reason is a solid one and indicates that the EPA is asking the right question: How much is the damage, rather than, is there any damage at all? And the fourth reason suggests again that a binary categorization of waste—Subtitle C (hazardous waste) versus Subtitle D (solid waste)—is inappropriate. The degree of care should reflect the degree of hazard and the size of the affected population.

every substance tested will harm some animal in some dosage at some time. This leaves the hazardous waste designators with a politically impossible decision—either declare almost every substance hazardous or ignore the Subtitle C criteria. Because the former path is the more politically and economically unpalatable—it would make too many relatively harmless wastes excessively expensive to generate and dispose of—the EPA is forced into a constant search for subtle ways to ignore the Subtitle C criteria. Semi–benefit–cost analyses intrude, even though Subtitle C does not permit them. "De minimis" wastes or waste generators may be exempted.[4] "Adequate" planning matters. Specific hazardous wastes get "delisted" after petition (see http://www.epa.gov/earth1r6/6pd/rcra_c/pd-o/delist.htm).

Hazardous Waste Trade-Offs

The government wants to encourage the use of fluorescent lamps because they are more energy efficient than the standard bulb. But each lamp does contain a few milligrams of mercury, and when mercury gets into the food chain, it poses a health risk to pregnant women and small children (Deutsch 1997; Warrick 1998). Until recently, the U.S. EPA classified all these discarded lamps as hazardous waste, which means business users could not just send the dead lamps to the nearest solid waste landfill. Rather, they had to spend something like $1 for each lamp to get it to a hazardous waste landfill. That's the private cost. But the longer trip also meant that more lamps broke, so the greater safeguarding of the hazardous waste landfill over a regular solid waste landfill is perhaps totally offset by the greater breakage and mercury emissions in the transport stage—a trade-off.

Of the lamp industry's Big Three—Philips, General Electric, and Sylvania—only Philips has so far achieved a lamp that is sufficiently low in mercury to pass the EPA tests, but the others are not far away. (GE is now marketing a fluorescent bulb that also passes the EPA tests. It still contains the same mercury, but by adding vitamin C to the bulb, GE neutralizes the mercury under testing conditions. Whether it neutralizes it in a landfill is an open question.) What has gotten lost in all this controversy about the permissible amount of mercury and the tests themselves is another trade-off. With fluorescent lighting, there is not only a possible energy–health trade-off, but also a mercury–mercury trade-off. Fluorescent lamps save electricity, and the major source of human-made mercury in the environment comes from coal combustion in electric power plants. Coal-fired electric power plants generate 20% of U.S. mercury emissions; fluorescent lamps, less than 1% (U.S. EPA 1997b).

When the EPA conducts a Cost and Economic Impact Analysis, it sounds as if it might be creeping up to a forbidden benefit–cost analysis. But look closely, for example, at the analysis of the listing as hazardous wastes of substances K141–K145, K147, and K148 generated by the Coke By-Products Industry (U.S. EPA 1992). After the EPA has listed these substances as hazardous wastes, it is permitted under HSWA to do an analysis of the "costs and economic impacts associated with these changes" (U.S. EPA 1992, 1). But what costs are they seeking? Not social costs but private costs! (The clue that private cost is the aim is seen from the fact that the operating and maintenance costs and the closure costs are all multiplied by one minus the firms' tax rate—if resource costs are paid by the government, they do not count.)

The purpose of the cost analysis is not to discover if the social benefits of the hazardous waste listing are worth the social costs but rather whether the listing adds private costs in excess of 1% of the sales revenues, which would be "a trigger indicating further analysis is called for to determine if the plant can recover profitability lost through price increases or would be forced to close" (U.S. EPA 1992, IV-16). In this case, the study found that "because the costs of compliance as a percent of the value of production are less than one percent, assuming no price increase, the coking industry is not significantly affected by the newly listed K wastes. A price increase of less than one-half of one percent would restore profitability to its precompliance level" (EPA 1992, IV-24)—ending "cost and economic impact analysis." (Because the annualized costs did not add up to $100 million, a fuller Regulatory Impact Analysis was not required and hence not done; such an analysis would have included estimates of the benefits as well as the costs.) Exceptions to the hazardous waste listing criteria seem to occur only if the cost to the industry would be onerous. But this is hardly a good reason for risking public health—should arson be permitted if pyromaniacs' happiness greatly depended on it?

As fast as the EPA finds ways around its legal obligation to list as hazardous waste anything that can be shown to be even the slightest bit toxic, science moves on to undermine the EPA's efforts. Ever better analytical techniques and ever sharper equipment means that ever smaller degrees of toxicity can be detected, requiring ever more substances to be declared hazardous. As long as the law precludes any balancing of costs and risks, the problem will not only exist but get worse.

In short, it is relatively easy to know what are the *potential* health risks of a substance—especially if all you need to know is whether they can or cannot be shown to be positive. It is much harder to know what the *actual* health risks are. And it is much, much harder to determine whether these risks are great enough that the substance warrants expensive treatment and disposal. And even if these much, much harder tasks were easy, Subtitle C of RCRA would not permit them to be used in the decision about whether a material was officially "hazardous" or not.

The RCRA attitude is that a pound of prevention is better than an ounce of cure. Consider the following two sentences:

> EPA estimates that the average cost of disposing of hazardous waste in compliance with RCRA regulations is about $90 per ton. The EPA estimate of the cost of cleaning up improperly dumped waste is up to about $2,000 per ton [these figures, if converted to 1997 dollars, would each about double]. (OTA 1983, 6)

The intended implication of juxtaposing these two highly different figures is that it is better to spend $90 than $2,000. Indeed, reasoning that way, it would be better to spend $1,990 than $2,000. But the $90 is spent on *every* ton of hazardous waste, whereas the $2,000 is spent *only* on those tons that cause problems. RCRA does not permit the EPA to ask, what is the probability that hazardous waste handling will lead to contaminated groundwater that will affect somebody's health? Although there are few close studies of the "prevention versus cure" trade-off, some show that the benefits of preventing groundwater contamination do not always exceed the cost of that prevention (Raucher 1986).

In fairness to the politicians, a law that openly stops short of zero risk is not easy to write. But if it were written, the EPA could fairly readily expand its analyses to reach more sensible decisions. Consider the steps required: (a) the various possible health and environmental exposures that can occur from treating a particular substance as ordinary solid waste must be identified; (b) the relationship between exposure and health effects must be estimated; (c) the population at risk of exposure, now and in the future, must be enumerated; (d) the expected (discounted) number of deaths, illnesses, and other environmental damages must be estimated; (e) these damages must be valued in (present valued) monetary terms; and (f) these dollar costs of failing to handle a particular waste material more carefully must be compared with the added costs of that more careful handling. The current EPA decision-making process provides the data for almost all of steps (a) through (d), and steps (e) and (f) could then be readily added. (There are even signs that the EPA is itching to add these final steps; see U.S. EPA 2000, 2001b).)

How to Handle Hazardous Waste

Hazardous waste is, well, hazardous. This in itself urges that it be monitored and handled more carefully than ordinary solid waste. But hazardous waste is the hardest pollutant to monitor. Most pollutants are generated in large quantities and appear in readily observable forms and places. Many hazardous wastes, conversely, are generated in small quantities and thus are harder to monitor.[5] This suggests the dilemma of hazardous wastes: Their toxicity calls for more costly treatment, but this greater cost invites illegal treatment. Evidence of such increased illegal dumping under RCRA comes from both econometric studies and sting operations (Sigman 1998a; Barbanel 1992).

Before RCRA in 1976, hazardous wastes were almost entirely regulated—or not regulated—by the states. Most hazardous wastes were "treated" on-site, and mostly by land disposal—drums, pools, lagoons, and so on (OTA 1983). RCRA initiated a system for tracking hazardous waste from its creation to its ultimate treatment or disposal, called cradle-to-grave tracking. The EPA specified precautions at each stage and established safeguards for the ultimate disposal. Command-and-control directions were utilized at every stage. Economic incentives were largely rejected on the grounds that mishandled hazardous waste would be so serious a danger that mere incentives would inevitably be inadequate. Nevertheless, the regulations had economic effects.

Hazardous waste landfills became much more tightly regulated and hence more costly to create and operate. Many, unable to meet the new regulations, closed, which increased the average transport distances for land-disposed hazardous wastes. The transport of hazardous waste, while tightly regulated in theory, is lightly regulated in fact, and transport accidents are frequent (Aizenman 1997). Finally, with the potential liability for environmental damages, landfill prices rose even further to cover insurance and risk. RCRA-induced price rises started the reduction of landfill disposal of hazardous waste, and HSWA in 1984 imposed a nearly total ban on such disposal. They succeeded—by 1990, a majority of these landfills had closed.

The details of the HSWA ban on landfill disposal of hazardous waste deserve attention. Congress, fearing that the EPA would move too slowly on ending landfill

---

The U.S. EPA's Evaluation of Its Hazardous Waste Regulations

During the period 1986–1990, the U.S. EPA itself conducted several "regulatory impact analyses" of its regulations for land disposal of hazardous waste. Adding up the costs incurred and cancers avoided by these regulations, the annual cost (converted to 1997 dollars) came to $2.0 billion, and the expected number of lives saved from cancer was 44 a year. The cost per expected life saved by the regulations studied was $46 million, an order of magnitude larger than most acceptable policies cost (Sigman 2000).

Although $46 million per expected life saved suggests excessively cautious regulation, it should be noted that the EPA studies overestimated costs by assuming that the regulated firms could make no short-term adjustments in their inputs or outputs to reduce the burden of the regulation; and that the regulations stimulated no new research that would enable firms to reduce their regulatory costs in the long term. Moreover, looking only at lives saved and exposures avoided neglects benefits to the environment in other dimensions—the regulations averted thousands of cases per year of noncarcinogenic toxic chemical exposure and thereby raised the welfare of those who had previously been endangered (Smith and Desvouges 1986).

---

disposal, listed specific substances whose landfill disposal would be automatically banned in so many months after the law passed unless the EPA had reached a determination before then that the landfill prohibition was not necessary to protect human health and the environment (U.S. EPA 1986b). This so-called hammer provision, though tempting as a way to push a sometimes recalcitrant bureaucracy, is far from ideal. It assumes that the EPA can always reach the correct decision within the specified time period. When that assumption is wrong, the hammer forces a perhaps incorrect disposal or treatment path on the specified substance.

The intention behind the closing of hazardous waste landfills was, of course, that safer means of treatment would be used in their place. In fact, this did happen to some extent. By 1999, less than 4% of the hazardous waste was disposed of in landfills or in surface impoundments (U.S. EPA 2001a). By 1999, nearly three-fourths of the hazardous waste was injected into deep wells or underground caverns. The higher landfill prices and increased bans led also to hazardous waste source reduction, as HSWA had also intended (Peretz et al. 1997). But two unintended and undesirable things also happened. First, some waste continued to go to landfills because they were still the cheapest means of disposal and because new treatment technologies were slow to get established—partly because landfills could not be closed until new treatment facilities were available, and partly from fear that access to the cheaper landfills would once again be permitted (Mazmanian and Morell 1992). And second, what waste continued to go to landfills went mostly to "existing" landfills

> ## Dioxin Disposal and Cancer Deaths
>
> In its regulatory impact analysis of dioxin disposal, the U.S. EPA considered two disposal paths, land disposal and incineration (U.S. EPA 1986b). Land disposal would cost $325 a ton—counting tipping fee and transport cost and assuming that private cost equals social cost. Incineration, with its nearly total destruction of the dioxin, would cost $1,800 a ton.
>
> For this extra cost of incineration, the lifetime risk to the maximally exposed individual would be reduced by 0.01. This is a big reduction, in magnitude it is roughly the equivalent of *removing* the lifetime possibility of highway death for such an individual. But the study also finds that the risk of cancer would be *increased* for the *average* exposed individual! How could this be? Incineration generates very small risk but puts every downwind person at that small risk; land disposal could leach to groundwater, seriously affecting only the few who use wells that contain that leachate. In short, the more costly incineration would actually increase the total number of expected cancers. Concern solely for the maximally exposed individual can lead to a wrong decision.

because the standards for them—that is, those built before the passage of HSWA—were less strict than the standards for "new" landfills, and their costs, prices, and hence care were lower (Ujihara and Gough 1989).

Besides banning much landfill disposal of hazardous waste, the law intends to make injection of hazardous waste into deep wells very difficult, because it is permitted only where the operator can prove that no injected hazardous waste will migrate to the groundwater in less than 10,000 years or that it will no longer be hazardous when it does migrate (Russell 1990). Notwithstanding that, nearly three-fourths of the hazardous waste is still disposed of in deep wells or caverns. But RCRA and HSWA between them should ultimately make it necessary to handle all hazardous waste by source reduction, recycling, or incineration.

There is good news and bad news about the incineration of hazardous waste. Incineration breaks down complex and toxic organic molecules into their constituent elements, thus detoxifying them. The process yields carbon dioxide; water; ash of carbon, salts, and metals; and some new organic compounds. But these new compounds, "products of incomplete combustion," are highly toxic dioxins and dibenzofurans. Hazardous waste incinerators are required to achieve 99.99% "destruction and removal efficiency" (called four nines) and must achieve 99.9999% (called six nines) for the most toxic compounds.

Again, better safe than sorry. But this may be carrying safety too far. These emissions limits are based on avoiding a lifetime cancer risk higher than 1×10^{-5} (i.e., 0.00001) for the "maximally exposed individual" (MEI), and an MEI is a person living at the point of maximum downwind exposure and living there all day, every day, for 70 years. How safe is this? If *every* American lived at maximum downwind exposure from a hazardous waste incinerator and stayed there all day, every day, for 70 years, each year 37 Americans would die of a cancer caused by this exposure. (This 37 is a very rough calculation, but it gives the order of magnitude involved; 37 expected deaths per year equals 0.00001 times 260 million Americans divided by 70 years.) By comparison, the lifetime highway-death risk of Americans is above 1×10^{-2} (i.e., 0.01)—more than 40,000 Americans die each year on the highway, and this is from average, not maximum, highway "exposure" (Porter 1999).

Even if we accept the approach that says better safe than sorry, the regulations that RCRA all but forced the EPA to adopt are far from ideal, for five reasons:

- Not all hazardous waste incinerators are surrounded by the same number of people. The social benefits of reducing toxic emissions are proportional to the affected population. Density around incinerators varies a lot. One EPA study found only 4,000 people living near one incinerator, but more than 1 million around others (Dower 1990). Economic efficiency, as well as common sense, dictate that greater marginal cost should be undertaken to avoid emissions where populations are greater. But the same EPA regulations apply to all hazardous waste incinerators, regardless of their location and the density of the neighboring population.
- Some hazardous wastes are more toxic than others. An EPA study of hazardous waste incinerators estimated that the most toxic incinerator was 100 *million* times more toxic than the least toxic one in the sample (Dower 1990). But the same EPA standards apply, regardless of what is being burned.
- There is one place where the same EPA incinerator standards are *not* applied: Small hazardous waste incinerators are held to less strict standards. There is no reason why size per se should matter for the standard. If it is more costly for small-scale incinerators to meet higher standards, then lesser standards make benefit–cost sense, but the step toward economic efficiency—where marginal benefit equals marginal cost—should be explicitly made. It may be better to apply the same regulations and force the more dangerous small incinerators out of operation.
- The concern for the MEI is inappropriate. We should be concerned for the *average* exposed individual. The expected number of cancers is the product of the probability of the *average* person getting cancer times the number of exposed people. Concern for the MEI could lead to wrong decisions (as is illustrated in the box titled "Dioxin Disposal and Cancer Deaths").
- The EPA goes beyond just telling incinerators how much of various substances they can emit. The EPA also tells them what technological means they must use

Lead in Car Batteries

Lead is highly toxic, especially to growing children. Lead is principally used in car batteries. Old car batteries are now more than 95% recycled, but the rest end up in landfills, where the lead can get into the groundwater, or in municipal incinerators, where the lead can get into the air (Whitford 2001). The conventional view is that greater recycling should therefore be encouraged, perhaps even mandated. Indeed, most states do ban the disposal of car batteries in landfills, but many still find their way there.

One study explores other means of reducing such disposal, by either reducing car battery production or increasing the recycling of the lead in old batteries (Sigman 1996a). Four policies are examined: (a) a tax on virgin lead extraction would reduce virgin lead output, would push up lead prices, encouraging increased battery recycling, and would push up battery prices, reducing their consumption; (b) a deposit-refund system on car batteries would end up having the same effects (recall, from Chapter 6, that a deposit-refund system means that consumers put down a deposit when they buy the product and redeem that deposit when they return the used-up product to a proper receiver; 12 states have mandated deposits on lead-acid auto batteries, and in the rest of the states, the private sector has created a similar deposit system, largely because of the value of the used batteries; U.S. EPA 2001b); (c) a subsidy for lead recyclers would also reduce virgin lead output and increase recycling, but it would lower battery prices and increase battery consumption; and (d) a minimum recycled content standard for batteries (issued through a marketable content system so that the chosen recycled level would be achieved at least-cost) would reduce virgin lead extraction and encourage recycling, but it might lower the price of batteries and encourage greater consumption. The first two policies are the preferred ones, because they promise to achieve all three goals—reduce virgin lead extraction, increase recycling rates, and reduce battery consumption—and hence reduce battery disposal in landfills.

Increasing lead battery recycling by 1 percentage point would mean 10 tons a year fewer airborne lead emissions from primary lead smelters and 45 tons a year fewer airborne lead emissions from municipal waste incinerators, while increasing the airborne lead emissions of secondary smelters (which do the recycling) by only 6 tons a year (Walker and Wiener 1995).

But wait. The two primary U.S. lead smelters are in Montana and Missouri, far from most children in the United States. The 50-odd secondary smelters are mostly located in eastern United States and California and Texas, in much more densely populated areas. By considering the density of the affected population as well as the tonnages of lead, the EPA estimated that *more* children would be damaged with a higher battery recycling rate. Accordingly, the EPA decided not to move forward on this front. Hurrah—the EPA looked at the average exposure of children, not the maximally exposed child. Indeed, it might have been useful to go on and think about what the effect of *reducing* battery recycling would have done.

to achieve the standards. This forgoes the potential for incinerator managers to find cheaper or better ways to do it (Russell 1988).

Having noted shortcomings in the actual U.S. approach to hazardous waste treatment and disposal, we should also note that there are hundreds of hazardous waste treatment facilities, not to mention perhaps thousands of on-site facilities, and individual benefit–cost analyses resulting in different rules for each would be expensive and time consuming, not to mention invitations to further time-consuming and expensive litigation. Uniform standards may be necessary in reality, and excessively safe standards for most such facilities may be appropriate. If people fear hazardous waste, and especially its incineration, more than, say, petrochemical plants, then is it wrong that we end up in a situation where a "hazardous waste incinerator, even one that occasionally makes minor deviations from its permit standards, is a much healthier neighbor than a petrochemical refinery" (Kopel 1993)?

Why do we bother to nitpick about best policies if their implementation is in reality impossible—for legal, political, or cultural reasons? Because we should keep the best in sight and mind whenever we stray from it into the realm of second best, lest we stray too far.

Pricing Hazardous Waste Disposal

Considering how much regulation there is of hazardous waste disposal, it is surprising that there is almost no regulation of hazardous waste creation (Russell 1988). Of course, the high price of its proper disposal is in itself a form of tax on its creation. Between 1976 and 1994, while the overall national price level roughly doubled, the cost of hazardous waste disposal in landfills rose from $10 to $250 a ton, and the cost of hazardous waste incineration was even higher, running as high as $2,000 a ton for some substances (Ujihara and Gough 1989; Gerrard 1994; Sigman 2000).

One might ask why a profit-seeking firm would ever send its hazardous waste to a more expensive incinerator. First, it may be sufficiently closer—the number of hazardous waste incinerators is growing while the number of hazardous waste landfills is declining. And second, a landfill may some day become a Superfund site—an area that requires expensive remediation that may have to be paid for by those who sent hazardous waste to it. With an incinerator, there is no future risk whatsoever. (Superfund is the subject of the next chapter.)

To the extent that the rises in disposal and incineration costs reflected the internalization of external costs, then the resulting reduction in the volume of hazardous waste creation was salutary. And there is extensive evidence that firms reduced their generation of hazardous waste by reusing, reducing, recycling, or finding substitutes (Peretz et al. 1997). Indeed, one econometric study estimated that the elasticity of hazardous waste generation with respect to its disposal cost is 15—that is, a 1% increase in the disposal cost of hazardous waste causes a 15% decrease in the volume of hazardous waste generated (Sigman 1996b).

If the optimal safeguarding of treatment and disposal facilities is achieved, and the cost of that safeguarding passed on to generators, then in principle that will provide the correct incentive not to produce too much of it. No further action on hazardous waste creation is needed.[6] In fact, however, many states do take further action. By 1995, 32 states taxed hazardous waste, either its generation or its disposal (Levinson 1999a). Most of them taxed landfill disposal of hazardous waste more highly than incineration, reflecting the perceived higher external costs of landfill disposal, and adding to the EPA incentives to turn to incineration.[7]

State taxes varied a lot. For chlorine solvent wastes, for example, state taxes ranged from $1 a ton up to $157.50 a ton. Eight states charged higher taxes if the waste was generated out of state, and one charged a lower tax if out of state (New Hampshire). Some states have charged tax rates that are equal to the rate charged by the exporting state and hence have tax rates that differ among states of origin. It is hard to believe that the external costs of hazardous waste disposal that remain after the EPA regulations have been adopted are so different across states. Moreover, if external costs do remain, they are presumably the same, regardless of the state of origin of the waste.[8]

It is worth noting in passing that *any* state taxes on hazardous waste disposal, whether generated in-state or imported, will be too high from a national viewpoint. One of the benefits to the state of such taxes is that they are collected (partly) from out-of-state businesses, whose loss of profit presumably doesn't count much in the state's objective function—but the revenue they generate does count. And a second benefit to the state of such taxes is that they encourage hazardous waste to be disposed of in other states, and the reduction of in-state pollution counts in the state's objective function, whereas the concomitant increase in other states' pollution presumably doesn't count much. For both of these reasons, a state tax will be set too high (Levinson 1999a).

In any case, state taxes for the most part don't add significantly to the cost of hazardous waste. Incinerator disposal costs more than $700 a ton, and state taxes average only $12 a ton. Of course, if the elasticity of hazardous waste generation with respect to disposal cost is indeed 15, then small increases in state taxes can have large effects. Consider a state tax that accounts for 2% of the total disposal cost. Doubling that tax would add another 2% to the total cost. With a price elasticity of 15, this would reduce hazardous waste generation by some 30% (i.e., 0.02 times 15). Indeed, empirical work supports this: State hazardous waste taxes *do* affect hazardous waste generation and shipments (Walters 1991; Levinson 1999a, 1999b). But all this simply begs again the question: If the command-and-control regulations on hazardous waste disposal are strong enough to correct the external costs, why are further Pigovian-like taxes needed?

Transporting Hazardous Waste

The transport of hazardous waste is controlled by the Hazardous Materials Transportation Act of 1976. Regulations control innumerable aspects of the trucks and packages, but they do not control routes. Presumably, optimizing private truckers will take the route that costs the least (or earns them the highest income), but this may pass through densely populated areas, and should an accident and hazardous waste release occur, many people are put at risk.

How serious is this risk? One study tried to answer this question (Glickman and Sontag 1995). It started with randomly selected pairs of state capitals and considered two routes between them, the minimum cost route and the minimum risk route, where the minimum risk route minimized the population exposed to a release accident during the trip. For each city-pair and for each route-pair, travel time and population exposure was calculated. On the average, the minimum risk route required 670 extra hours of driving per person-impact avoided. Assuming that trucks average 43 miles an hour on minimum-risk routes and that trucks cost $1.10 a mile to operate, these 670 hours become $31,691 per person-impact avoided. Further assuming that 1–5% of these impacts would prove fatal, that translates into $0.6–3.2 million per life saved, a cost that our society is almost always willing to pay to save an expected life.[9]

All this is hypothetical, but it is suggestive of two things. First, the obvious, hazardous waste transport also generates external costs that may need attention. And second, there are benefits that we have not as yet noticed to on-site treatment of hazardous waste.

Prevention of Careless Treatment or Illegal Disposal

As it is more and more recognized that hazardous waste must be handled carefully, the costs of handling it have risen. Today, the costs of an ordinary municipal solid waste landfill are usually less than a tenth the cost of a proper hazardous waste facility.[10] An unknown but undoubtedly large number of hazardous waste generators probably succumb to the temptation to take the less costly, albeit illegal, path. The EPA once estimated that a seventh of all toxic waste generators discarded wastes illegally (U.S. EPA 1983; cited in Sullivan 1987).

The best policy—as always, in theory—is to tax the illegal dumping, with the tax being equal to the marginal external damage that it causes. Unfortunately, that is impossible in fact, and second-best approaches must be examined. We will look briefly at four:

■ *Legal liability*. The discovery of illegal dumping brings on two costs: for optimally remediating the extent of the external damage; and for neighbors who continue to be affected by the unremediated external damage. (In principle, the cleanup operation should be expanded as long as the benefit of additional cleanup—i.e., the reduced external costs imposed on neighbors—exceeds the marginal cost of that cleanup. We will say much more on this in the next chapter.) Making the perpetrator of the illegal dumping legally liable for all of both kinds of costs does, in theory, lead potential perpetrators to abstain if all illegal activities are detected, if all perpetrators can be identified, if all costs can be measured, and if all the liable damages can be collected (Calabresi 1961; Shavell 1980; Alberini and Austin 1999). Merely listing the "ifs" makes it clear that one cannot rely on liability and courtroom redress when it comes to the illegal handling of hazardous waste.

■ *Subsidizing legal disposal*. Such subsidies remove the temptation to illegal disposal, but not without cost. First is the deadweight loss of the added taxation needed to finance the subsidies (or the forgone benefits of other government programs that are curtailed to provide the funds). But more than that, the reduced *private* cost of hazardous waste disposal encourages firms to generate more of it (Sullivan 1987). If hazardous waste as an input in productive processes is highly substitutable for labor, capital, or nonhazardous raw materials, the subsidy could greatly, and wrongly, increase the generation (as well as the disposal) of hazardous waste.

■ *Deposit-refund system*. The theory is straightforward for beverage containers and the prevention of litter (as was discussed in Chapter 6). With hazardous wastes, the prevented "litter" is not just an eyesore, but a serious health hazard. The problem with all such deposit-refund systems is that they require a second waste collection system that largely duplicates the first. But with many hazardous wastes, a second, separate collection (and treatment) system may be desirable, so the duplication of

Period	Inspections (average annual number)	Violations (average annual number)	Violations per inspection (percent)	Table 14-1
1986–90	22,554	8,781	39	Hazardous Waste Disposal Inspections and Violations
1991–95	28,764	16,682	58	

Source: Stafford 2000.

waste services may not be inefficient. Another problem does arise here. Many hazardous wastes have undergone chemical transformation, so it is not always clear on what purchase a deposit should be levied. If the deposit can be evaded and the redemption collected, then we are back to the subsidy discussed just above.

■ *Monitoring, enforcement, and punishment.* The theory here is also straightforward and long known (Becker 1968). With adequate enforcement effort and fines, the government can make the expected cost to the illegal dumper equal to the external costs the dumping imposes on society. In principle, the enforcement effort should be small, because it costs real resources, and the fine should be high, because it does not. But this runs into the risks that the huge fines will be successfully evaded in the courts, that the legal costs of collecting the fines will be excessive, or that the convicted firms will escape through bankruptcy. The empirical literature on this approach has generally, though not always, found that tighter enforcement does indeed lead to greater compliance with disposal regulations (Magat and Viscusi 1990; Gray and Deily 1996; Laplante and Rilstone 1996; Helland 1998; Stafford 2000).

The EPA has largely settled on the enforcement approach to hazardous waste disposal. Firms that generate hazardous waste must notify the EPA that they do so, and the EPA then conducts periodic inspections to ensure compliance. Notice that, with this approach, there is little but the ethics of the generator and the fear of an internal whistle-blower to forestall nonnotification and illegal dumping. Generators are expected to self-report any accidental noncompliance.

In the 1980s, there were complaints that the EPA set low and inconsistent penalties for hazardous waste disposal violators, and the policy was revised in 1990. Penalties were increased to 10–20 times the previous levels, and penalties became dependent on the probability of harm, the extent of the deviation from the disposal regulations, and the firm's past record of compliance (Stafford 2000). The data on inspections and violations show the changes (Table 14-1).

In the five years after the policy change, the average number of inspections increased by a fourth, and the average number of violations nearly doubled. But one must be careful in interpreting these data. It could be that disposal violations just plain increased despite the stricter policy. Or it could be that the more and more careful inspections discovered violations that were missed before and that the true number of total violations might have decreased as a result of the stricter policy.

How much inspection should the EPA undertake? On the one hand, if the EPA inspects almost no one, then self-interested firms will not comply and the external costs of illegal disposal will be very high. On the other hand, if the EPA inspects everyone and frequently, illegal disposals will be few, but the resource costs expended on the inspections will be very high (Stafford 2000). Clearly, there is an optimum between nothing and everything.

Small-Quantity Generators

Although hazardous waste is hazardous regardless of who generates it, the U.S. EPA has always been reluctant to get into the cumbersome and expensive business of monitoring the disposal of hazardous waste when it involves many generators and small quantities. Accordingly, there has always been a distinction between large-quantity and small-quantity generators (SQGs), with the break between the two at about one ton a month. Not until the HSWA in 1984 were SQGs regulated at all.

Although SQGs generate barely 1% of the nation's hazardous waste, there is nothing small about the problem they pose. There are a quarter-million SQGs in the United States, mostly vehicle repair shops (Abt Associates 1985). Many are located in densely populated urban areas, and many produce some of the most highly concentrated and toxic wastes. But the difficulty of monitoring so many small firms means they are subject to less onerous regulation (U.S. EPA 1995c).

The final contributors to the U.S. hazardous waste problem are the millions of households, which average five pounds a year of household hazardous waste (HHW), almost all of it into the MSW stream (Glaub 1993). HHW consists mostly of batteries, used motor oil, and such household maintenance products as paints, cleaners, polishes, and pesticides (Duston 1993). Although the volume per household is small, the total is significant because of the large number of households and because much of our worry about landfills and incinerators for MSW stems from the HHW they contain.

Currently, most cities do little more than offer a drop-off possibility for HHW, though increasing numbers are beginning to offer curbside pickup, although usually only once or twice a year. Why is

---

"Nothing Dry about Dry Cleaning"

There are 30,000 dry cleaners in the United States, and each year they use 150,000 tons of a toxic fluid, perc (short for perchloroethylene), and much of that is released in the process into the air and sewage (Montagu 1995; Papenhagen 1995). Perc can cause nervous system disorders, infertility, and cancer. There is an alternative dry cleaning technology available, "multiprocess wet cleaning," which uses only water, soap, steam, and heat. Unfortunately, from the viewpoint of private profit, perc is the cheaper process.

The obvious Pigovian answer to the perc external costs is a tax. Because a large equipment changeover would be required, and this would prove expensive for small businesses—which many dry cleaners are—perhaps the best policy is a small tax at first, with announcement of future tax increases, and perhaps ultimately a ban on the use of perc (though a high enough tax would be the effective equivalent of a ban).

However, the EPA has chosen the regulation route. Dry cleaning regulations have been gradually tightened until, by 1996, the EPA required "dry-to-dry" cycles, where garments go in and come out dry and almost all the perc is recaptured. (Most of what is not recaptured ends up in the used filters, which once were pitched into the municipal solid waste but now must be disposed of as hazardous waste—at a cost of nearly twice their original price.)

Because of the now higher cost of using perc, many dry cleaners have already switched away from perc, yet not to multiprocess wet cleaning but rather to petroleum cleaners, the preferred technique before perc. Petroleum does not cause cancer but it does release smog-causing particles and it yields clothing that can catch on fire. Moreover, the machinery costs more than that for multiprocess wet cleaning. Why is it chosen? Because with wet cleaning, the clothing has to be ironed, and that means big added labor costs (Kravetz 1998b). If wet cleaning is the socially least-cost technique, then a Pigovian tax on the petroleum cleaners is also needed.

---

there not more and more frequent curbside pickup? HHW collection programs are expensive, and the participation rate of households is low, typically less than 10%. As a result, curbside HHW collection costs are high. Some estimates say the cost is $100 a household visit or $18,000 a ton of HHW collected and processed (U.S. EPA 1986c; Duxbury 1992; Miller 1995b).

Are there ways to collect HHW without curbside pickup? Mandatory deposit-refund schemes are an obvious possibility, but to my knowledge no state has tried this for HHW. Some come close. California and Rhode Island tax some products that become HHW and use the tax revenues to fund programs to reduce and recycle HHW (Temple et al. 1989; Merrill 1996). But remember, deposit-redemption programs have their own costs.

One as yet untried but interesting idea for reducing the harmful disposal of HHW is to use the financial strength and know-how of the firms that deal in HHW by-products. Consider, for example, waste oil, which is now about 30% recycled. Suppose the major oil refiners were mandated to increase their recycled oil by, say, 2% a year for the next 10 years—this would get the recycled oil up to 50%, and more important get a lot of oil out of the MSW stream. Specialists who collected and recycled oil would get coupons from the EPA, which they could sell to major refiners, which in turn could count these toward their new 2% higher quotas. Households would end up being paid enough that they would choose to meet the higher recycling targets.[11]

An alternative to preventing dangerous disposal of HHW is to reduce the volume of hazardous substances that reach households. Consider mercury thermometers. Not only do they generate 18,000 calls a year to poison control centers; once discarded, they contribute four tons a year of mercury to U.S. landfills (Neus 2000). One state (New Hampshire) and a few cities have responded by banning the sale of mercury thermometers (McMullen 2000c). San Francisco goes a step further, exchanging free new digital thermometers for old mercury thermometers. Many discount and pharmacy

An Industry Program to Collect Discarded Nickel–Cadmium Batteries

One of the most rapidly growing segments of a rapidly growing household battery market is the rechargeable nickel–cadmium (Ni–Cd) battery. These batteries now contribute most of the cadmium in the municipal solid waste stream. The environmental release of cadmium poses a serious potential health threat. Through landfill leachate, incinerator ash, or the food chain, it can affect human health, particularly by causing kidney, lung, blood, and reproductive damage.

City recycling programs have been reluctant to add these batteries to their recyclables collection because of the high cost, $2,000–5,000 per ton (Miller 1995b). So states instead began requiring producers themselves to take back Ni–Cd batteries for proper disposal. To meet these mandates without exorbitant cost, most firms in the industry joined together in 1995 to form the Rechargeable Battery Recycling Corporation (RBRC; for more information, see http://www.rbrc.com). This nonprofit company collects, transports, and recycles discarded Ni–Cd batteries, with its activities financed by a self-imposed fee on domestic battery sales of its members. Retailers and communities that participate in the RBRC program must set up costly special collection procedures, but they are then spared the expense of shipping their batteries themselves or suffering the health problems of casual disposal (Fishbein 1997).

Within a year, the RBRC was collecting one-fourth of the discarded Ni–Cd batteries, and it anticipates taking back three-fourths within five years. The average cost in its first year was $2,000 per ton of batteries—expensive, but not an outrageous figure for so hazardous a waste. (Recall from Chapter 2, and the analysis of producer take-back responsibility there, that this kind of re-collection *is expensive*, but for very hazardous products, the benefits may well be worth the high cost. We are not talking paper and plastic here.) Communities that now operate collection systems for their own household hazardous waste face comparable cost per ton. If consumers cooperate—a big "if" because households have no financial incentive to participate—the RBRC system should be able to bring the average cost down, but the major plus is that the cost is now borne by battery purchasers and not by all those who pay local property taxes.

Medical Waste

Traditionally, medical waste was treated as ordinary solid waste. When, however, syringes began washing up on New Jersey beaches, at the very time of growing fears of hepatitis and AIDS, the Medical Waste Tracking Act of 1988 (Subtitle J of RCRA) was enacted (U.S. EPA 1990). Most hospitals reacted by installing incinerators. Although the incinerators sterilized the waste, they also raised fears of dioxins and toxic metals.

Because most hospital incinerators were relatively small (relative to electricity-generating waste-to-energy incinerators), most hospitals planned to close their incinerators rather than install new, expensive air pollution controls. Even those that had already upgraded their pollution-control equipment often were pressured by the neighborhood to close their incinerators (Acohido 2000c; McMullen 2000a). The usual alternative was an "autoclave," whereby hospital waste was sterilized, shredded, and then landfilled. The switch is expensive. Not only is autoclaving more costly than burning, the hospital lost the electricity and/or steam heat that the incinerator produced, and the landfill volume was greatly increased. The EPA estimates that the overall nationwide cost increase per unit of waste treated would be $150–300 per ton (http://www.epa.gov/ttnuatw1/factsns.html).

Not surprisingly, after decades of moving toward throwaway medical supplies because of their sterility and convenience, hospital administrators are once again looking for ways to reduce, reuse, and recycle. The major U.S. hospitals have promised the EPA to seek 50% reduction of their overall waste by 2006.

chains have discontinued the sale of mercury thermometers, and many hospitals and clinics have pledged to become mercury free. This sounds great—but wait. These thermometers contribute only 1% of the mercury that goes into our landfills, and the new digital thermometers now cost more than twice as much—typically $8.00 (and up) versus $3.50 (see http://www.epa.gov/glnpo/bnsdocs/hg/thermfaq.html). Of course, the cost of digital thermometers will go down as time goes by and more are produced (Goodstein and Hodges 1997; Harrington et al. 1999). In the end, however, there is no getting around the fact that reducing the hazardous waste generated in small and varied quantities by SQGs, very-small-quantity generators, and households is going to be expensive for someone.

Environmental Justice

During the past decade, more and more people in the United States have expressed concern that the poor suffer unfairly by having to live and work near hazardous waste facilities. This concern for environmental justice involves two very different questions. First, are hazardous waste facilities in fact predominantly located near the residences of the poor? Second, if this is true, what should we be doing about it? Let's look at each question.

Much research has been done on the first question, on the proximity of hazardous waste facilities and the residences of the poor.[12] Unfortunately, much of this research seems to bear out the adage "Believing is seeing." Two examples (GAO 1995): A study by the National Association for the Advancement of Colored People and United Church of Christ Commission for Racial Justice found that, for a sample of 530 commercial hazardous waste treatment facilities, minority populations are more likely than nonminorities to live in the ZIP code areas where such facilities are located. Another study, financed by Waste Management Incorporated, and the Institute for Chemical Waste Management, found that, for a sample of 454 commercial hazardous waste treatment facilities, there was no consistent association between the location of the facilities and the percentage of minority people living nearby.

Although one can find studies with findings on all sides of this question, a majority do conclude that hazardous waste sites and minority populations tend to coincide (Michaels and Smith 1989; Hamilton 1995; GAO 1995; Gayer 2000). Then the question is, why? There are two plausible answers. First, minorities tend to be poor, the poor tend to live where land is cheap, those siting hazardous waste facilities also seek out cheap land, and hence such facilities tend to get sited where minorities live. A variant of this reasoning is that hazardous waste sites appear where the neighbors, being poor, have too little political clout to forestall them (Hamilton 1995).

Second, hazardous waste facilities are sited first, more or less at random, but their presence drives down the prices of the surrounding land, from which the rich then flee to escape the risk and to which the poor then flock to acquire cheap land for their houses. That housing prices would be lower when hazardous waste is nearer seems obvious, but the empirical evidence is not clear; some studies find support for the hypothesis, but others do not (Adler et al. 1982; Schulze et al. 1986; Michaels et al. 1987; McClelland et al. 1990).

There is evidence that both answers are right some of the time (GAO 1983; Bullard 1983; Been 1994). And this ambiguity of causation is awkward for policy-making. To the extent that the first causation exists, the environmental injustice can be prevented by laws forbidding hazardous waste facilities from locating where the surrounding population is predominantly poor and/or minority. But to the extent that the second causation exists, wherever one sites hazardous waste facilities, the poor will eventually come to live. Laws about siting cannot easily stop such market dynamics. But sizable "host fees" can. Such host fees will slow the outmigration of the rich, slow the decline of land prices, and hence slow the in-migration of the poor.

This brings us to the second big question: Should we try to prevent this juxtaposition of hazardous waste and minority residence? The official answer to this is a resounding yes. President Clinton signed Executive Order 12898 in 1994 directing each federal agency to develop a strategy for "identifying and addressing ... disproportionately high and adverse human health or environmental effects of its programs, policies, and activities on minority populations and low-income populations in the United States" (http://www.epa.gov/civilrights/docs/eo12898.html). As a result, new hazardous waste facilities must prove to the EPA that they are not in predominantly minority neighborhoods.

There are, however, two very distinct sets of arguments that the answer should be no. First, from libertarians, who see no good reason for interfering with the market here—or, indeed, anywhere. They have a point. The poor often get mistreated by the invisible hand of the marketplace. That's why they're poor. Nobody suggests that the fact that the poor get less broccoli than the rich urges a federal broccoli distribution policy. Yet health is not broccoli, and we often collectively choose to step in to ameliorate the disadvantages of poverty when it comes to health, education, and family.

The second "no" argument is more surprising—at first—because it comes from the residents who live near actual or potential hazardous waste facilities. These facilities not only bring risk to the area but also bring property taxes, jobs, and sometimes sizable host fees. These benefits may be perceived as outweighing risks, and laws preventing facility siting may not be welcomed. Indeed, the environmental justice movement may have already done the job of such a law—big firms that generate health risk may not want to locate in inner cities, even if welcomed, because the areas are largely populated by African Americans and the firms fear getting hit with the racism label.

Final Thoughts

This has been a long chapter, but on a serious and important subject. Next to the U.S. income tax code, the RCRA hazardous waste system may well be our most complex regulatory program, fraught as it is with exemptions and exceptions. I keep coming away from thinking about Subtitle C of RCRA with the conclusion that its complexity simply masks an inability or unwillingness to ask how we should handle hazardous wastes to save lives cost-effectively.

Notes

1. There are now, of course, many state laws as well. In fact, most hazardous waste control programs are largely run by the states, after U.S. EPA approval—the EPA directly runs the programs in only 10 states. Other federal laws regulate the production and use of toxic materials, for example, the Toxic Substances Control Act of 1976 and the Federal Insecticide, Fungicide, and Rodenticide Act of 1947. So these acts also, albeit indirectly, affect the quantity and quality of toxic materials that enter the waste stream, but we will focus more narrowly on just the treatment and disposal side of hazardous waste. RCRA regulates both hazardous waste (under the section, Subtitle C) and ordinary solid waste (under Subtitle D).

2. If hazardous waste is hazardous, isn't the optimal quantity zero? But remember, hazardous waste is a by-product of industrial processes, and those processes produce useful products. Zero hazardous waste would mean either that these products were no longer produced or that they were produced at higher costs as nonhazardous inputs were substituted for the hazardous inputs.

3. Indeed, the EPA's own efforts have interpreted hazardous waste reduction to mean hazardous waste minimization—with reports titled *Minimization of Hazardous Wastes* and *Setting Priorities for Hazardous Waste Minimization* (U.S. EPA 1986a, 1994b). Only occasionally in such reports, the EPA "recognizes that factors other than hazard may also be important," with these factors including "economic feasibility of waste minimization alternatives" (U.S. EPA 1994b, 3-25).

4. In EPA regulation, de minimis, meaning trivial, applies in three possible ways: wastes that cause a lifetime cancer probability of less than 10^{-4} (i.e., 1 in 10,000) need not be regulated; wastes that contain a very small percentage (less than 0.1%) of cancer-causing substances need not be regulated; or waste producers who produce relatively little hazardous waste (called small-quantity generators or very-small-quantity generators) may be exempted or regulated less strictly.

5. Consider just the acutely toxic materials (i.e., the "P code"). In 1997, the EPA listed 205 such materials, and the United States generated a total of 80,451 tons of materials that fell (only) into these categories (U.S. EPA 2001a). That's an average of 392 tons a year for each of these substances; to put it differently, for the total of the 205 substances, it is about nine ounces per capita a year.

6. A caveat on that statement (Watabe 1992): Firms that generate hazardous waste and treat it on-site choose two things, how much to generate and how carefully to treat it. When treatment costs rise through increased legal liability—this induces them to generate less hazardous waste. This, however, may have a perverse effect on safeguarding. Because the firm is generating less waste, it is running less risk of causing, and becoming liable for, environmen-

tal damage, and profit-maximizing considerations could lead it to reduce the safeguarding of what hazardous waste it does generate.

7. Some states, for example California, progressively tax hazardous waste—that is, higher tax rates on larger generated volumes (U.S. EPA 2001b). It is hard to understand why progressive taxation is used. Each unit of a particular hazardous waste presumably generates the same amount of external cost regardless of how many other units of that hazardous waste accompany it.

8. Aren't these discriminatory taxes unconstitutional? Yes, said the U.S. Supreme Court of Alabama's $40 per-ton tax on in-state hazardous waste and $112 per-ton tax on imported hazardous waste. But other states' discriminatory taxes continue to exist, arising about as fast as the Supreme Court strikes them down. Some states have switched to large per-ton transport taxes, which must be paid on all hazardous waste that is generated in or imported into the state. Such taxes do not discourage export of hazardous waste but do discourage its import (Levinson 1999a).

9. The minimum risk route, being longer, does add danger to the driver, but very little because well-protected big-truck drivers are rarely injured in their accidents. In theory, though, should we count these added driver risks as an offset to the reduced exposure of the general population? No, because the risks to the driver are presumably already incorporated into the driver's wage—which we have already counted in the operating costs of the truck.

10. Solid waste incinerators are also destinations of hazardous waste (Lipton 1998). Recall that incinerators need to run at capacity and that increased recycling has left many with inadequate solid waste. Some of these incinerators have begun soliciting industrial waste, in which is often mixed hazardous waste that ought not be burned in an ordinary solid waste incinerator. Incinerators even have an advantage over landfills for illegal dumping because the evidence goes up in smoke.

11. Look closely at the effect of the coupon and the mandate. Let P_{newoil} be the price of new oil, and assume for simplicity here that this price is unchanging as coupons are introduced. Then, because new and recycled oil are essentially substitutes, the price of new oil would equal the price at which recyclers buy waste oil from households ($P_{wasteoil}$), plus the cost of recycling oil ($C_{recycoil}$), minus the value of the attendant coupon (V_{coupon}); or $P_{newoil} = P_{wasteoil} + C_{recycoil} - V_{coupon}$.

Suppose $C_{recycoil}$ went up, discouraging recycling. Without any coupon system, $P_{wasteoil}$ would go down, and households would bring in less waste oil to be recycled. But with a law mandating ever more oil recycling, V_{coupon} goes up instead, enough to raise $P_{wasteoil}$ high enough to meet the mandated higher recycling rates. The system is efficient in that it gets the recycled oil collection process into the hands that can do it most cheaply. But don't be fooled into thinking everybody wins—collecting and recycling oil may be worth it for the environment, but it is still expensive in comparison with virgin oil. Indeed, the magnitude of V_{coupon} tells us how much more expensive it is to produce oil from recycled waste oil than from virgin crude oil.

12. The worry about environmental justice also extends to lesser hazards, such as solid waste landfills. But a government study found that "the percentage of minorities and low-income people living within one mile of nonhazardous municipal landfills was more often lower than the percentage in the rest of the county" (GAO 1995, 20). This makes sense if you think about it. Municipal dumps were originally sited well out of town, where land was cheap and people were few. While these old dumps were growing into today's landfills, middle-income and rich people were moving out of town into new suburban developments. Some of these developments were near landfills, and some actually *on* former landfills. Because the poor have been unable to participate in this suburban sprawl, they have been largely spared from living in proximity to municipal landfills.

Superfund

We all know it doesn't work—the Superfund has been a disaster. All the money goes to lawyers and none of the money goes to clean up the problem it was designed to clean up.

—President Bill Clinton, in a speech to business leaders
at the White House, 11 February 1993

Until the Resource Conservation and Recovery Act, there were few restrictions on how hazardous waste generators could handle their waste. Owing to ignorance or avarice, most dumped such waste into open lagoons or put it into barrels that were buried shallowly. Fortunately for us, most of these casual disposals have not provided long-lived dangers to our health. But many have. With the advent of Subtitle C regulations of current and future hazardous waste came the discovery of past hazardous waste disposals that posed current and future health and environmental dangers.

Accordingly, in 1980, Congress enacted the Comprehensive Environmental Response, Compensation, and Liability Act (CERCLA), commonly known as Superfund, which empowered the U.S. EPA to respond to existing hazardous waste sites. (CERCLA was amended by the Superfund Amendments and Reauthorization Act, or SARA, in 1986. For readable summaries of CERCLA and SARA, see http://www.cnie.org/nle/leg-8/j.html#_1_8 and http://www.epa.gov/superfund/action/law/sara.htm.) CERCLA also created new taxes to generate some of the revenue needed for these remediations, but the principal source of revenues was intended to be the companies that generated the hazard in the first place, the "responsible parties."

CERCLA authorized the EPA to investigate suspected hazardous waste sites and to place those found indeed to be hazardous on a National Priorities List (NPL). The EPA would then undertake any immediate cleanup necessary and would seek out "potentially responsible parties" (PRPs) to help finance the ultimate remediation process. The number of NPLs was not small, and the cleanups were not quick and not cheap. (States also have Superfund-type programs. These programs pick up what doesn't get on the NPL. There are many sites, but most are very low-cost relative to NPL cleanups.)

No one knows how many NPLs will ultimately be located. The EPA inventory of possible sites is currently around 40,000—more than 1,300 of which have already been placed on the NPL—but the General Accounting Office has guessed that there may ultimately be more than 400,000 suspect sites to be investigated (GAO 1987). (If you are interested in knowing more about the NPL sites near you, go to http://www.epa.gov/superfund/sites/query/queryhtm/nplfin.htm, which has information on where the NPL sites are, what was wrong there, what has been done, and what is still planned.)

The remediation process is not quick. On the average, the time from a site's initial notice to its designation on the NPL is nearly four years. From NPL designation to the completion of a list of possible remedies is nearly two more years. From there to the determination of the proper remedy is another three years. And then it takes nearly four more years to complete the remediation and "delist" the site from the NPL. Overall, on the average, from start to finish, the process takes 12 years (Acton 1989). Only half of the NPL sites have completed their remediation.

The remediation process is not cheap. The average Superfund cleanup process costs around $30 million in total public and private costs, excluding litigation and administration costs (Viscusi and Hamilton 1996). And the litigation costs are high, too, as the PRPs fight against liability. Not without reason has CERCLA sometimes been called the Comprehensive Employment for Regulators, Consultants, and Lawyers Act (Ray and Guzzo 1993).

In this chapter, we are going to explore the big Superfund questions: (a) Which hazardous waste sites should we clean up? (b) How much cleanup is appropriate? (c) When should we clean up toxic sites? (d) Who should pay for the cleanup? We shall find it not at all obvious

Love Canal

More than a century ago, William Love began to excavate a canal southeast of Niagara Falls, New York, in order to generate electricity and provide ships with a bypass around the falls. The project was abandoned after one mile of canal had been dug. By 1920, the land had become a disposal site, principally for chemical wastes of the Hooker Chemical Corporation. In 1953, Hooker covered the waste dump and sold it to the city's Board of Education for one dollar, with a warning that chemical wastes were buried there and with a full shift of any chemical liability to the board. Shortly thereafter, the board built a school and sold land to developers, who installed sewers and excavated basements for a growing residential community. By 1978, more than a thousand families lived on or near the chemical waste dump.

In that year, the *Niagara Gazette* began a series of articles on suspected hazardous waste problems around Love Canal, as the area had come to be known. Anecdotal evidence and preliminary epidemiological studies suggested risk to pregnant mothers and small children. President Jimmy Carter declared the Love Canal neighborhood an emergency area and provided funds to permanently relocate the families who lived closest to the dump. More families were later temporarily relocated, and eventually more than 1,000 families received money from the federal government and Hooker (now Occidental) Chemical Corporation. The widespread publicity of the Love Canal crisis gave rise in 1980 to the Superfund legislation.

One can still find great disagreement about how serious the health crisis *really* was. Dozens of studies have been done, and studies of the studies. One recent detailed examination concluded that the "scientific study of Love Canal looks more like a prize fight than a search for truth" (Mazur 1998, 192). Love Canal may have triggered a long-overdue response to hazardous waste legacies, or it may have triggered an excessively costly overreaction to them. Consensus appears no closer now than 20 years ago on Love Canal.

What has happened to Love Canal? It was listed under CERCLA for remediation in 1983, and a variety of cleanup actions soon commenced. In 1988, the area was declared again habitable, and houses that had not been demolished during the cleanup were again being sold to new residents. The deeds contain a clause stating that if the new owners become sick, are harmed, or die as a result of exposure to the Love Canal wastes, the city, state, or federal government will not be responsible. This clause, ironically, is similar to the "Hooker Clause" in the earlier land transfer. The area is now known as Black Creek Village.

that the usual EPA answers are the right ones—the usual EPA answers being: (a) all, (b) total, (c) now, and (d) PRPs.

Which Sites Should We Clean Up?

It would be nice if we could answer that question by simply saying that we should fully remediate everything that poses any threat at all to human health or the environment, now or at any time hereafter. Indeed, that is almost what the original Superfund law says—the government should step in whenever

> any hazardous substance is released … [or] there is a release … of any pollutant or contaminant which may present an imminent and substantial danger to the public health or welfare … or the environment. (CERCLA 1980, Section 104a)

In fact, we probably lack the resources, and we certainly lack the budgetary will, to *fully* clean up *every* potentially hazardous legacy that lingers anywhere across our landscape. If only we could recognize and admit that, we would be half way to answering the question, which sites should we clean up?

We could then ask, what is so hazardous that it deserves our limited resources for cleanup? For each potentially hazardous site that we identify, we could estimate the cost of cleaning it up (C) and the present value of the benefit of cleaning it up (B). The cost estimate is usually straightforward, although unexpected changes in costs may later occur if and when an actual cleanup is undertaken. For most hazardous sites, various degrees of cleanup are usually possible, but we will leave the question of *how much* cleanup for later in this chapter. Note that the relevant cleanup costs for a benefit–cost analysis are those *not yet* undertaken. The costs of studying the site are already expended, and such sunk costs should not count. The benefit of a cleanup consists largely of the present value of the future expected lives saved—there are also environmental and health benefits of a less-than-life-saving nature, but they are almost always relatively small.

Estimating the number of expected lives saved by a cleanup is not easy. For most potentially hazardous sites, there are a number of chemicals of different degrees of toxicity, but we will for now assume that there is just one. The excess cancer risk of

---

Glossary and Chronology of Superfund Actions

A reader's guide to Superfund's terms, timings, and numbers may be useful before we bury ourselves in the economics.

When sites are identified as possibly hazardous, the first step is a Preliminary Assessment (PA) of the risk. If the PA suggests serious risk, a more careful Site Inspection (SI) is undertaken, with some of the SIs leading to a numerical estimate of a Hazardous Ranking Score (HRS). If the HRS is high enough, the site is placed on the National Priorities List for eventual remediation. (At any time, an emergency Removal Action may be undertaken if a site poses an immediate health threat.)

For all NPL sites, the next steps are Remedial Investigation and Feasibility Studies (RI/FS), which produce detailed risk assessments and analyses of possible remediation procedures. After a period of public comment on the RI/FS, the U.S. EPA issues a Record of Decision (ROD, pronounced "rod"), which explains what cleanup remedy has been chosen and why. For the chosen remedy, there is then a detailed Remedial Design (RD), and the Remedial Action (RA) actually undertakes the cleanup. When that is completed, the site is delisted from the NPL.

How many sites are we talking about? Tens of thousands of sites have undergone PAs. More than 1,300 have been placed on the NPL. The RDs and RAs have been completed for about half of the NPL sites, and these sites have been delisted.

---

many materials depends on cumulated exposure, but we will assume for now that the risk is a simple probability. Degrees of exposure usually vary, across people of different ages, residences, and occupation, but we will assume for now that each affected person has the same risk of exposure. The number of affected neighbors may change over time, but we will assume for now that the number is fixed. Finally, any actual cleanup is less than complete, so that there is almost inevitably some lingering toxicity and risk of exposure to it, but we will assume for now that the cleanup removes all such risk.

To estimate the expected lives saved, we need estimates of three things: the probability that the toxic chemical of the site will cause cancer if a person is exposed (p); the probability that a site neighbor will be exposed to this toxic material at some time during a year if the site is not cleaned up (q); and the number of site neighbors who would suffer this excess cancer risk (N). We also need a political consensus about the "value of life" (V_L)—that is, the amount we are willing to spend to save one expected life—and about the appropriate (real) rate of discount (i). We can then ask whether the present value of the life-saving benefits are greater or smaller than the cleanup costs—in symbols, whether

$$pqNV_L/i \gtreqless C \qquad\qquad (15\text{-}1)$$

(Recall that a flow of benefits, constant every year forever after, has a present value of $1/i$ times that annual benefit.) By multiplying inequality (15-1) through by i, we can phrase the question slightly differently, whether the *annual* benefits of the cleanup are greater or smaller than the *annualized* costs of the cleanup— in symbols, whether

$$pqNV_L \gtreqless iC \qquad\qquad (15\text{-}2)$$

Obviously, comparing (15-1) and (15-2), we see that the two are just different ways of asking the exact same question. Indeed, there is yet a third way of looking at inequality (15-1). Divide both sides of inequality (15-1) through by the present value of expected lives saved (pqN/i) and you get the following inequality:

$$V_L \gtreqless \dfrac{C}{\left(\dfrac{pqN}{i}\right)} \qquad\qquad (15\text{-}3)$$

The right-hand side of (15-3) is the cleanup cost divided by the present value of expected lives saved, or the cost per expected life saved. Asking whether the policymakers' valuation of lives exceeds this cost per expected life saved is the third way of asking the same question. This third way is the way we ask the question in the next section. (If any of these ideas—present value, discounting, annualized costs, or value of life—bothers you, look back at Appendix B of Chapter 1.)

Given how highly we value lives and the rate at which we discount, inequality (15-1) tells us that we should be more anxious to clean up a potentially hazardous site the greater the toxicity of the hazardous chemicals, the greater the likelihood that neighbors will be exposed, the greater the number of neighbors at risk, and/or the lower the cleanup cost. There are a number of complexities that we will soon have to deal with, but in its simple form, inequality (15-1) shows us the way to answer the question, what sites should we clean up?

What the EPA Actually Does

The EPA actually starts out along the path of inequality (equation 15-1). It estimates each of the key factors—p, q, and N, as were defined just above. But it does this separately for each of four toxic pathways (i.e., groundwater, surface water, soil exposure, and air migration), then it awards "points" for each of p, q, and N on each pathway, adds these points up for each pathway, and finally combines them into a Hazardous Ranking Score (HRS) by means of the formula,

$$\text{HRS} = 0.5(S_{gw}^2 + S_{sw}^2 + S_{se}^2 + S_{am}^2)^{1/2} \tag{15-4}$$

where S_i refers to the score for the ith pathway and the pathways include groundwater (gw), surface water (sw), soil exposure (se), and air migration (am). Each of the S_i calculations is limited to being between 0 and 100, so the resulting HRS is similarly limited. If the HRS is 28.5 or higher, the site is considered dangerous and is added to the NPL and scheduled for remediation.

What started off sensibly has diverged steadily away from a meaningful estimate of the expected lives that could be saved by a cleanup. It finally arrives at an HRS for which no sensible interpretation can be offered. It certainly has no cardinal meaning—a site with an HRS of 57 cannot be considered twice as dangerous as one with an HRS of 28.5—and even its ordinal ranking value is suspect. For a particular site, a higher HRS would seem to indicate a more toxic site, but the HRS cannot logically be used to compare the toxicity of different sites.

There are many other difficulties with using the HRS as even a rough indicator of the toxicity of a site; two are sufficiently critical to warrant mention. First, when calculating the toxicity of a particular pathway, the U.S. EPA utilizes information on only the *most toxic* substance. From logic alone, this means that a pathway with many mildly toxic substances will appear less hazardous than another pathway with only one, but very toxic, substance—even though the pathway with many toxic substances may, if properly evaluated, be much more hazardous. And second, the HRS cumulates toxic information from all the observed potentially hazardous pathways, so the HRS will usually rise with the amount of time and money spent studying the site. (I say "usually" because later studies sometimes find the original studies overestimated the toxicity of the site, and this can result in revising the HRS downward.) Sometimes, studies go on just long enough to get the HRS up over the NPL threshold of 28.5—then they stop, so we don't know how far above 28.5 the HRS would have gone if more studies had been done.

For the EPA, the critical value of the HRS is 28.5. HRS = 28.49, site not hazardous; HRS = 28.50, site hazardous. (If the HRS < 28.50, the site does not necessarily go untouched. It may still be added to a state's "Superfund" list.) Why 28.5? The original CERCLA specified that the EPA should find at least 400 NPL sites—a number suspiciously close to the number of Congressional districts. The EPA's choice of a 28.5 point produced 406 sites. The requirement to find 400 sites was removed in the 1986 SARA, but the 28.5 has remained.

To estimate how many lives can be expected to be saved by a cleanup, one needs (among other things) estimates of the *average* exposure of neighbors to toxic substances and *average* rates of excess cancer deaths from exposure. (In principle, it is also necessary to estimate what average exposure and average rate of excess cancer deaths remain *after* a proposed cleanup, but the U.S. EPA rarely makes these estimates—or at least rarely publishes them. One study of 50 Superfund remediation decisions found that only *six* of the decisions reported such estimates; Doty and Travis 1989.) In both its HRS calculations and its later remediation decisions, however, the EPA continually puts greater reliance on such concepts as reasonable maximum exposure and the maximally exposed individual. The EPA can get incredibly high estimates of excess cancer risks when it multiplies a high estimate of exposure duration times a high estimate of ingestion rate times a high estimate of contaminant concentration times a high estimate of substance toxicity (Burmaster and Harris 1993; Nichols and Zeckhauser 1986). How much too high? One sample of 150 NPL sites examined the differences between the average cancer risk of all the sites with the EPA's "conservative" risk estimates (Hamilton and Viscusi 1999a). For the entire sample, they found the *average* lifetime excess cancer risk to be 0.00047, whereas the *conservative* risk estimate averaged 0.013—*roughly 25 times as high.*

Using maximums instead of averages does not "just" distort the expected cancer death estimates and excessively scare any of the site neighbors who trouble to read the EPA documents. Because the EPA *always* cleans up sites where the measured lifetime excess cancer risk is greater than 10^{-4}, unnecessary cleanups will occur.[1] It may also distort the apparent cleanup priorities because there is no reason for thinking that the ratio of the maximum risk to the average risk is the same across sites. Using maximum probabilities instead of average probabilities can lead to very curious results when the EPA starts adding up the

EPA Procedure and the Dominant Error That Ensues

The process of examining a site to see if it qualifies for the NPL also yields potential error. Because the investigations are not all perfect, in the end some sites will be listed on the NPL that ought not to be there, and other sites will be declared innocuous that ought to be on the NPL. Let's examine how the U.S. EPA procedure affects the percentage of each type of error.

Look at the first two steps, the Preliminary Assessment and the Site Investigation. If, in the PA, the site is found to be nonhazardous, it is removed from further consideration. If it is found to be potentially hazardous, it goes on to an SI. Again, if found nonhazardous, it is removed. If it is found to be potentially hazardous, it goes onto the NPL.

Consider a group of S sites (OTA 1989a). A fraction, h, are in fact hazardous and should be cleaned up. The probability of getting a correct evaluation at each stage is c. This means that $(1 - c)hS$ sites are *incorrectly* removed from NPL consideration at the PA stage (and $c(1 - h)S$ sites are correctly removed from further analysis). chS sites are correctly passed on to the SI stage, but at the SI stage, $(1 - c)chS$ will be *incorrectly* not given NPL listing. The total number of hazardous sites that are incorrectly ignored is $[(1 - c)hS + (1 - c)chS]$, which, after terms are collected, equals $(1 - c^2)hS$. Meanwhile, innocuous sites are getting onto the NPL. At the PA stage, $(1 - c)(1 - h)S$ are *incorrectly* passed on to the SI stage, and at the SI stage, $(1 - c)^2(1 - h)S$ get *incorrectly* listed on the NPL. In the end, the fraction of the innocuous sites that get incorrectly listed as NPL is $(1 - c)^2$, and the fraction of the hazardous sites that fail to get listed as NPL is $(1 - c^2)$.

Suppose $c = 0.99$. The proportion of innocuous sites that get on the NPL is 0.01%, whereas the proportion of hazardous sites that fail to get listed on the NPL is 1.99%—a percentage *nearly 200 times as high.* (Some might object that all this assumes the probability of incorrectly listing a harmless site is the same as the probability of incorrectly listing a harmful site. I do assume that in the text, to keep the algebra simpler. But the asymmetry remains even if the probabilities are different.) Common sense suggests that this is backward. Listing a harmless site on the NPL means spending a little more money on it—no big deal—but failing to list a real hazard means not only keeping the surrounding people at possibly serious risk but also keeping knowledge of that risk from them.

probabilities across different cancer pathways—in one case, it arrived at an overall estimate of the neighbors' cancer probability of *5.1*—apparently, the EPA failed to recall that a probability cannot exceed one (Hamilton and Viscusi 1999a).[2]

A final problem with using maximum risks rather than average risks is that seasoned Superfund observers become cynical about the EPA studies. Many have ended up dismissing all Superfund risks as trivial. This is a shame because they are not all trivial, by any means. In the study of 150 NPL sites referred to above, more than 2,000 potentially carcinogenic pathways were considered by EPA (Hamilton and Viscusi 1999a). The lifetime excess cancer probabilities were greater than 0.1 for 40 of these pathways and greater than 0.01 for another 70. These are big risks and very worthy of Superfund notice. Of the risks we can do something about, only the lifetime death rates from smoking and driving are higher than 0.01 in the United States.

A final inconsistency in EPA's actual handling of potentially hazardous sites concerns the affected neighborhood population. The EPA *does* consider the size of the affected population when it calculates the HRS and decides whether a site should or should not be put on the NPL. But then, in its remediation decisions, it no longer explicitly considers the size of the population at risk (Gupta et al. 1995; Travis et al. 1987). The EPA is concerned to reduce the maximum exposure to some reasonable level, and this implies no concern for the number of exposed individuals. Thus, many NPL sites receive extensive remediation even though very few people live near them. In the study of 150 NPL sites mentioned above, there were about 700 expected future cancers, but 90% of these deaths would occur at *one site alone* (Hamilton and Viscusi 1999a, 1999b). At only 10 of the sites would there be more than 1 expected cancer if the site were left uncleaned. For many of these sites, the authors concluded that the hazards to the cleanup workers were probably greater than the hazards to the nearby residents.

Furthermore, much of the cancer risk that cleanup removes is from *future* populations that are assumed to migrate into the NPL neighborhood, inexorably and exogenously. Only 18 of the 150 NPL sites in the Hamilton and Viscusi sample have *any current neighbors at all*. This raises several questions:

- Should we not at least take present values of lives saved, just the way we take present values of dollars? Expected lives saved decades or centuries from now are not the same as expected lives saved today (OTA 1989a).
- Can we not prevent the neighbor population around NPLs from expanding? Governments can zone land areas to prevent residential growth. Even where there is now a local population, the best way to "remediate" an NPL site might be to subsidize the neighbors sufficiently that they are willing to move elsewhere, as has actually been done in a few cases (Nossiter 1996).
- If a site is known to be seriously contaminated, and hence shunned by people, where are the benefits (to people) of cleaning it up? If no one would live there if the site were unremediated, then the benefits of remediation should *not* count all the lives of those who would have been saved if they had moved in.
- If the affected population is small but destined to grow, despite common sense and government control, might it not make sense to postpone the remediation until the population is large enough to warrant it? We will come back to this "when" question shortly.

In fairness to the EPA's decision to ignore the size of the affected population in reaching its cleanup choice, we should look more closely at a concept of environmental equity: that each American has an equal right to environmental protection regardless of the number of his or her neighbors. It is a tempting credo, but it utterly fails to notice the tremendous differences in the risks faced by urban and rural people. Nobody claims that cities should have air just as clean as rural areas, because the cost would obviously be huge (perhaps infinite) to achieve that. Nobody claims that farmers should have just as close access to hospitals as city-dwellers, again because of costs. There is no logical reason for suddenly discovering an equity issue when it comes to NPL cleanups.

In this long list of failings in the discovery and remediation process, I have continually referred to "the EPA" as if it were single-handedly responsible for the shortcomings. In fact, many people in the EPA are fully aware of all the problems I have been discussing. But the CERCLA and SARA legislation discourages the EPA from doing anything like a proper benefit–cost calculation. The EPA has to obey the law! The point here is not to levy blame but to point out the gap between what ought to be done and what is in fact being done.

The Cost per Expected Life Saved

The EPA cleans up some very serious life-threatening hazardous waste legacies, but it also cleans up many sites where few lives are at risk. It is time to ask, has Superfund been a cost-effective program?

One early attempt to answer that question proceeded as follows (OTA 1989a). The average Superfund cleanup is of an NPL site surrounded by 5,000 at-risk neighbors. The average cleanup reduces the lifetime excess cancer risk from 10^{-3} to 10^{-6} (i.e., from one chance in a thousand to one chance in a million). The average cleanup cost for an NPL site is about $30 million (brought up to 1997 dollars). The study then estimated the cost per expected life saved as the $30 million cost divided by the (roughly) five expected lives saved to get a $6 million cost per expected life saved.[3] These numbers suggest that Superfund is a slightly more expensive life-saving program than we usually undertake, but not much more. Recall from Appendix B of Chapter 1 that programs that save lives at a cost of less than $1 million we almost always

Benefit–Cost Analysis and Environmental Justice

Noneconomists often consider benefit–cost analysis as just another way for the rich to gain at the expense of the poor. An analysis by Hamilton and Viscusi (1999a) shows just the reverse.

At the most cost-effective cleanups—those that save expected lives at a cost of less than $5 million per life saved—the minority percentage of the population is higher than at the average NPL cleanup. Put differently, the most expensive cleanups per life saved are mostly undertaken to save white lives. If benefit–cost analysis rather than politics and bureaucracy were to decide cleanup priorities, minorities would be better treated by Superfund activity.

This should come as no surprise. Cleanup decisions ignore the size of the neighbor population. Minorities are poor and live in more densely populated neighborhoods. Given the cost of cleanup and the toxicity of the waste, those NPL sites where more lives are affected are the ones that should be cleaned up first. Benefit–cost analysis incorporates that thought. U.S. policy does not.

The absence of a benefit–cost test of cleanup possibilities also leaves the EPA more open to political pressure, and that pressure is more readily applied by wealthier communities. One study found that higher income neighborhoods get significantly faster NPL listing, though not faster cleanup (Sigman 2001). Other studies have found that higher income neighborhoods get more extensive cleanup, which probably explains why the cleanups are not faster (Hamilton and Viscusi 1995; Gupta et al. 1995).

NPL Sites, Cleanups, and Property Values

Another way to look at the benefits of cleaning up a toxic waste site is to look at the impact of NPL designation and later cleanup on property values in the neighborhood. After all, property values are the capitalized values of people's willingness to pay to live in a particular house in a particular neighborhood. So changes in property values should tell us something about how much people are willing to pay to *not* live near toxic waste.

In fact, property values are not very helpful when it comes to toxic waste. We do not always know where to start looking at changes. The date when a site is placed on the NPL may be long after residents and prospective buyers all know of the hazards of the neighborhood. Indeed, the date when it is put on the NPL may herald that, at long last, the U.S. EPA will start doing something about it, and as a result, property values might even begin to rise when the NPL listing is announced.

Nevertheless, there is much evidence that property values are depressed by proximity to toxic waste and are raised by cleanup of that waste (Kohlhase 1991; Greenberg and Hughes 1992; Hamilton and Viscusi 1999a). On the basis of statistical regressions of property values on many housing and neighborhood characteristics and on distance from the nearest NPL site, one study (of the Houston area) found that the average house, if it were situated at an NPL site, would cost $140,000, but if an identical house were six miles away from the nearest NPL site, it would cost $165,000 (Kohlhase 1991). (The study was done in 1985, but the prices of the original study have been inflated to 1997 prices. Houston, by the way, is a particularly fertile statistical area for Superfund study, because it is located in Harris County, which has one of the highest incidences of NPL sites in the United States.) This difference, $25,000, converts at an annual real interest rate of 4% to a household willingness to pay $1,000 per year to move six miles from an NPL site. Assuming 3.3 persons per household on the average, that means an annual per capita WTP of $300.

What does that $300 mean? Suppose that being near an NPL site carries an annual excess cancer risk of 10^{-5}, whereas being six miles away from the nearest NPL site carries an annual excess cancer risk of only 10^{-8}. (These are roughly the annualized excess cancer probabilities of NPL sites that are not cleaned up and are cleaned up, respectively; OTA 1989a.) Then the $300 shows a willingness to pay by the residents of $30 million per expected *own* life saved (i.e., $300 divided by [0.00001000 − 0.00000001]).

Do people really value their lives this highly? Other studies of life-saving activities find much lower values (see Appendix B of Chapter 1). Or are people unable to comprehend how small are the average NPL risks? Or should we just remain skeptical of studies of property values in which very small but very serious risks are involved?

embrace, those that cost between $1 million and $5 million we sometimes undertake, and those that cost more than $5 million we usually reject.

But—did you notice?—the above Office of Technology Assessment study is flawed. It compared the present value of the costs with the undiscounted lives saved. Those expected lives that would be lost if an average NPL site were left unremediated would be lost gradually during a stream of time long into the future. Convert the expected cancer probabilities of the preceding paragraph into annualized cancer probabilities, and you find that the cleanup reduces the annual cancer rate from roughly 10^{-5} to 10^{-8} (see Appendix B of Chapter 1). For the population of 5,000, this translates into roughly 0.05 cancer deaths avoided each year forever after. The present value of 0.05, with a 4% discount rate, becomes roughly 1.25 expected, present-valued lives saved. The present-valued cost per present-valued expected life saved is about $24 million—much higher than we are usually willing to undertake.

As Superfund experience has grown, more detailed studies have become possible. The most recent and extensive is that of Hamilton and Viscusi (1999a). In their sample of 150 NPL sites, they found that only 3% of all Superfund cleanup expenditures were made at sites where the cost per expected life saved was less than $5 million per life. Indeed, only 29% of all Superfund expenditures were made at sites where the cost per expected life saved was less than $100 million per life. Even using the EPA's own conservative cancer-risk probabilities and counting undiscounted lives saved, they find the *median* cost per expected life saved across their 150 sites to be almost $400 million.

What studies like this are telling us is that, despite the social importance of some Superfund cleanups, many cleanups are very expensive and affect very few lives. Hamilton and Viscusi put it neatly:

> Allocating resources strategically [i.e., doing first the cleanups with the lowest cost per expected life saved] could ... eliminate 99.5 percent of the expected cancer cases from hazardous wastes with only 5 percent of the expenditures .... EPA spends 95 percent of remediation resources to eliminate under 1 percent of the risk. (Hamilton and Viscusi 1999a, 125)

How Much to Clean Up

So far, we have been thinking of cleanup as an either–or thing: Either the NPL site is completely remediated, or it is left completely untouched. In fact, there are usually choices to be made as to *how much* to clean up. For each site placed on the NPL, the possible remedies are spelled out in the EPA's Remedial Investigation and Feasibility Studies, and one of these possible remedies is chosen in the Record of Decision (see the box above titled "Glossary and Chronology of Superfund Actions"). Typically, the RI/FS options range from very low cost (e.g., simply containing the toxic materials within impermeable walls or removing it to a hazardous waste disposal facility) to very high cost (e.g., extracting the soil or groundwater, treating it until it is made harmless, and then returning it to the original site). In the Hamilton and Viscusi sample of 150 NPL sites, the remediation cost of the 30 lowest cost sites averaged less than $4 million a site, whereas that of the 30 highest cost sites averaged more than $40 million a site (Hamilton and Viscusi 1999a).

In theory, it is easy to answer the question, how much to clean up? Recall, from earlier in the chapter, inequality (15-1): $pqNV_L/i \geq C$. There, we were implicitly assuming that there was only one way to clean up a site and that the cleanup completely removed any probability of exposure. Now suppose there are choices to be made about the extent of the cleanup. The U.S. EPA could clean up a little, lowering the probability of exposure to the toxic substances by q_1 at a cost of C_1. Or the EPA could conduct a more extensive remediation, lowering the probability of exposure by q_2 (where $q_2 > q_1$) at a cost of C_2 (where $C_2 > C_1$). Applying inequality (15-1), suppose we find that *both* of these possible cleanups pass, that is, $pq_1NV_L/i > C_1$ *and* $pq_2NV_L/i > C_2$. Now we have a new problem: not just whether to clean up, but also how much to clean up.

The decision should rest on whether the *extra* benefit of the more complete cleanup is worth the *extra* cost of the more complete cleanup, that is, whether

$$\frac{p(q_2 - q_1)NV_L}{i} \gtrsim C_2 - C_1 \qquad (15\text{-}5)$$

We should choose the more careful remediation if the *marginal* value of the expected lives saved exceeds the *marginal* cost. This criterion readily extends to more than two possible cleanup techniques. Suppose inequality (15-5) shows that the type 2 cleanup is better than doing only type 1 cleanup. If there is a third, even more complete, means of remediation, we have only to ask whether this type 3 cleanup is better than type 2 cleanup, that is, whether $p(q_3 - q_2)NV_L/i \gtrsim (C_3 - C_2)$. And so on, if there is a still better type 4 cleanup possibility.[4]

The economic approach to how much cleanup to do is straightforward. Alas, the Superfund legislation did not adopt such criteria. Worse than that, it did not adopt *any* criteria (Landy et al. 1994). Indeed, the original CERCLA did not mention costs of remediation, and the SARA went even further—it also did not mention costs, *and* it stated a preference for "permanent remedies" (i.e., more expensive remedies) to site toxicity. The EPA had to react to this SARA preference. For NPL sites with groundwater problems, practically all sites involved containment without treatment before 1984, but after 1984 four-fifths of them involved the much more expensive treatment.

But what was the poor EPA to do? The law said "permanent solutions to the maximum extent practicable" and did *not* say anything about the costs of these solutions. People living near NPLs wanted—nay, demanded—complete cleanups. After all, they lived there, and they weren't paying for the cleanup. Consultants and contractors profited from more complete and expensive remediation. Where cleanups are labor intensive, laborers and their unions benefit from bigger cleanups. And the bureaucratic loss function was very, very asymmetrical. Saving the government a few million dollars earns an EPA official very little professional prestige, but being identified as the one whose penny-pinching cleanup decision cost lives could mean a serious setback to one's career. But even "right-minded" EPA officials feared that there were many intangible or hidden benefits to cleanups that would be missed by benefit–cost analysis and hence that such analysis, even if permitted, would fail by recommending too little cleanup.

The law and all its attendant incentives urge the EPA to excessively clean up NPL sites. Again, one's first reaction is that it is better to be safe than sorry, even if it does waste some cleanup resources. But it also wastes lives. Within the limited remediation resources available, when the EPA spends more than it should on the sites it does clean up, it means that other sites must wait for the beginning of their cleanup. Many receive some emergency remedial action, but then must wait a long

---

"The Last Ten Percent"

The question of how much to clean up has been called the problem of "the last ten percent" by Stephen Breyer (Breyer 1993). He argued that too often 90% of the money was spent fixing up the last 10% of the problem. He tells of a case in his court (before he became a Supreme Court justice) where one party balked at spending $9.3 million to remove and incinerate the last little bit of contaminated dirt on a site that was already mostly cleaned up. The case, which consumed 40,000 pages and 10 years, seemed to have achieved agreement among all parties that this incineration would have made it safe for small children playing on the site to eat dirt there for 245 days of the year instead of only 70 days. But there were no children near the site because it was a swamp; and small children were not ever likely to be living and playing in or near this swamp. Moreover, half of the contamination would probably evaporate in a few years. "To spend $9.3 million to protect nonexistent dirt-eating children is what I mean by the problem of 'the last ten percent'" (Breyer 1993, 11 f).

time for more permanent solutions. By 1990, a full decade after the passage of CERCLA, only 29 of the more than 1,000 NPL sites had been cleaned up, verified, and delisted from the NPL (Church and Nakamura 1993). A decade later, by 2000, about half have been cleaned and delisted.

Budget restrictions and excessive care and cleanup bedevil each stage of the Superfund process. On average, it requires 43 months from the EPA's first notice of a site to its placement on the NPL, then 20 months more to the start of its RI/FS, then 38 months more to the issuance of the ROD, then 18 months more to design the cleanup (the RD), and then 25 months more to complete the cleanup (the RA). The entire process, from first notice to completed remediation, averages 96 months—eight years.[5]

The timing is even worse than this appears. Because the most hazardous sites have the most toxic substances to study and the greatest complexity of remediation, they are the slowest to move through the process. Far from cleaning the worst first, there is evidence that we clean the least bad first (Church and Nakamura 1993). The increasing political emphasis on speeding up cleanups encourages the EPA to spend its limited budget on the sites that offer quick solutions, and these are not usually the most serious health-risk sites (Hird 1994; Guerrero 1995; Czerwinski 1996). Nevertheless, many feel that the current laws and institutions make the problems intractable. Some have suggested that the federal government should just get out of the Superfund business, simply give money to the states, and let the states decide what and how much to clean up (Landy et al. 1994). Indeed, the states might just pass the money on to the neighbors of NPLs and let them decide what to do—where, of course, the neighbors would have to provide their own money if they wished to exceed the state's subsidy.

When to Clean Up

People, even economists, often think that a Superfund decision is either–or: Either it is serious enough to clean up, or it is not. The corollary of that kind of thinking is that if it should be cleaned up, it should be cleaned up now. Actually, there are three choices: Clean it up now, clean it up later, or don't clean it up at all. The theory is straightforward, if not entirely obvious (Porter 1983c).

There are many reasons why it might be socially sensible to clean something up later, but do nothing (or very little) now. The neighboring population might be growing, so that at some future time there will be enough people at risk to justify a cleanup.[6] People's willingness to pay for a risk-free neighborhood might be growing, so that at some future time their willingness to pay will exceed the cost of the cleanup. Cleanup technology is always improving, and postponement of a cleanup may mean that it can be done easier, faster, better, or cheaper at a later time.

Cleaning the "worst first" sounds like good advice, but if budgetary or labor constraints require prioritizing NPL sites for cleanup, the "worst first" may not yield the right priority. Think about the benefits and costs of postponement. When we postpone a cleanup, three things happen. First, expected health and environmental damage occurs during the postponement—and that's bad. Second, cleanup costs are postponed and hence have a lower present value today—and that's good. Third, the cost of cleaning up a site later may change—and that can be good or bad, depending on whether the cleanup cost rises or falls (relative to the overall price level) in the interim. The worst-first criterion focuses on the first of these three impacts and ignores the second two.

Who Pays for Cleanups?

Although it is nowhere near the cost of controlling air pollution, Superfund has been a very expensive part of the U.S. effort to reduce human assaults on our own environment. The total costs of cleanup have been averaging more than $2 billion a year for the past two decades, and many estimate that the ultimate costs will run well past $100 billion (Acton and Dixon 1992; Dixon et al. 1993; Probst et al. 1995). Although the public perceives toxic waste sites as seriously dangerous and deserving of expensive remediation, Congress has not perceived the public as being willing to pay the high cost through increased taxation or reduced public programs elsewhere.

Congress' answer to the dilemma of increased expenditure without the necessity of greatly increased government budget was to make those responsible for the creation of Superfund sites also responsible for their remediation—the "polluter pays principle." Simultaneously with the identification of NPL sites, the EPA was instructed to identify the potentially responsible parties and either to collect the costs of the remediation from them or to direct them to undertake adequate remediation themselves.

Unfortunately, identifying the PRPs and getting them to perform the cleanup is not usually easy. The stereotypical NPL site is a chemical plant that used to toss its toxic wastes in rusting barrels into shallow trenches in back of the plant and that now can be made to correct this hazardous disposal. But relatively few NPLs follow this stereotype. There may be many—sometimes very many—contributors to the toxicity of a particular NPL. More than 10% of the NPL sites have more than 100 PRPs (Probst et al. 1995). It may be difficult to discover the responsible party or parties. Many sites are far from the plants that produced their wastes. The responsible party may not have sufficient financial resources to underwrite the remediation costs. Or the PRP may simply no longer exist.

Accordingly, it has long been recognized that Superfund cleanups would require money when PRPs couldn't be found if such "orphan sites" were going to be cleaned up. This money is collected in several ways. The federal government makes general fund allocations. Superfund also gets revenue from a general corporate tax. Special taxes were levied on the petroleum and chemical industries on the grounds that these have been the leading contributors to the creation of NPL sites. (Confounding any economic logic, the Superfund tax rate on imported petroleum was initially higher than that on domestically produced petroleum.)

Potentially Responsible Parties

A potentially responsible party is anyone the EPA can show contributed *any* toxic waste to a Superfund site. The PRP then becomes liable for the costs of whatever cleanup is selected. PRP liability has three important features. First, liability is "strict" in the sense that it is *not* mitigated by the fact that others were also negligent or that the toxic disposal was at the time legal, accidental, or not known to be hazardous. Second, it is "joint-and-several" in the sense that *any* contributor to the NPL is fully liable for the entire cost of the remediation. Third, it is "retroactive" in the sense that the liability in no way depends upon how long ago the disposal occurred nor on how legal it was at the time. The idea is not to assign guilt or achieve fairness but simply to make the EPA's job of gathering cleanup funds as easy as possible.[7]

Think for a moment about guilt and fairness. PRPs are (mostly) firms that at some often long-ago time used cheap but ultimately hazardous toxic waste disposal techniques. Who really benefited from this cheap disposal? Basically, the answer is our parents, grandparents, and great-grandparents. As customers of these firms, they paid lower prices. As stockholders, they earned higher dividends. As managers, they earned higher salaries. As workers, if well unionized, they may have earned higher wages. They benefited at our expense. Fairness might require a retroactive intergenerational transfer that we can no longer make.[8] And it might not, because those generations also bequeathed to us the added capital and technology that lets us live at a much higher standard of living than they did, even after allowing for the Superfund costs we also inherited.[9] As for guilt, only a few of those who benefited knew that there was wrong being committed, and most of those few are now beyond the reach of human law. Superfund liability could not do much with guilt and fairness even if it tried.

The real issue with Superfund liability is whether it is an effective way of raising cleanup money. There are problems.

The major problem, as you might guess, is that fingered PRPs do not go gentle into unlimited cleanup liability. They fight such designation in the courts, and the litigation costs can be high. Studies vary greatly, but some conclude that roughly a fourth to a half of all Superfund expenditures go not to cleaning up toxic sites but to litigation over who should pay and how much (Anderson 1989; Dixon 1995).

The U.S. EPA well knows this and attempts to keep litigation costs down by going after one responsible party for the entire cleanup cost. This party of course must have deep pockets, because cleanup costs can be high. Although this keeps the EPA's litigation costs down, the legal battles do not stop with its successful conviction of one PRP. A convicted PRP with deep pockets can keep its costs down by suing shallower pocketed PRPs for a share of the cost. And smaller PRPs sue still smaller PRPs (Dixon 1995). Because fixed costs are attached to each of these pyramiding trials and the variable costs of trials are presumably related to the sums involved, the EPA system of going after the deep-pocketed PRP, though it saves the EPA costs, ends up adjudicating many more dollars than if the EPA had arranged to try one case involving all PRPs (Tietenberg 1989). Although the EPA's goal should not be to minimize total adjudication costs, neither should it be to maximize them, and the current system may come close to that.

PRPs not only sue other PRPs; they sue their own insurance companies. Before CERCLA introduced strict, joint-and-several, retroactive liability, most big firms insured against only the damages due to "sudden and accidental" toxic releases (Revesz and Stewart 1995). The new CERCLA rules made it unclear how much of the PRP's liability should be paid by their insurer (i.e., by the insurer at the time of the Superfund-causing releases). Indeed, Superfund may have been single-handedly responsible for the almost complete disappearance of the "environmental impairment liability" insurance market. The insurance industry relies on its ability to predict and mitigate risk, but the new Superfund system meant that it could not. Insurers cannot predict what or when the EPA will suddenly declare a new liability. Moreover, the insurer has no control over firms it does not insure, but for whose environmental damages it may become liable under joint-and-several liability (Katzman 1989).

The timing, care, and cost of NPL cleanups are also affected by CERCLA liability. Litigation usually stalls the pace of cleanups. The EPA's full recovery of cleanup costs is less likely if it does the cleanup itself and later tries to recover the cost from a PRP, so the EPA tends to postpone the cleanup whenever it is planning to seek PRP responsibility (Sigman 1998a). When PRPs have to pay for cleanup, they may be able to exert bargaining leverage either directly on the EPA or indirectly on the political process to achieve a less than optimal remediation. Or the EPA may feel that, when its own budget is not being spent, it should choose a larger than optimal remediation. One study found that the EPA chooses less extensive remedies when PRPs are expected to bear a heavy part of the costs (Sigman 1998a).[10] As for cost, PRPs may have information or incentives that the EPA lacks that will permit them to design more cost-effective remediation.

Finally, there are incentive problems with the CERCLA liability system. It might seem at first that there would not be. Why should a sudden, unexpected, once-and-for-all liability for some long-past activity affect a firm's future activity? It can have three kinds of effects:

1. Having been stung once, firms may redouble the care with which they handle current hazardous waste. This may sound like a good thing, encouraging more careful handling. But if the Resource Conservation and Recovery Act has already achieved the correct treatment of hazardous waste, then even more careful handling will be excessive.
2. The limit on a firm's hazardous waste liability is its equity. Big firms risk a lot, and small firms only a little. Unlimited liability may lead to the growth of inefficiently

small firms or lead big firms to contract out their hazardous waste handling to small firms, possibly to fly-by-night firms that cut the corners of RCRA regulations because their maximum loss if caught is capped by their modest assets.[11] The government can offset these incentives to some extent by requiring all firms, large or small, to maintain sizable bank balances or insurance coverage so that no one can escape environmental liability.

3. Superfund increases the cost of capital to PRPs, as investors feel there is either a probable loss of assets or an increased risk of future liabilities (Garber and Hammitt 1998). This reduces investment there and increases prices of their products. But there is no good social reason to make these products more expensive now just because they caused external damage in the past. Just because past consumers got the products too cheaply is no reason to make future consumers pay too much, if the firms are now optimally abating external costs and the costs of that abatement are already incorporated into the price of the product (Katzman 1989). Despite this conceptual worry, most empirical studies of Superfund liabilities as a fraction of total value added suggest that such price increases would be very small (Probst et al. 1995).

Given these problems, and especially the immense waste of resources in legal expenses, maybe we should think again about an end to retroactive liability and a beginning of strictly government financing of Superfund activities.

Government Responsibility for Cleanups

If the government were to assume the responsibility for future NPL cleanups, it would mean an immediate and complete end to the very costly legal and transactions costs of Superfund. The budget allocation for CERCLA could be spent entirely on identifying and remediating NPL sites. This is a huge plus. But such government assumption of complete responsibility is not without its own problems (Probst and Portney 1992; Probst et al. 1995).

The first problem is with the transition. We cannot go back to 1980 and discuss which of PRP or government responsibility is the better choice. We have already lived with 20 years of PRP responsibility. Switching now would mean that many PRPs have already fulfilled their responsibilities, whereas others that have not yet been identified or have dragged their feet successfully would escape payment. Not only would this be seen as unfair, but also it would create incentives in all future waste issues for private parties to litigate and lobby in the hopes of a later change in the rules. To handle this problem, many proposals for ending PRP liability include the idea of reimbursing past PRP expenditures through tax credits (Probst and Portney 1992; Probst 1995). This of course is the equivalent of going back to 1980 and having the government finance the entire Superfund cost right from the start. But that is exactly what Congress refused to do then because of the high budgetary cost, and now that the costs of Superfund are known to be higher than originally thought, such tax credits have probably become even more unthinkable.

The purpose of government responsibility is to speed up the pace of cleanups. To the extent that it ends legal costs, it does that by freeing up the EPA budget. The flip side of this, however, is that the EPA no longer can use PRP monies for the cleanups. The loss of PRP money far exceeds the gain in avoided legal costs. One careful study estimated that eliminating PRP liability would require a *doubling* of annual EPA expenditures (Probst et al. 1995).

Suppose Congress did not increase the EPA budget at all. Superfund cleanups would effectively go at half the current pace. The average length of time for a Superfund cleanup would roughly double. This doubling of the average cleanup time is not inevitable, of course. The EPA could make less extensive cleanups than it currently does (or than it currently requires PRPs to do). As our discussion above of "how much to clean up" suggested, most current cleanups are excessive, and reduced expenditure per site would probably be a good thing. But wishing would not make it so—the same political and bureaucratic forces that make for excessive cleanups would still be there unless there were basic changes in the incentive structure of CERCLA. Furthermore, even if these incentives could be altered, the impact would only gradually emerge. For many NPL sites, the remedy has already been chosen, and suddenly cutting it back would be very difficult.

Although less complete remediation of NPL sites would probably be a good thing in itself, and especially so if the total amount of monies for Superfund cleanups were reduced, it is not at all clear that a switch from PRP responsibility to government responsibility would lead to less extensive and less costly cleanups. It might mean that the EPA would slide into regular "Cadillac cleanups" without the expertise and cost-cutting enthusiasm of involved PRPs (Probst and Portney 1992). Currently, private-sector cleanups are one-fifth less expensive than government cleanups, and one study estimated that shifting all cleanups entirely to the EPA would add $3.6 billion (in 1997 dollars) to the total cost of cleaning up the existing NPL sites. It is interesting that this is almost exactly what would be saved in legal costs by eliminating PRP responsibility. Between the two effects, lower legal costs and more costly cleanups, entirely eliminating private responsibility for Superfund cleanups would save a mere 4% of total social costs (Probst 1995).

Final Thoughts

Superfund has *not* been a disaster, as President Clinton claimed. Not *all* the money has gone to the lawyers. It *has* worked. It has probably saved many American lives and illnesses. The problem is that it has done its work at very high cost. The reason for this high cost is clear. The EPA essentially has been told to make everything perfectly clean again, and do it soon, but it has been given a very limited budget to achieve this. The limited budget is understandable—to make everything clean might well require trillions of dollars, and few voters or representatives are willing to pay so huge a fraction of our national output for this purpose. When told to do the impossible, the EPA is left with two choices: Stall, or pretend that it is being done. The EPA has done some of each.

What the EPA and the public need to recognize is that everything cannot be made perfectly clean right away. Priorities must be established. The limited budget should be spent where it will do the most good, saving the most (present value of expected) lives. Cleanups at some sites not only must be postponed, perhaps forever, but many *can* be postponed at a low cost in expected lives lost. For other NPL sites, partial remedies may save most of the endangered lives, and may suffice.

All this requires that the EPA frankly introduce the concept of the cost per expected life saved. With that concept out in the open, the EPA can begin to make sensible decisions as to what to clean, how much to clean, and when to clean.

APPENDIX

Social Benefit–Cost Analyses of Two Michigan Superfund Remediations

The real world is often a special case.

—Horngren's Observation

The approach to doing a benefit–cost analysis of Superfund remediation is outlined in the text. Here it is applied to two Michigan remediations, one of which surely passes a benefit–cost test, and one of which is a close call. Benefit–cost analyses, even when as simple as those presented below, add something to our understanding of the social urgency of cleanups. This is evidenced here by the fact that the site with the lower HRS score, the fewer affected people, and the greater cleanup cost is the one that more clearly deserves the full cleanup that both in fact received. (Abstracts of the RODs for these and all other NPL sites are available at http://www.epa.gov/superfund/sites/rodsites, and one copy of any full ROD can be ordered at that address.)

In both the cases, and in the analyses below, full remediation was adopted. Before starting on the details, a caveat is in order: Even if complete cleanup passes its benefit–cost test, some less complete technique might have been even better; and even if complete cleanup fails its benefit–cost test, some less complete remediation technique might have passed.

Rose Township Dump

In the 1960s, Tucker Ford, a waste hauler, dumped unknown amounts of solvents, paint and battery sludges, PCBs, oils, and grease on a 12-acre plot of Howard Wilson's land, perhaps with the landowner's permission. Some of this waste was buried in drums, and some bulk waste was simply dumped into pits and lagoons or scattered on the surface (Kafka 1997). Only a dozen residences adjoin this site, but 4,600 people live in surrounding Rose Township, and almost all of them rely on wells for drinking water.

Dumping was halted by 1971, and 5,000 drums were removed by the State of Michigan in 1980. In 1983, the site was added to the NPL with an HRS of 50.92. In 1985, the area was fenced off, the drums were sampled, and some soil was removed. The names of PRPs were gathered; and in 1988, 12 of them agreed to implement and pay for the cleanup. This involved excavating, incinerating, and returning 25,000 cubic yards of soil; installing and operating a groundwater extraction and treatment system; and monitoring the groundwater on and around the site for the next 30 years. The EPA and the PRPs considered less permanent remediation techniques, but in the end the most expensive—and the only one considered "permanent"—was selected. The present value of the total cost was $47.4 million. (This cost, which has been converted to 1997 dollars, counts only the costs during EPA's oversight, and not the costs of the initial work done by the state.) There were many different carcinogens at this site, and the path of contact could be either groundwater or soil. The ROD gave many estimates of the overall lifetime excess cancer probability of each Rose Township resident if this site were not cleaned up. These estimates ranged between 0.01 and 0.70. That's the equivalent of an *annual* excess cancer probability of between 0.00013 and 0.01494. The ROD of course stopped there, but we don't have to. If the population of Rose Township stays constant at 4,600, these probabilities mean that cleaning up the site will save somewhere between 0.6 and 68.7 lives a year, for each year forever after. Valuing each life saved at $3 million (see Appendix B to Chapter 1), we get an annual benefit of the cleanup of $1.8–206.1 million. With a discount rate of 4%, the annualized cost of the cleanup is 0.04 times $47.4 million, or $1.9 million. The annualized benefits almost certainly exceed the annualized costs, even though the estimate of lives saved has a wide range.

(As we have seen in the text of this chapter, the U.S. EPA's procedures lead to an overstatement of the probability of cancer death, and hence an overstatement of the benefits of cleanup. But we have also omitted many other benefits of this Rose Township cleanup—mainly, reduced child exposure to lead and improved recreation in the area—and these omissions mean we have understated the benefits of cleanup.)

Spiegelberg Landfill

From 1966 to 1977, a subsidiary of Ford Motor Company dumped paint sludge and other wastes into the abandoned Spiegelberg gravel pit, contaminating the soil and threatening an ever larger area of groundwater below it. The 18,000 people who lived within three miles of the site mostly draw upon these shallow aquifers for their drinking water. The contamination from the paint sludge posed both carcinogenic and other health risks to them (Drucker 1997).

In 1983, the Spiegelberg landfill was added to the NPL, with an HRS of 53.61. Ford, the sole PRP, agreed to remove and incinerate (off site) the paint sludge deposits and to pump, treat, and reinject the groundwater. The remediation was completed in 1997 at a total cost of $27.3 million (in 1997 dollars).

This remediation completely removed the cancer threat to the people living nearby, a threat that the ROD estimated to be a lifetime excess cancer probability of between 0.00036 and 0.00190 for each resident. That's the equivalent of an annual excess cancer probability of between 0.000005 and 0.000024. Should the threatened number of residents have remained at 18,000, these probabilities mean that fully cleaning up the site will save somewhere between 0.08 and 0.43 lives a year, for each year forever after. At a value of $3 million per life saved, there is an annual lifesaving benefit of $0.2–1.3 million. With a 4% discount rate, the annualized cost is $1.1 million—right smack in the benefit estimate range.

One could argue either way on this cleanup. Despite the large number of people potentially affected, the probabilities were very small and the costs very high. If one were to use a slightly higher interest rate, the benefit–cost test would be more likely to fail. If one were to apply a slightly higher (or lower) value of life, the benefit–cost test would be more likely to pass. If one were to incorporate into the benefits the reduced odors, hypertension, and birth defects, the benefit–cost test would be more likely to pass.

Notes

1. If the EPA's lifetime excess cancer risk estimate is greater than 10^{-4}, the site must be cleaned up; if the risk is between 10^{-6} and 10^{-4}, then cleanup is at the discretion of regional EPA officials; if the risk is less than 10^{-6}, cleanup is not generally warranted, though exceptions can be made (U.S. EPA 1991c). Just how big is a cancer probability of 10^{-4}? The average American faces a lifetime risk of fatal cancer of about one in four, 0.25. If proximity to an NPL site raises that by 10^{-4}, the person's cancer risk rises to 0.2501—that is, it rises by 0.04% (Walker et al. 1995).

2. How can anyone get a probability of five when a probability cannot exceed one? The probability that you will eat breakfast tomorrow is say 0.80; the probability that you will eat lunch is say 0.70; and the probability that you will eat dinner is say 0.90. The EPA would say that the probability that you will eat at some time tomorrow is 2.40 (i.e., 0.80 + 0.70 + 0.90 = 2.40). In fact, the probability that you will eat tomorrow is *not* additive in the probabilities. The true probability that you will eat tomorrow is equal to one minus the probability that you will *never* eat tomorrow, 0.994 (i.e., $1 - [1 - 0.80][1 - 0.70][1 - 0.90]$).

3. More precisely, $(10^{-3} - 10^{-6})(5,000) = (0.001000 - 0.000001)(5,000) = 4.995$ expected lives saved.

4. This neat sequential marginal approach breaks down if it is not true that $q_1/C_1 > (q_2 - q_1)/(C_2 - C_1) > (q_3 - q_2)/(C_3 - C_2) > \dots$, but these inequalities will almost always be fulfilled for actual cleanup schedules. The first, cheapest possible cleanup will usually reduce exposure risk a lot per dollar of cost, and the marginal risk reduction per marginal dollar of cleanup expenditure will decline as more complete cleanups are contemplated. This is just a variation on the principle of diminishing returns. If the inequalities given just above are not fulfilled, a straightforward but more cumbersome set of calculations is required.

5. If you did the arithmetic carefully, you might have noticed that the average time for the complete Superfund process (i.e., 96 months) is less than the sum of the average times for each stage. This is not bad math. A little hypothetical arithmetic shows why. Suppose Superfund started 36 months ago. At that time just three sites (I, II, and III) were specified, and there were just three cleanup stages (A, B, and C). Then the times of each site at each stage (in months) would be: Site I: stage A, 12; stage B, 12; stage C, 12. Site II: stage A, 18; stage B, 18; stage C, not yet complete. Site III: stage A, 36; stages B and C, not yet complete. The averages of the sites that have completed each stage would be 22 for stage A, 15 for stage B, and 12 for stage C.

The average times at each stage sum to 49 months. But the average time to completion of the (only one) completed cleanup is only 36 months. In short, a slow-moving site gets counted in (some of) the stage averages but not in the overall average because its overall time is not yet known (Church and Nakamura 1993).

6. When is cleanup warranted? Recall inequality (15-2): $pqNV_L \geq iC$. There, we assumed N was fixed now and forever. Suppose N is growing exogenously at rate g. Let the current population be N_0. If you postpone the cleanup from now until next year, the cost is the value of the lives that are lost to cancer because the site was not cleaned up this year. That value is pqN_0V_L. The benefit of not cleaning up this year is the interest earned on the cleanup expenditures that are not incurred this year—of course, it is not the interest per se that we are earning from the postponement but the benefits of other projects to which we can devote more of this year's budget. That is iC. Suppose $pqN_0V_L < iC$. Then it pays to postpone the cleanup. How long? Until such time (T) as the population has grown to the point where $pqN_0(1 + g)^T V_L = iC$.

7. One past EPA administrator openly admitted that he saw nothing wrong with a system that punishes the innocent as well as the guilty (Connolly 1991). And one judge (in *U.S. v. Cannons Engineering*) pointed out that "CERCLA is not a statute that places a great premium on fairness" (quoted in Clay 1991).

8. Levying special taxes primarily on the petroleum and chemical industries is an administratively expensive way of attempting this impossible intergenerational task. Such taxes raise the current price of these products, not the price our forebears paid.

9. A few people have even argued the "wastebasket theory" of toxic waste sites. You don't run to the garage with every bit of household waste. You drop it into a wastebasket or garbage pail and empty that only when it is full. The in-house waste is unsightly and perhaps even unhealthy, but instead of spending your time dashing to and from the garage, you are enjoying other much more productive or interesting activities. The problem with applying this theory to toxic wastes is that proper disposal would have been very cheap compared with the cost of today's NPL site cleanups. Almost no marginal investment our forebears could have made would have earned enough to cover the very costly cleanups that we are now undertaking.

10. Recalling the EPA's tendency to overclean NPL sites makes it tempting to suggest a positive force for the CERCLA liability system, that the influence of the PRPs is to force the cleanup choice back toward the optimal. But it also delays the cleanup—it takes on average two years longer to complete an NPL cleanup if PRPs are involved (Sigman 2001).

11. This small-firm effect of unlimited liability has been evidenced in many areas. In states with strict liability for hazardous waste pollution, small firms are responsible for a larger share of spills involving toxic chemicals (Alberini and Austin 1999). As liability for worker illness became more extensive, small firms began to grow more rapidly in industries where workers are subject to disease exposure (Ringleb and Wiggins 1990). Within a year of the Oil Pollution Act of 1990, which greatly expanded oil-spill liability, nearly half of the oceangoing tankers serving the United States were owned by single-ship companies, and many large shippers were transferring oil to small, independently owned ferry ships before entering U.S. waters (Sullivan 1990; Ketkar 1995).

Radioactive Waste

Put all your eggs in the one basket, and—WATCH THAT BASKET.
—Mark Twain, *Pudd'nhead Wilson*, 1894

We come now to the disposal of the most dangerous of all human wastes: radioactive waste. It may well be that the problem, to most people, completely transcends economics—and that disposal calls strictly for a scientific and political solution, implemented at any cost. To some extent, as you will see, I agree. But there are degrees of hazard to radioactive waste. For much of it—what is called low-level radioactive waste (LLRW)—economics may sensibly play an important role in the solution.

What is low-level radioactive waste? LLRW consists basically of all the nuclear waste other than that generated as spent nuclear fuel in electric power plants and as a by-product of weapons manufacture. It comes mostly from electric power generation, but also from medical and academic research, medical diagnostic and therapeutic services, industrial research, and manufacturing processes.[1] Categorizing radioactive wastes as "low-level" does not imply that they present no safety risk, just that the risk is much lower than that of high-level waste.

For this LLRW, there is much scope for economic analysis. The U.S. Nuclear Regulatory Commission (NRC) "believes storage can be safe over the short term as an interim measure, but favors disposal rather than storage over the long term" (http://www.nrc.gov/NRC/NUREGS/BR0216/br0216.html#_1_16). Then the question becomes, what is the optimal number and location of LLRW disposal facilities in the United States? By "optimal" we will mean the plan that minimizes the present value of the social cost of storing, transporting, burying, and monitoring (for at least a century) the nation's LLRW. (The word "optimal" is put in quotation marks because all we are trying to do here is handle the existing flow of LLRW cost-effectively. For a true optimum, we also should be asking how much LLRW we should be generating.)

The optimal number and location of sites will take account of two opposing cost concerns. On the one hand, these disposal sites exhibit great economies of scale, which suggests the establishment of a small number of large facilities. On the other hand, the transportation of LLRW is expensive, which suggests the establishment of several, small, dispersed, regional disposal sites. Only after the lessons from LLRW disposal have been learned will we turn, somewhat more briefly, to the much more difficult problem of high-level nuclear waste disposal.

Low-Level Radioactive Waste

Figure 16-1 shows the annual volume of LLRW received by commercial disposal facilities since they began accepting waste in 1962. (Figure 16-1 shows the total volume of LLRW *shipments* from generators to commercial disposal facilities; there are no data on LLRW *generation*; DOE 1995, 1998a. What is shipped and what is generated will differ to the extent that the volume of on-site temporary storage changes.) The steady upward trend ended in the early 1980s when disposal fees began to rise, from $20 per cubic foot (in 1983, converted to 1997 dollars) to at least $500 per cubic foot today; fees now vary greatly in different parts of the United States. LLRW generators have reacted to the more costly disposal in several ways, by using less radioactive material, by building on-site storage facilities, and by using incineration and compaction (Ring et al. 1995; Berry and Jablonski 1995; Malchman 1995).

The problem with LLRW is no longer what to do with it. There has long been agreement that permanent on-site storage of LLRW is not appropriate, and ocean dumping and reprocessing (i.e., recycling) have more recently been rejected. There is now a consensus for land burial, with careful monitoring for hundreds of years. What has been bothering policymakers for the past 20 years is how many LLRW disposal sites there should be, and where they should be located.

There is no theoretical solution to the question of how many sites. If transport costs are high, then many regional sites are appropriate. If there are large fixed permit and construction costs that are independent of the size of the disposal site, then few—perhaps only one—disposal site is appropriate. We must look at the empirical magnitudes of these costs.

Figure 16–1

The Volume of Low-Level Radioactive Waste, 1962–1997

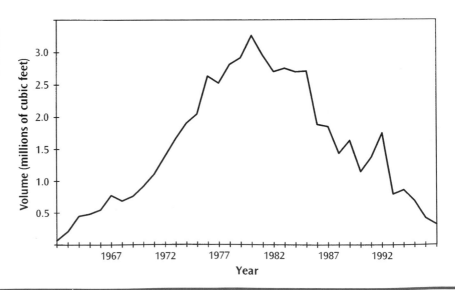

LLRW Transport Costs

The total LLRW transport cost depends multiplicatively on the volume shipped, the cost per mile, and the average transport distance. Let us look first at the distances. Although we do not have data on all the sites that generate LLRW—and eventually ship it somewhere for burial—we do have data on the state of origin of LLRW shipped and each state's volume of LLRW shipped. For a rough estimate of distances, let us assume two things: All of a state's LLRW is generated in its biggest city; and if an LLRW disposal site is located in a state, it will be located in the state's biggest city. Using the five-year 1992–1996 average LLRW shipments from each state, we can calculate the average distance LLRW must travel for a variety of numbers and locations of disposal sites (Kearney and Porter 1999). (For simplicity, we calculate the shortest route to the nearest disposal facility, even though the shortest route may not be the fastest or the safest route. Alaska and Hawaii are omitted from this exercise, and the District of Columbia is included.) It is then an easy programming exercise to find, for any given number of disposal sites, where they should be located to minimize the total travel distance of U.S. LLRW. The least-transport-distance sites and the average distances are shown in Table 16-1.

The bottom line of Table 16-1 is that, though the average LLRW transport distance declines as the number of optimally located facilities increases, it declines very slowly by the time we have reached three or four sites. Given the great economies of scale that we shall soon encounter in the construction of LLRW facilities, it becomes quickly clear that the optimal number of LLRW disposal sites is not likely to be large.

Transport costs for LLRW run about $3–5 per truck per mile (Feizollahi et al. 1995; McDaniel et al. 1995). Because LLRW is usually shipped on a truck that handles 640 cubic feet, this means that for average distances (as in Table 16-1) of 234–898 miles, the transport cost of each cubic foot of LLRW runs less than $7—not

Number of sites	Average distance (miles)	Host state(s)	**Table 16-1**
One	898	Arkansas	Location and Average Distance for Low-Level Radioactive Waste Disposal Facilities
Two	523	Kentucky	
		Oregon	
Three	355	Oregon	
		Pennsylvania	
		Tennessee	
Four	281	Illinois	
		Oregon	
		Pennsylvania	
		Tennessee	
Five	234	Colorado	
		Georgia	
		Illinois	
		Oregon	
		Pennsylvania	

very expensive, as we shall see when we look at the facility costs. (We are also ignoring, but just for a while, any external costs of either transport or disposal of LLRW.)

LLRW Disposal Facility Costs

Because no major LLRW disposal facility has been constructed in the United States in more than 25 years, there are no recent, hard data that can be used to estimate the future costs of location, construction, operation, and closure of LLRW disposal facilities. The estimates that do exist are tentative. Because of this, we will compare cost estimates developed by several different sources (DOE 1993, 1994; Coates et al. 1994; Baird et al. 1995).

Basically, there are four principal cost components of an LLRW disposal facility:

- *Pre-operations.* These are the costs associated with site selection, site characterization, licensing, land purchase, as well as the actual construction costs of the facility and its supporting structures. Typically, these costs are incurred between 1 and 7 years before operations begin.
- *Operations.* These are the costs of the receipt and disposal of LLRW, environmental monitoring, cell construction, and operating equipment. The lifetime of a facility is usually assumed to be between 20 and 40 years.
- *Closure.* These include the cost of covering the disposal site, removal of any buildings, site protection, and operation during the typical 5-year closure period.
- *Postclosure.* These primarily include environmental monitoring and security during the necessary postclosure period. Different sources use different periods, ranging from 100 to 300 years.

The cost estimates below will be broken down into these four categories.

Table 16-2 shows estimates of the present value of total cost (PVTC) for an LLRW disposal facility with a capacity of 1.5 million cubic feet, filled at the rate of 50,000 cubic feet a year for 30 years. The various estimates differ widely, from a PVTC of $91 to $218 per cubic foot of capacity. Although we should not be comparing

Table 16-2

Various Estimates of the Present Value of Total Cost (PVTC) of a 30-Year, 1.5-Million-Cubic-Foot Facility (millions of 1997 dollars)

PVTC measure	DOE (1993) estimate[a]	DOE (1994) estimate[a]	Coates et al. (1994)[b]		
			Low estimate	Medium estimate	High estimate
Pre-operations	62.8	80.3	73.5	147.0	220.5
Operations	65.7	204.1	61.3	80.5	99.6
Closure	5.0	9.1			
Postclosure	2.3	0.2	4.6	5.7	6.8
Total PVTC[c]	135.9	293.3	139.4	233.1	326.9
PVTC/ft^3 ($/ft^3)[d]	91	196	93	155	218

[a]DOE is the acronym for the U.S. Department of Energy.
[b]Coates et al. (1994) give three estimates and do not separate the estimates for closure and postclosure costs.
[c]Totals may not add because of rounding.
[d] PVTC per cubic foot of capacity, in dollars per cubic foot.

Table 16-3

Various Estimates
of the Present Value
of Total Cost (PVTC) of a
30-Year, 900,000-Cubic-
Foot Facility
(millions of 1997
dollars)

| PVTC measure | DOE (1993) estimate[a] | DOE (1994) estimate[a] | Coates et al. (1994)[b] | | |
			Low estimate	Medium estimate	High estimate
Pre-operations	61.7	76.7	73.5	147.0	220.5
Operations	61.2	146.7	36.8	48.3	59.8
Closure	5.0	7.7			
Postclosure	2.3	0.2	4.6	5.7	6.8
Total PVTC[c]	130.3	231.3	114.8	201.0	287.1
PVTC/ft³ ($/ft³)[d]	145	257	128	223	319

[a]DOE is U.S. Department of Energy. The DOE 1994 estimate is for 35,000 cubic feet a year.
[b]Coates et al. (1994) give three estimates and do not separate the estimates for closure and postclosure costs.
[c]Totals may not add because of rounding.
[d]PVTC per cubic foot of capacity, in dollars per cubic foot.

present-valued dollars (PV) to un-present-valued cubic feet, it suffices to make the point that it is high relative to transport costs. To convert PVTC per cubic foot to PVTC/PV per cubic foot for a 30-year steady flow of LLRW and a discount rate of 4%, you simply multiply the former by 1.73; i.e., by 30 times 0.04, divided by $(1 - 1.04^{-30})$. Table 16-3 repeats this exercise, for a smaller facility used at a slower rate during the same number of years. The PVTC per cubic foot of capacity ranges from $128 to $319. The higher cost per cubic foot estimates in Table 16-3 than in Table 16-2 indicate that the PVTC rises much less than proportionately as capacity increases. Cost estimates for still larger facilities show even lower estimates of cost per unit of capacity (DOE 1994; Coates et al. 1994; Baird et al. 1995; Rogers and Associates Engineering 2000).

Combining Transport and Facility Costs

Transport costs of LLRW are low—less than $7 per cubic foot—for any plausibly chosen distance and the cost per mile. Facility costs of LLRW are high, usually more than $100 per cubic foot for large facilities and even more than $200 for smaller facilities. It should come as no surprise that, when detailed calculations are done, the cost-minimizing number of LLRW disposal facilities is *one*. Only if transport costs were several times higher than they are would a second disposal facility become optimal.

So far, we have ignored external costs. Could they alter the result that one facility is optimal? It's hard to see how. It would require that external costs be very high at the transport stage *and* very low at the facilities stage.

LLRW transportation involves the possibility of leakage accidents that affect the health of people around the accident site. The probability of a radioactive leak is the probability per mile of an accident times the probability of a leak given that there is an accident times the probability of a harmful radioactive leak given that there is a leak of some kind. All three of these probabilities are very low. Trucks are safe vehicles on highways—at least for their own drivers and cargoes. From June 1974 to March

1993, there were only 17 reported accidents involving the highway shipment of LLRW. In only 4 of the accidents was there a release of radioactive material, and these releases were never above background levels of radiation (Kearney and Porter 1999).

Facility external costs—possible damages imposed on the neighboring community—seem more worrisome. There have been no efforts to directly estimate the expected damage of an LLRW disposal facility, but there are three indirect methods of approaching such a measure. The first is to examine the extent of the decline of nearby property values after an LLRW disposal facility has been announced or opened. Although there have been no studies specifically of property values around LLRW disposal facilities, studies of other sites contaminated (or potentially contaminated) by radioactive materials have shown no effect of proximity on residential housing values (Gamble and Downing 1982; Payne et al. 1987). Second, because many waste disposal facilities make contributions to the neighborhood, the magnitude of these host fees can be interpreted as a measure of the community's willingness to accept the facility and hence a measure of the expected damage from it. Host fees, however, are few (so far), and the amounts vary greatly.[2] And third, the history of LLRW disposal facilities is fraught with environmental damage. Of the six such facilities that ever operated in the United States, all have experienced some leakage and overflow of radioactive rainwater, and two had to be closed because of ongoing environmental damages.[3]

One may be the optimal number of LLRW disposal facilities, but one is not the number toward which we are headed. The number being considered is greater than a dozen. It may be interesting to see how this huge deviation from least-cost disposal has come about.

How Not to Calculate the Efficient Number of LLRW Facilities

One effort to estimate the optimal number of LLRW disposal facilities concluded that there should be *five* (Coates et al. 1994). How do they find that number? They estimate the fixed cost of siting each facility and the variable cost of interring LLRW there—the costs reported in Tables 16-2 and 16-3. Next they guess at the maximum price for LLRW disposal that they think "political pressure" will permit (about $230 per cubic foot when converted to 1997 prices). They are then able to calculate the maximum number of sites that can be established within the constraint that facility revenues must cover total facility costs. That number is five.

They claim that their solution is "efficient." But what is so efficient about achieving zero profit (or loss)? Given the volume of LLRW to be disposed of, efficiency requires its disposal at the lowest present value of social cost. Indeed, they claim that their solution is "equitable." But what is so equitable about having five times the efficient number of people worried about their proximity to an LLRW disposal facility?

U.S. Policy on Siting LLRW Disposal Facilities

Of the six different LLRW disposal facilities that opened in the 1960s and 1970s, only two remain active today, in Barnwell, South Carolina, and Richland, Washington. If you look back at Table 16-1, you will notice that the distance-minimizing location of two facilities was in Kentucky and Oregon, not much different from the actual locations of those remaining two, in South Carolina and Washington—the average distance shipped to the actual sites (if each state shipped to its nearest site) was only 65 miles greater than the distance to the optimally located two sites.

Almost nobody would claim that the invisible hand of the market had managed to locate almost the right number of disposal sites in almost the right locations. But once the visible hands of federal and state governments began to move, any semblance of optimality began to fade.

It all began in 1979, when the governors of Washington and Nevada temporarily closed their LLRW facilities, citing safety concerns after several accidents (Peckinpaugh 1988; the Beatty, Nevada, site has since closed permanently). Probably more important, the governors expressed a fear that their states were unfairly becoming the only and the permanent LLRW disposal sites of the entire United States. South Carolina joined in this fear, cutting in half the amount of waste that its Barnwell facility would accept. (It is interesting to note how dramatically this sentiment has changed in the past 20 years—the Barnwell facility is now seeking new sources of volume to offset rising disposal costs; Hayden 1997.)

As a result of this impending LLRW disposal crisis, three working groups were created to search for a solution that would be acceptable to all the involved parties—the National Governors' Association, the U.S. Department of Energy (DOE), and the State Planning Council set up by President Jimmy Carter. All three task forces strongly urged that the responsibility for LLRW disposal be given to the states themselves.

This state-based approach was chosen for a variety of reasons. First, many states were bothered by the poor safety history at the established disposal facilities and felt that state control would be more effective. Second, many felt that communities would be more willing to establish new LLRW facilities if they were not forced upon them by the federal government (Kemp 1992). Third, political equity seemed more likely to emerge from interstate negotiations (Paton 1997).

To implement this state-based approach, Congress passed the Low-Level Radioactive Waste Policy Act in 1980. This act contained three major provisions governing the disposal of commercially generated LLRW: Each state was made responsible for the disposal of the commercial LLRW generated within its borders; states were encouraged to form regional com-

What About Recycling LLRW?

Some LLRW—that labeled Class A—is so low in radioactivity that constant exposure to some of it for a year would be much less harmful than a chest X ray. As a point of comparison, the average American is exposed each year to natural background radiation from sun and earth that is the equivalent of about 30 chest X rays. The question arises regularly, why not recycle such slightly radioactive scrap (Chen 1996)?

Almost accidentally, in the Superfund Recycling Equity Act of 1999, Congress has encouraged such recycling by making the recyclers exempt from liability in their use of radioactive material—another example of very indirect encouragement to recycling with much scope for cup–lip slippage. If they bury it or dispose of it, they would be liable under Superfund, so recycling, whether otherwise profitable or not, is a neat way to avoid liability (Greenwire 2000b).

Even without this encouragement, the case for recycling is strong. LLRW disposal currently costs up to $1,000 a ton. But the metal—ignoring its slight radioactivity—is worth some $200 a ton as scrap. So, for each ton that is recycled instead of buried, society gains $1,200. Moreover, if there are external costs to mining and smelting virgin materials, these are avoided by recycling, and that should also be added into the social benefit.

At what cost? The radioactive effects on workers who handle the scrap, on workers who turn it into new products, and on consumers who ultimately use the cars, refrigerators, and hammers made of that scrap. Even if these effects are low, they still may be unacceptable to Americans for two reasons. First, natural radiation is unavoidable, whereas radioactive hammers are avoidable. Second, chest X rays are taken voluntarily, whereas any radioactivity in a hammer is foisted on us without our permission (and usually without our knowledge). Said one observer, "I can't think of another time when the steel industry, the steelworkers union, and the environmentalists have all been against a proposal" (quoted in Fialka 1999, A16). Nevertheless, the NRC is still considering permitting such uses (Duff 2001a).The NRC is also considering "recycling" low-level-radioactive soil from nuclear plants by selling it for use as fill in yards and playgrounds (Schneider et al. 2000). For a sampling of public reaction to such radioactive recycling, see http://www.citizen.org/cmep/radmetal/radioactive_recyclingindex.htm.)

pacts to dispose of such waste; and states were granted the authority to exclude waste not generated by fellow compact member states after 1985 (Glicksman 1988).

By 1985, most states had made some progress toward the formation of interstate compacts, but they had made little progress toward developing new disposal facilities. In response to this inactivity, the Low-Level Radioactive Waste Policy Act Amendments were passed by Congress in 1985. These amendments were much more detailed than the 1980 law and contained six key provisions: (a) the seven pending compact agreements were adopted; (b) the date when states could begin excluding out-of-compact waste was delayed by seven years, to 1993; (c) a series of milestones were established that states were required to meet to be considered in compliance with the regulations; (d) surcharges on LLRW disposal fees were added for states not meeting the milestones; (e) states unable to provide for disposal by 1996 would be required to "take title" to—that is, accept all liability for—the LLRW generated within their borders; and (f) yearly volume limits were set for the three existing LLRW disposal sites.

Even with these seemingly firm milestones, only one new LLRW disposal facility has been developed (DOE 1996; Hayden 1997). Much of the states' inactivity toward siting new disposal facilities stems from the removal of the "take title" provision in the 1985 amendments when in 1992, the Supreme Court ruled that this provision was an unconstitutional intrusion on states' rights.

> ## The Midwest Compact Failure
>
> Look, for example, at the history of the Midwest Compact. The Illinois government had entered preliminary agreement to join the Midwest Compact, but when it became apparent that it would be chosen as the first host state for the compact's disposal facility, Illinois withdrew and formed the Central Midwest Compact with Kentucky so that the total volume of LLRW a facility must absorb would be significantly reduced.
>
> The Midwest Compact then determined that Michigan would be the first host state. Michigan set siting criteria that made every inch of the state unacceptable for LLRW disposal. The Midwest Compact then voted to revoke Michigan's membership.
>
> The remaining members of the Midwest Compact designated Ohio as the host state. But Ohio has expressed reluctance to operate the site and has made no real progress toward selecting a site (Coates and Munger 1996; Tompkins 1997).

Some of the lack of progress toward the development of new facilities can also be attributed to a free-rider problem. Compact membership is obviously a valuable commodity. However, although every state wants to be a compact member, no state wants to be the host state.

Currently, nine states either have an operating facility or are planning to construct facilities for the "permanent" disposal of LLRW. Four states have halted their plans because they currently have access to the Barnwell facility. In addition, a number of states are not planning to construct any disposal facilities but are examining other options, including permanent storage at the place where the LLRW was generated—called on-site storage. How many disposal sites we end up with is obviously very uncertain, but if U.S. policy provides any guidance in the end, there will be more than one—and perhaps more than a dozen.

The Cost of the Compact System

Aside from the cost of the seemingly interminable political and bureaucratic squabbling over membership and host states in the compacts, the anticipated compact system would be very expensive, if implemented. First of all, having many disposal facilities would not reduce transport costs nearly as much as one might at first think. The so-called regional compacts include some strange bedfellows. The most extreme is the regional compact that joins Texas, Vermont, and Maine. If the 13 (as of 1999) host facilities were all opened, and if every state shipped its LLRW to the nearest disposal facility, the average distance that LLRW would travel would be reduced to 141 miles—about a sixth of the average distance with one optimally located disposal site. But with every state shipping only to its own compact's facility, the average distance would be 257 miles—more than a fourth of the average distance with one optimally located facility. Indeed, the average transport distance with the 13 operating or proposed compact facilities is roughly the same as having 4 or 5 optimally located sites (with each state shipping to its nearest site).

The 13 compact sites currently envisioned would mean a transport saving of 641 miles (i.e., 898 minus 257), which for the average of the 1992–1996 LLRW (i.e., 900,000 cubic feet a year), at $5 per truck-mile (i.e., the high end of our earlier estimate), in 640-cubic-foot trucks, would mean an annual saving of just under $5 million. To earn this $5 million transport-cost saving requires building 13 disposal facilities instead of 1. The present value of the nonoperating costs of an LLRW disposal facility runs about $80 million (plus another $5.5 million for each million cubic feet of capacity) (Kearney and Porter 1999). Thus, 13 facilities are going to cost almost $1 *billion* more than one facility, even though the 1 facility is 13 times as large. Annualizing that $1 billion yields a cost of the compact system of about $70 million a year (assuming a 30-year facility life and a discount rate of 4%).

Spend $70 million per year to save $5 million per year? The present value cost of the compact system, relative to a single optimally located facility, is more than $1 billion, not even counting what has already been spent on developing the system. Is it really impossible to find a state, and a locality within it, that would agree to become *the* national LLRW disposal site for the next 30 years if it were accompanied by a host fee of something up to $65 million per year? All of these calculations assume that the volume of LLRW continues at its 1992–1996 average of about a half-million cubic feet a year. If prices were raised at *the* national disposal site, not only would less LLRW be generated and the site last longer, but also a greater part (or all) of the host fee would end up being paid by LLRW generators (and their customers).

High-Level Nuclear Waste

Existing U.S. nuclear power plants will generate 80,000 tons of spent nuclear fuel during their lifetimes. This fuel will remain dangerously radioactive for 10,000 years. Currently, most of this fuel is stored at the plants themselves. This on-site storage is, however, generally thought to be an interim measure. (When the spent-fuel pool of a utility becomes full, the utility has three choices: It can re-rack the spent-fuel rods, moving them closer together; it can encase the rods in casks of steel, concrete, and lead, and store them in reinforced concrete bunkers; or it can ship the spent fuel to a "monitored retrievable storage" facility.) The problem the United States faces, with ever increasing urgency, is making decisions about *how* and *where* to dispose of this high-level nuclear waste (HLNW).[4]

The first question we should ask is, how did the country get into a position of only now trying to decide what to do with its HLNW? No sensible family would buy a house and take up residency without buying toilets, wastebaskets, and garbage pails and arranging disposal service. But, as one observer of HLNW nondisposal has noted, from "the beginning of the nuclear age, governments neglected wastes as best they could" (Lenssen 1992, 47). To the extent that there was any thought given to the disposal issue in the 1950s, when the first nuclear power plants were constructed, the industry and government expectation was that the spent fuel would be reprocessed. By the 1970s, however, it had become clear that this would not be safe and cost-effective in the United States, and other alternative disposal means were sought.[5]

Several low-cost options were rejected on scientific or political grounds, and by 1980, there was a near-consensus in the world's scientific community that the best way to deal with HLNW was to isolate it in a deep underground repository. This consensus was not reached in haste. Scientific preference for a deep underground repository dates back to 1957 and has been reiterated several times, most recently in 2001 (NAS 1957, 2001; DOE 1980). The Nuclear Waste Policy Act of 1982 wrote this belief into law and instructed the Department of Energy to locate the HLNW repository. Time passed … and passed. As the cost of "characterization" (i.e., the assessment of the scientific suitability, security, and safety) grew at each potential site, and estimates of the development and construction cost of any selected site also grew, so also did the expected date of actually opening the repository fade into the twenty-first century (see Table 16-4). It seems that the more money is spent, the less clear become the answers to where and when our HLNW will be cared for.

So far in this section, there has been very little to do with economics. Indeed, in the few times when economists have attempted to offer some advice on the HLNW disposal process, their analyses have become so politicized that they hardly deserve being called economics

Uranium Mine Tailings

For every ton of uranium ore, barely five pounds of uranium is extracted. The rest is finely ground "tailings," which go from the mill as a slurry to tailings ponds and eventually, when dry, into a tailings pile, usually near the mill. There are more than 200 million tons of tailings in the United States, mostly in New Mexico, Wyoming, Colorado, and Utah. Left untended, the radioactivity is carried off by water and wind. The U.S. EPA has estimated that the lifetime excess lung cancer risk of a person living near a bare tailings pile is 0.02—greater than the lifetime risk of the average American of dying on the nation's highways—and that 500 people would die in the next century if no countermeasures were taken (Carter 1987; Diehl 1999). In some cases, the dried tailings have been used for the construction of houses; the EPA estimates the lifetime excess cancer probability of the residents of these houses to be 0.04. This kind of disposal is no longer considered acceptable.

The current practice is to cover the tailings with a thick layer of earth and monitor the groundwater, but past tailing piles will continue to present ongoing problems. In more serious cases, it is necessary to move the tailings to a site where fewer people live and where the pile can be better monitored (the times involved are mind-boggling—uranium-238 has a half-life of nearly 5 *billion* years). More than a half-billion cubic feet of such tailings have already been moved.

Future tailings are not likely to be a problem in the United States. Uranium prices declined in the 1980s, and most U.S. uranium is now imported. Even the U.S. uranium enrichment facility, which was privatized a few years ago, is nearing bankruptcy under the triple whammies of declining warhead production, closing nuclear electricity-generating facilities, and falling prices due to Russian imports that use improved technology (Wald 2000b).

(Porter 1983a). It is not that the scientists and politicians have conspired to keep helpful economics out of the picture. There are good reasons why economics can be only marginally helpful.

Aspect of cost (billions of 1997 dollars)	1982 estimate	1986 estimate	1989 estimate
Characterization (per site)	0.1	1.4	15.0
Development and construction	1.6	12.6	32.6
Expected year of repository opening	1998	2003	2010

Table 16-4

Estimates of the Cost of Locating and Constructing an HLNW Repository

Source: Keeney and von Winterfeldt 1994.

Is There Any Scope for Economics Here?

You may have noticed that economics is at its most useful when we are dealing with a small, immediate problem. As consequences get more distant in time—and 10,000 years *is very distant*—the economist's handling of different-time comparisons through discounting becomes ever more suspect. And as problems get bigger—and 80,000 tons of dangerously radioactive material *is* a very big problem—the neat and (I think) convincing partial-equilibrium analyses that are applied economists' stock in trade become ever less applicable. The proliferation of relevant variables, the need to extrapolate functions into unknown regions, plus problems of ignorance, uncertainty, and equity, overwhelm our simple attempts at economic analysis, and ever more complex attempts become quickly debatable if not incredible.

Think, for a minute, about discounting. Even if we use a very low (real) interest rate, the present value of $1 accruing (as benefit or cost) in T years gets very small for large values of T. At a discount rate of 2%, $1 accruing in 100 years has a present value of only $0.14, in 500 years only $0.00005, and in 1,000 years only $0.000000003. If we were to use simple discounting in a benefit–cost analysis of repository care, we would end up preferring to save $1 in repository cost today even if we *knew* that this saving would cause a $400 *billion* radioactive disaster in the year 3000. A recommendation based on such a calculation would not survive the laugh test.

Economists have been trying for a half-century to find some way around this problem with long-term discounting (Strotz 1956). The dilemma is this. If we use even lower discount rates—say 0.0002% a year—in an effort to give weight to distant events, too many public investments with very marginal and very distant benefits would be declared profitable. The present generation would starve providing the savings for such investments. If we use not-too-low interest rates for near-term investments and very very low interest rates for distant investments, then our benefit–cost analyses become dynamically inconsistent. Analyses could end up telling us today to do something tomorrow, when tomorrow it will tell us not to do it. The advice could change even if there were no change in the information on which it is based.[6]

Even ignoring the discounting problem, no sane economist would dare estimate the benefits or costs 10,000 years from now of any activity undertaken today. Indeed, ecological historians who regularly deal with millennia find 10,000 years to be "an almost unimaginably long time" (Flannery 1994). Demography changes— 10,000 years ago, nobody lived in what is now Ireland. Economic structure changes—10,000 years ago, there was no settled agriculture. Technology changes— 10,000 years ago, the Stone Age had not yet ended. Tastes change, too—10,000 years ago, humankind probably thought bliss was survival for another year. In short, economics is far from well-placed to settle the questions of when, where, and how to handle HLNW. But it can offer some advice.

Pricing Civilian Reactor Waste

Since 1983, the federal government has collected a tax of one tenth of one cent per kilowatt-hour on nuclear-generated electricity and placed the revenues in the Nuclear Waste Fund. The total revenues so far collected are greater than $15 billion, some of which has been spent and much more of which will soon be spent, because the courts have determined that the Department of Energy is now liable to "take title" to spent reactor fuel, even though it currently has no place to put it (Fialka 1998, 2000).

For ordinary businesses, introduction of a tax on one of its inputs would both reduce the use of that input and reduce output. Most electricity generation, however, is not an ordinary business, but rather a regulated regional monopoly. The regulators permit the monopoly to pass on any taxes into the electricity prices, so there is no direct discouragement of nuclear generation. Moreover, no nuclear electricity-generating plant would have been built anyway in the United States in the last two decades, for reasons that have little to do with waste or taxation.

Even if the tax has no effect in practice, it is interesting to think about what its purpose is. It is clearly not a Pigovian tax on the external costs of nuclear plants, because we haven't a clue how big those are. The fact that the revenues of the tax are earmarked to finding, building, and operating a permanent HLNW disposal facility suggest that the purpose of the tax is to prevent any hidden subsidy in the disposal process—those that use electricity produced by firms that have nuclear facilities will pay higher prices for their electricity, thereby paying the cost of the fuel's final resting place. But the special tax will ultimately bring in something like $30 billion, whereas the HLNW disposal will cost at least *twice* that much (Manning 1998).

The special nuclear-electricity tax will not entirely remove the hidden subsidy that will ultimately apply to HLNW disposal. It is not even clear that it is a fair way to try to allocate those disposal costs. Neither the electricity companies nor their customers knew about the tax or the disposal costs when the decision to build the nuclear plants occurred. One could make a good case that *all* electricity should be taxed to pay for the HLNW repository. Or one could make a good case that, because almost everyone uses electricity in the United States, we should not bother at all with a special electricity tax.

Where to Site the Repository

Most of the search for a site for an HLNW repository in the United States has followed what might be called an overlay process (Porter 1983a). This process goes like this: You put a map of the United States on the wall and begin pulling see-through plastic sheets down over it. Each sheet blacks out the areas that do not meet a specific criterion. The first sheet blacks out areas with too dense a population, the second sheet blacks out areas with too much groundwater risk, and so on. After all the overlaid plastic sheets have been pulled down, the potential repository sites are those that can still be seen; that is, those that have not been blacked out on any of the criteria. The obvious inefficiency inherent in this process is that a site may be chosen that barely passes all the criteria and another may be rejected that passes all but one of the criteria by a wide margin and fails that one criterion by a small margin.

Why has this process been followed? Because the U.S. government wants to maintain the fiction that the final site will be *perfectly* safe. So we fudge a little on each criterion, declaring something to be perfectly safe when it is in fact slightly less than that. Consider, for example, population density. A repository is perfectly safe only if *no one* lives near it. We fudge and declare the criterion to be that it is perfectly safe if fewer than x people per square mile live near it. Then, following the overlay method, we can in the end declare—wrongly of course—that the areas unblackened on any criterion are perfectly safe.

In short, the entire process has not been a search for the best site, but for a site that passes all of several arbitrary criteria. Cost has never been a worry—nearly $7 billion has already been spent examining the proposed Yucca Mountain site in Nevada, and the DOE estimates that the final total cost (when it is completed, no earlier than 2010) may reach $75 billion (Fialka 1998; Grunwald 1998; Grove and Manning 2001). And the extra cost per expected life saved has never intruded into the comparison of sites, because all sites are supposedly equally—and totally—safe. This assumption— that the long-term risk of deep underground storage is zero—even received the imprimatur of the U.S. Supreme Court when it ruled unanimously that the NRC did not act capriciously in declaring it so (in *Baltimore Gas And Electric Co. v. Natural Resources Defense Council*).

An Attempt to Site the Repository by Host Fee

The Nuclear Waste Policy Amendments Act of 1987 authorized the President's Nuclear Waste Negotiator (NWN) to negotiate with state governors and Native American tribes to host a high-level nuclear waste permanent repository within their jurisdictions. In 1991, the NWN invited governors and leaders of Native American tribes to apply for grants to assess the conditions under which they might consider hosting such a facility. Within a year, the NWN had received 20 applications for grants.

Ultimately, the Mescaleros Apaches expressed a willingness to accept the facility. Asked why, their tribal leader said, "the Navajos make rugs, the Pueblos make pottery, and the Mescaleros make money" (Hanson 1995). However, the final NWN decision was never made. In 1994, Congress ended the funding for the NWN office (Bacot et al. 1998).

With the Nuclear Waste Policy Act of 1982, Congress asked the DOE to find three possible sites for the first U.S. repository. As the search for sites proceeded, actual and estimated costs rose rapidly, and Congress stopped the search by designating Yucca Mountain in Nevada as the sole candidate in the Nuclear Waste Policy Act Amendments of 1987—sometimes called the "Screw Nevada" bill.[7] Studies that have been done during the past few decades do not give us a clue about whether Yucca Mountain is the best possible site, in the sense that the extra expected lives saved by choosing it are worth the extra costs. (For an exhaustive list of the titles of the Yucca Mountain studies, see http://www.ymp.gov/doclist.htm; for the most recent study, see DOE 2001.) But we can ask, is Yucca Mountain a good choice? Unlike all other possible U.S. sites, Nevada has the "advantage" that the state is already hopelessly contaminated with radioactive legacies. Many years of underground nuclear weapons testing have left radioactive cavities with no barriers to prevent water from flowing toward public water supplies (Tsipis 1994).

The only real question about Nevada's suitability is transport costs. Nevada is not the least-transport-distance site for spent U.S. nuclear fuel. The shipment of all U.S. HLNW to Yucca Mountain would mean transporting it through all but seven states—a "mobile Chernobyl," Nevadans warn us.[8] But transport dangers seem slight, from past experience. During the past 20 years, thousands of shipments totaling more than 1,000 tons of HLNW have been transported within the United States, with only eight accidents and without a single radiological release (Holt 1998; Fialka 2001a).

Many Nevadans are beginning to think about negotiating the size of the host fee (Kanigher and Manning 1998; Fialka 2001a; Sebelius 2001).[9] The problem with host fees here is that, studies suggest, willingness to accept *nuclear waste* is not much affected by the size of a host fee (Kunreuther et al. 1990). Interviews with local neighbors (in Beatty, Nevada) reveal that the only host fee they want out of Yucca Mountain is some jobs there (Kuz 1998).

When to Move HLNW to the Repository

In principle, we should be making simultaneous optimizing decisions about what to do with HLNW, where to put it, and when to put it there. In fact, the United States has been making these decisions sequentially. It officially decided on the technique of deep underground storage in the 1980s, and it has—almost—decided on Yucca Mountain as the place for that storage. The decision about when to move the HLNW has never been explicitly considered. Presumably, it is being assumed that it is optimal to move it there as soon as possible—some would say it was optimal to move it there decades ago. Let us examine that implicit decision.

Suppose the Yucca Mountain repository were up and receiving right now. Then the relevant question would be, should spent nuclear fuel be moved there now or later or not at all? (The word "now" is shorthand for "soon"—in reality, spent fuel and other HLNW should not go to a permanent repository for at least a decade, during which its radioactivity would decay greatly; see Carter 1987.) If the answer is "not at all," then it begs the question, why did we go to all the trouble and expense of characterizing, planning, and constructing the Yucca Mountain repository? Presumably, the very search for a single permanent repository indicates that we have, I hope wisely, decided that the risks of Yucca Mountain (including any risks attendant to moving the HLNW to Yucca Mountain) are smaller than the risks of having many small (mostly temporary and on-site) repositories. The question, now or later, is also easily answered. If it makes sense to build the repository, what sense could there be to not use it as soon as possible? If that is not obvious, plow through the algebra of the box titled "The Algebra of Moving HLNW to Permanent Storage Now or Later."

All the above makes it sound like disposing of radioactive waste, even long-lived waste, is a simple proposition—economically at least. But it is not just the politics that makes it hard. There are economic problems as well, beyond the obvious one that the relevant cost comparisons are filled with data uncertainties and long horizons. If technical advances are just around the corner, it pays to wait on any decisions; unfortunately, we have no way of knowing when or whether such technical advances will appear, and continued on-site storage is risky and costly. If reprocessing technologies are about to surface as socially profitable, then burying the HLNW irretrievably in a deep underground repository would be a mistake; again, we have no way of knowing when or whether that will happen (Lake 2001). To some people, the answer to such uncertainties is monitored retrievable storage, which is safer than on-site storage and more flexible than deep burial.

Final Thoughts

Whatever we do with radioactive waste, whether high-level or low-level, there are potential risks. Part of the whole radioactive waste disposal problem is our impossible assumption that, if we look hard enough and long enough, we can find a perfectly safe repository. It means that we are doomed to look forever. But not doing anything may be even more risky—LLRW stored in barely guarded sheds and HLNW shallowly buried in canisters on-site at soon-to-be-closed nuclear plants.

Or is it more risky? We may be overreacting to the risk. According to the Department of Energy's own estimates, the cost per expected life saved in radioactive waste management activities is about a half-billion dollars (with that figure converted to 1997 dollars from earlier estimates; DOE 1979, 1982). Indeed, one physicist has estimated that *randomly* burying all the spent nuclear fuel of the existing electricity generating plants with simple precautions would cause a total of less than 100 expected U.S. deaths (Cohen 1986).

The repository we are hoping to build will in the end cost about $60 billion—and spending this on the repository rather than on other needs (including health) could cost more than 10,000 expected U.S. lives (see Appendix B of Chapter 1). Even if that repository prevented every single one of the 100 expected deaths from on-site burial, it would do so at a cost of more than a half-billion dollars per expected life saved—two orders of magnitude above what we are usu-

The Algebra of Moving HLNW to Permanent Storage Now or Later

A few parameters: The expected social cost of transporting HLNW to its permanent repository is X. The social cost of continued on-site storage is Y per year. The cost of characterizing, planning, and constructing the repository is F (assume, for simplicity, that it happens all in one year). And the social operating cost of the repository is V per year. We start with the assumption that building the permanent repository was a smart move, which means that the present value of the social cost of permanent repository storage ($PV_{reposit}$) is smaller than the present value of the social cost of permanent on-site storage ($PV_{on\text{-}site}$). In symbols, the two PVs can be written out as (recall that i is the real discount rate; $[1 + i]/i$, or $[1 + 1/i]$, is the present value of a flow of one per year starting now and going on forever):

$$PV_{reposit} = X + F + \frac{(1+i)V}{i} \qquad (1)$$

and

$$PV_{on\text{-}site} = \frac{(1+i)Y}{i} \qquad (2)$$

By assumption, $PV_{reposit}$ is smaller than $PV_{on\text{-}site}$, which means that

$$X + F + \frac{(1+i)V}{i} < \frac{(1+i)Y}{i} \qquad (3)$$

Now the question is when to move the HLNW to the permanent repository. The present value of waiting one year to move it (PV_{wait}) consists of three terms: first, a benefit of postponing for one year the social cost of the transport of the HLNW to the repository ($+ iX/[1 + i]$) (the PV of the cost of transporting the HLNW this year is X; the PV of the cost of transporting it next year is $X/[1 + i]$; the difference is $iX/[1 + i]$); second, a benefit of avoiding the social operating cost of the repository this year ($+V$); and third, a cost of having to store the HLNW on-site this year ($-Y$). Waiting is socially profitable only if

$$\frac{iX}{(1+i)} + V - Y > 0 \qquad (4)$$

But, by multiplying both sides of inequality (3) by $i/(1 + i)$, we see that the inequality in (4) must be negative, which means that waiting is unprofitable. If it makes sense to build the repository, it makes sense to use it as soon as possible.

ally willing to spend to save an expected life. But if on-site is the preferred disposal method, why are we spending so much time and money "characterizing" an inferior alternative?

In any case, all this may be moot. Recently, Senator James Jeffords switched parties, and the new Senate majority leader, Tom Daschle—before dashing off to the Bali Hai Golf Club for a dinner that was expected to raise a half-million dollars in campaign money—announced that Yucca Mountain "is dead" (Moller 2001). What is live is not yet clear. Nienberg's Law ("progress is made on alternate Fridays") may be overly optimistic.

Even more recently, the 11 September 2001 terrorist attacks have cast new doubts on the security of accessible nuclear fuel. Most of the spent nuclear fuel being stored at plants is housed in fairly ordinary buildings with no special antiterrorist safeguards—the Union of Concerned Scientists calls them "Kmarts without neon" (Grunwald and Behr 2001, A01).

Notes

1. LLRW is typically contaminated protective shoe covers and clothing, rags, mops, filters, reactor water treatment residues, equipment, tools, luminous dials, medical tubes, swabs, syringes, and laboratory animal carcasses and tissues. The most intensely radioactive wastes are typically found in water treatment residues, discarded parts from nuclear reactors, and small gauges containing radioactive material. LLRW is classified as either Class A, B, or C waste, depending on the level of radioactivity of its contents. Approximately 90% of the LLRW is Class A, which has a very short half-life and will decay to safe exposure levels in less than 100 years (and often in a matter of only hours or days). Class B carries stricter waste requirements and can pose health hazards for up to 300 years. Class C makes up only 2% of the volume of LLRW but the largest amount of radioactivity; it will not decay to safe levels for 500 years (Vari et al. 1994; MILLRWC 1985; OTA 1989c).

2. In Pennsylvania, the search for a volunteer host community for an LLRW disposal facility involved a host fee that included a promise to spend $4 million a year hiring local workers, to give preference to local materials suppliers (up to $12 million a year), to offer tax preferences and home sale price guarantees to homes within 2 miles of the facility, to provide college grants and student scholarships, and to make a direct payment of $300,000–$600,000 a year (Burk 1996). A proposed Massachusetts host fee for its neighbors includes payment of property taxes, annual payments of $150,000 for the first three years, additional annual payments of $240,000 in each year of operation, and $3 million in an up-front "impact" payment (DOE 1994).

3. The two that were closed for environmental problems were at West Valley, New York, and Maxey Flats, Kentucky (Gershey et al. 1990; Gerrard 1994; DOE 1996). In the 1970s, at all six facilities, the method of disposal was shallow land burial—drums of LLRW were placed in shallow, excavated trenches and, once full, covered with a layer of dirt. In addition to leakage and overflow, workers at one facility were discovered taking tools (that had been accepted as LLRW waste) from the facility and selling them. Cleaning up West Valley has already cost more than $1 billion and is expected to cost up to $8 billion more before completion (Greenwire 1999b). Here, as with Superfund sites, an ounce of precaution would indeed have been worth at least a pound of cure.

4. There is also much HLNW from the U.S. nuclear weapons program of the past half-century. It is being handled separately from civilian reactor waste—or *mis-handled*. The Department of Energy now estimates that nearly a third of the plutonium used in weapons making has been released into the soil because it was buried in "flimsy containers" (Greenwire 2000a; Wald 2000a). For much of the rest, the government has developed a 2,150-foot-deep, salt-

cavern, underground Waste Isolation Pilot Plant near Carlsbad, New Mexico, to which waste from weapons production sites is being transported (Jensen 1999). The Department of Energy estimates that the complete cleanup of U.S. weapons sites may end up costing as much as $250 billion (DOE 2000). Less, or no, cleanup is of course less costly—requiring only the "iron fences" to keep people away—but the legacies of such inaction have been ominously labeled "national sacrifice zones" (NAS 2000).

5. Whenever disposal is very costly or virgin material mining generates large external costs, recycling makes more sense. But four factors more than offset these forces for nuclear waste. First, reprocessing turned out to be expensive, something not noticed when it was applied only to bomb making, where cost hardly mattered. Second, reprocessing safety was questioned, a worry that was fueled by the environmental disaster at the West Valley, New York, reprocessing facility. Three, uranium was becoming increasingly cheap and imported, lessening U.S. concerns for uranium mining externalities (Carter 1987). Fourth, although reprocessing reduces the volume of HLNW, it does not reduce its heat content or remove its fission products, so reprocessing does not remove the problem of locating an HLNW waste repository (NAS 2001). In any case, *private* reprocessing of spent fuel was banned in the United States in 1979, although some scientists continue to promote it and it is being done elsewhere (Fialka 2001b). France routinely reprocesses its own spent nuclear fuel and finds that process no more expensive than using fuel once and then storing it indefinitely (Lake 2001). British Nuclear Fuels Limited in northern England imports spent fuel and reprocesses it, charging $1 million per metric ton for the service.

6. Consider the following example. An investment costs $1 in the year it is initiated, and it then yields $2 of benefit in each of the two years following. Suppose we always use 5% as the discount rate for the coming year and 1% as the discount rate thereafter. The present value of undertaking the investment now is $[-1 + 2(1.05)^{-1} + 2(1.05)^{-1}(1.01)^{-1}] = [-1 + 0.95 + 0.94] = +0.89$. But the present value now of waiting to do it next year is $[-(1.05)^{-1} + 2(1.05)^{-1}(1.01)^{-1} + 2(1.05)^{-1}(1.01)^{-2}] = \{-0.95 + 0.94 + 0.93\} = +0.92$. The PV now is higher if you postpone the investment for a year, but next year it again will be higher if you postpone it a year. This profitable investment will never get done.

7. Not only Nevadans complained about the abrupt choice. One (non-Nevadan) author concluded that "in retrospect, the choice of Yucca Mountain for the nation's first high-level nuclear waste repository appears to have been neither systematic nor the result of an organized program to identify the best site …. the choice … had little to do with the ability of that site to isolate radioactive materials from the biosphere" (Jacob 1990, xv, 172). It is amusing, and perhaps ominous, that Yucca Mountain sits between Little Skull Mountain to the east and the Funeral Mountains to the west. By the by, the "Screw Nevada" bill did *not* pass the Senate 98–2, as was widely rumored; the vote was 61–28.

8. The city of Las Vegas commissioned Louis Berger & Associates to determine the risk on the 90 miles that HLNW would travel near Las Vegas (Sahagun 2000). Berger estimated that the probability of a "nuclear accident" would be 0.000000003 a mile and that there would be 44,250 shipments of HLNW before the repository would be filled. The probability of a Las Vegas accident is therefore one chance in 90 (i.e., $1 - (1 - 0.000000003)^{(90)(44,250)}$, which equals 0.011, or 1 chance in 90).

9. The Russians are also thinking about it. There are parts of Russia already heavily radioactively polluted from decades of sloppy handling of HLNW—e.g., Muslyumovo is said to be more radioactive than 20 Chernobyls (Glasser 2001). Taking in HLNW from abroad at something like $1 million a ton would earn Russia the money to dispose of its own HLNW and begin to clean up its own nuclear pollution. Currently, Russian HLNW either is stockpiled awaiting budget for disposal or is *sold* to the United States. Why do we *pay* for HLNW when everywhere else in the world countries *get paid* for accepting it? Because the U.S. government fears that the radioactive materials of dismantled Soviet nuclear weapons might otherwise fall into the wrong hands (GAO 2000).

PART 4

Final Thoughts

What Have We Learned?

Life can only be understood backwards, but it must be lived forwards.

—Sören Kierkegaard

When writing about any economic policy, it is fashionable to claim that it is all wrong. U.S. waste policy is clearly *not* all wrong. Waste of all kinds can create hidden, long-lived external costs if it is not handled carefully, and recent U.S. policy has forced the more careful handling of waste. The changes in policy during the past half-century were overdue and in the right direction. The questions about waste policy are twofold. First, have our policies created properly functioning waste markets? Second, have these policies achieved their results at the *lowest possible* social cost? We have seen, time and again in the preceding chapters, that the answers to these two questions are too often *no* and *no*.

Why have U.S. waste policies failed both to achieve proper handling of waste and to do so at least cost? I think our review has taught us that two basic failings have undermined our waste policy efforts. First, we have regularly maintained that the goal of waste policy is to achieve an unattainable zero-risk environment. Second, we have not worked hard enough to get waste pricing right.

A Zero-Risk Environment Is Unattainable

Some Americans enjoy taking real risks, such as parachuting, hang gliding, or auto racing. Some enjoy taking monetary risks, such as playing the lottery, buying gold-mine stocks, or choosing boom-or-bust occupations. For most of us, however, risk is something to be avoided. We look both ways before crossing streets, we look for jobs promising promotion and permanence, and we carry insurance of various kinds to reduce our losses when disasters strike. Because we spend so much of our life trying to avoid serious risk, is it not logical that we should also insist that risks to our health from our environment should also be minimized—especially because these risks are largely outside of our individual control?

No, it is not logical. The key word in the preceding paragraph is "minimized." To minimize risk means to get it as near zero as possible, at any cost. We don't do this in our private lives. It would mean looking 10 times before crossing streets, or perhaps never crossing them at all. It would mean taking out a special lightning-insurance policy every time we planned a picnic.

We do not, however, approach environmental risk with this attitude. There, zero risk is the publicly pronounced target. Landfills must never damage groundwater. Incinerators should emit no air pollution. Everything should be recycled. Superfund sites should be cleaned to the point where a maximally exposed individual, living downwind for an entire lifetime, should suffer no health risk at all. High-level nuclear waste should be disposed of in such a way that it generates zero probability of environmental damage for at least ten millennia. The problem with setting an unachievable target is not just that it is impossible and hypocritical, but that it gives us no guidance as to where to stop in our subsequent attempts to reduce environmental risk.

Saying we cannot remove all risk does not mean that we should remove no risk. Once we admit that zero risk is unattainable, we begin to face explicitly the problem of how far we should go to reduce environmental risk. We need to admit that not all environmentally caused diseases, illnesses, and injuries could or should be prevented. More than that, we need to admit that not all environmentally caused deaths could or should be prevented.

In short, we need to do more explicit benefit–cost analyses and cost-effectiveness analyses on environmental risk reduction. Doing benefit–cost analysis on waste problems means simply putting the benefits of various possible actions in dollar terms and comparing them with the dollar costs of those actions. Because the benefits usually involve lives saved by the action, measuring them in dollar terms means deciding how much we, collectively, are willing to pay to save an unidentified expected life. Once we admit that we are not willing to spend infinite amounts on saving any expected life, the rest is easy. Most of the data on costs, affected populations, and probabilities of life-threatening exposures are already collected, calculated, or estimated by the U.S. EPA in the course of formulating policies. All that is needed is that these data must be brought together in a fashion that permits explicit comparison of expected life-saving benefits and costs.

Explicit benefit–cost analysis is not for everyone, and we should not expect that the EPA's calculations will go unchallenged, especially by those who live near what they perceive as great hazard. But benefit–cost analysis will at least change the focus of their arguments. Currently, the benefit–cost data remain buried in mountains of little-read documents, and the affected neighbors and their political representatives cry out for the most complete remediation of the hazard. With an explicit benefit–cost analysis, those who disagree with its conclusions must challenge the data or the calculations themselves. This benefit–cost analysis will hardly ever solve policy differences, but once its results are on the table, a much more useful debate will ensue.

Still, we must remember that benefit–cost analysis only adds up the benefits and costs to get a single total. There are winners and losers. But a slight extension of benefit–cost analysis can specify the size of the net losses to the losers, and this gives a good idea of the host fee, or other compensation, that will make the losers into winners as well. For any project or policy that passes its benefit–cost test, the gains of the winners are large enough that the losses of the losers can be fully compensated. (Another way of saying this: If the maximum host fee that the winners can offer is *insufficient* to compensate the losers, the project does *not* pass its test.)

Making ourselves willing to put a value on an expected life saved does not quite solve all our benefit–cost problems. There are other benefits for which it is difficult to place a value. How much are we willing to pay for better health or a cleaner environment? For some waste policies and projects, these intangible or hard-to-value benefits may be of overwhelming importance. In some of these cases, there may be a consensus that benefit–cost analysis is quite inadequate and hence inappropriate for helping solve an environmental problem. Does this mean that we must forget economics and seek a purely political solution? No. There is still cost-effectiveness analysis. Cost-effectiveness analysis starts with a socially accepted target without trying to defend that target through benefit–cost analysis. It then looks only at the costs of various possible ways of achieving that target and asks which way incurs the lowest social cost.

EPA Landfill Regulations Revisited

Just to remind you of the gap between proper benefit–cost analysis and what the U.S. EPA actually does, recall the EPA evaluation of their landfill regulations—discussed in Chapter 4 (U.S. EPA 1991a, 1991b). The EPA estimated that the new "final rule" for landfill regulation would save a total of 2.4 expected lives during the next three centuries and cost $450 million a year for the next 30 years (all converted to 1997 dollars). Not only are these numbers not brought together and compared, they are presented in such a way as to make it hard to compare them. We made the effort of doing the proper present-valuing, and we found that the EPA's own data told us the new rules were going to cost $32 *billion* per expected life saved—four orders of magnitude higher in cost than we permit for most U.S. policy.

The EPA also devoted much of its analysis of their new regulations to trying to determine if any particular groups would be seriously affected by the new regulations, calculating the increased cost per household and municipality for various kinds and locations of landfills. There is nothing wrong with this kind of analysis—it may be appropriate for triggering some side-payment subsidies to those whose welfare is seriously hurt by the new rules. But such analysis ought not be used as an excuse for weakening the rules. If some activity is causing, say, deaths and stopping that activity would save lives at a cost of less than a few million dollars per expected life saved, then the activity should be stopped, regardless of who is undertaking the activity. Only when the marginal cost of saving expected lives is equalized over all those undertaking an activity will cost-effectiveness be achieved. The EPA should recognize this for waste regulations. When we make arson illegal, we do not stop to analyze whether arsonists are poor or how important watching buildings burn is to their well-being.

LLRW Disposal Sites Revisited

Just to remind you of the gap between cost-effective and cost-ineffective policies, recall the process by which the United States is planning its future disposal sites for low-level radioactive waste—discussed in Chapter 16. Each of a self-selected collection of state "compacts" is (supposedly) building its own LLRW disposal facility. More than a dozen such compacts have been formed. Although the cost of such a facility is not known with certainty, because so few have ever been constructed (and none recently), we were able to explore the questions of how many facilities should be constructed and where they should be located to minimize the social cost of disposing of U.S. LLRW. Our answer was quite clear: One facility minimizes cost.

Building and operating 13 facilities instead of 1 will cost the United States more than 1 billion dollars (in present value and in 1997 prices). Transport costs would be somewhat reduced with 13 facilities, but the costs of permitting and constructing so many LLRW disposal facilities would be greatly increased. To my knowledge, no concern for cost-effectiveness has been shown by anyone during the past 20 years while the compacts have been mandated and formed (and deformed).

In our cost-effectiveness analysis, we did not ask how much LLRW the United States should be generating each year. We would need to know that to ask about an optimal overall LLRW policy. But that is a much harder question to answer. So we took the quantity of LLRW as a given and asked only the easier question, how should we dispose of it? Cost-effectiveness avoids the tough issues and tries to minimize costs with respect to the more readily controlled variables. Where life and fairness and justice are involved, as in many waste decisions, it may be impossible to decide what overall policy is optimal, but there are always parts of the problem that can be examined for cost-effectiveness. *Whatever* we do, we should do it at least cost.

External Costs Must Be Reduced

The costs that are relevant to both benefit–cost analysis and cost-effectiveness analysis are social costs, which means private costs plus external costs. A private cost is usually easy to measure. Because someone is paying it, and usually with money, it is not hard to put a dollar figure on it. An external cost is being suffered by someone else, someone who has not given permission for this imposition and who is not being compensated for it. It is often hard to know who is suffering an external cost, and it is always hard to put a dollar value on the suffering. Just because it is hard to enumerate the victims and to measure the damage does not mean that we should ignore external costs. After all, most of the waste problems that we have been talking about involve external costs—from noise and litter, from air and groundwater pollution, from toxic and radioactive endangerment.

We do not want to ignore external costs. Neither do we want to eradicate them. As we have seen at various places throughout this book, usually there is an optimal reduction of an external cost, and the optimal reduction means some reduction, maybe much reduction, but not total elimination. Groping toward an optimal reduction of external costs involves thinking marginally.

Thinking marginally often means incremental policymaking—starting with a small change in policy and later adding to it if the first step seems wise. It is often the only politically feasible way to make changes. But it is also a sensible way to grope toward an optimum in an uncertain world. After a small change is made, data emerge that let us test whether the benefits from that change exceed the costs. If so, a further change should be considered.

For many waste policies, however, moving incrementally is not an option. If either the environmental and health damages are so great as to require immediate nonmarginal action or the policy will require large fixed investments that cannot be later modified to meet higher standards, then a nonmarginal policy action must be taken. But this does not remove the need to *think* marginally. As stricter regulations are applied, marginal costs usually rise and marginal benefits usually fall. Only by thinking about various policy actions in terms of the marginal benefits and marginal costs of ever stricter rules can we make sensible estimates of when to stop.

In waste policymaking, we have leaned heavily toward regulation as the means of reducing external costs. Some of these regulations have proven very expensive. In many cases, where the EPA has attempted some sort of benefit–cost analysis of its policies, its cost estimates are very high. And other researchers have found even higher cost estimates. One economist estimated that 30% of the drop in U.S. industrial productivity in the 1970s and 1980s can be traced to regulations imposed by the EPA and by the Occupational Safety and Health Administration (Gray 1987). Although not all economists agree that the cost has been anywhere near this high, most agree that it has been too high (Goodstein and Hodges 1997).

Right Prices Help

The theory of the optimal choice between regulating by price and regulating by quantitative restriction is fairly complex. If we knew the exact marginal cost of safeguarding something and the exact marginal benefit of that safeguarding, then we could use either price or quantity tools to regulate the degree of safeguarding. In practice, of course, we never know these things exactly, and so either regulation by price or regulation by quantity may be the better choice. All we know for sure is that, if an activity generates external costs, it suggests regulation by one means or the other (or both).

For most consumer products and their packages, neither the producer nor the consumer pays anything for their disposal. And neither the producer nor the consumer is quantitatively constrained on how much product and package waste they can generate. As a result, they generate too much waste, reuse products too little, and recycle too little used material. None of these failings create social costs so large as to urge cumbersome quantitative regulation of product and packaging size and composition, but correct pricing can help a lot.

Recall the trash pricing scheme developed in previous chapters. Producers must be given incentives both to reduce the size of their products and packaging and to make them more easily recyclable. One way to do this is to levy an advance disposal fee (ADF), equal to the marginal net recycling cost of the product and its package. (By net recycling cost is meant the social cost of collecting the material and processing it in a MRF, minus the revenue earned from its sale. If this net recycling cost is less than the marginal social cost of collection and disposal—by the cheaper of landfill or incinerator—then recycling is socially profitable.)

Households must also be given incentives to generate less trash and recycle more of what can be recycled. To do this, we should introduce a trash collection charge, equal to the excess of the landfill or incinerator cost over the net recycling cost. Neither charge nor subsidy is appropriate for recycled materials. Because the ADF is usually passed on in the price of the product, this means that households end up paying the marginal social cost of the landfill or incinerator for what they dispose of and that they pay the marginal net recycling cost for what they recycle. The trash collection charge induces them to reduce, reuse, and recycle. The ADF induces manufacturers to make products and packages that are more cheaply recycled.

No pricing scheme is perfect—because no realistic scheme can overcome the temptation to illegally dispose—and this scheme's trash collection charge even adds to this temptation. There are other problems. The ADF must be uniform across the country, for we cannot know where products will be sold, and this means that the ADF will be too high where recycling is cheap and too low where recycling is expensive (or not possible). The trash fee must be uniform across products in any municipality because charging for trash by its specific contents would be very costly, and this means that the trash fee will be too high for hard-to-recycle items and too low for easy-to-recycle items when they are put instead into the trash. Despite these problems, the pricing scheme takes a big step toward making producers and consumers aware of the disposal costs of their products and packages.

Second-Best Policies Are Second Best, at Best

When we fail to price trash, we are forced to adopt second-best policies to induce a socially more desirable amount of recycling. We have come across many such second-best policies in the preceding chapters, and they almost always have unintended and undesirable side effects.

A few examples as a reminder of this. Landfill taxes levied on businesses do indeed have the ability to internalize any external costs of landfill disposal, but landfill taxes on municipalities do *not*, because the municipality usually absorbs this tax in its general fund budget. Individual households receive no *marginal* impact of the tax and hence are unlikely to alter their trash behavior. Recycling subsidies, of whatever kind, do indeed encourage municipalities to recycle more and also encourage firms to use more recycled materials. But if recycling subsidies are not passed on to households, they do not affect household willingness to recycle. Moreover, recycling subsidies reduce the cost of products and packaging and may increase their volume. Producer take-back responsibility makes producers clearly aware of the disposal or recycling costs of their products and packages, but it does so at the cost of greatly reducing the efficiency of the waste collection system. There is no need to go on.

In general, when an activity generates external costs, the activity should be taxed; how much depends on the size of the external damages. This tax gives the external cost generator an incentive to reduce these external costs. Taxing something else that is "sort-of" related to the activity with the external costs is, at best, second best. And worse, subsidizing something that is sort-of substitutable for the activity with the external costs is not only second best; it also adds budgetary problems to the allocation inefficiencies.

Banning something "bad" or mandating something "good" is almost always second-best policy. Usually, the optimal amount of any bad activity is not zero, and usually, it is not optimal to insist that everyone always participate in a good activity. Before turning to such heavy-handed policy instruments, we should be looking for ways to alter prices so as to induce people to undertake socially more desirable activities.

Final Final Thoughts

We have come to the end of a long journey, which I hope has not been a "waste" of time. During the past third of a century, U.S. policy toward waste has moved a long way, from near laissez-faire to quite stringent regulation to safeguard our health and the environment. The direction of this policy movement has clearly been correct. Studies by the EPA and others have shown that expected lives would have been lost that fairly cheap and simple regulations could avert.

The real question, as we have been asking over and over, is whether we have gone too far or not far enough. Many of the benefit–cost analyses, by the EPA itself and by other researchers, suggest that U.S. environmental legislation and its EPA implementation regularly err on the side of excessive regulation, where the cost of the marginal safeguarding exceeds the marginal benefits in expected life-saving. The word "suggest" appears in the previous sentence because environmental law permits and the EPA undertakes so few explicit benefit–cost analyses, and it is often difficult for other researchers to cheaply reshape the EPA data into benefit–cost format. The Superfund study on which we relied heavily in Chapter 15 estimated expected life-saving benefits for just 150 of the some 1300 Superfund sites (the National Priorities List), and it took the two principal researchers and a "large number" of research assistants three years to complete that study (Hamilton and Viscusi 1999a; 15 research assistants are thanked by name in their preface).

We have also seen that environmental policy has not gone far enough in other respects. Neither businesses nor households are charged for much of the waste they generate, even though the disposal or recycling of that waste is costly, both in private cost to the municipality that collects it and in external cost to health and the environment. There is still much scope in the United States for advance disposal fees and trash collection charges.

Americans seem to go extremes with waste policy. Perhaps this is a corollary of the sound-bite society—in a few seconds, one can argue zero or one, but it is impossible to come to grips with a fraction (Scheuer 1999). Health must be perfectly protected. The proper marginal charge for trash is zero. Everything or nothing should be recycled. In a sentence, what I have been trying to point out is that the optimum where waste is concerned is not usually all or nothing at all.

As a final, final thought, let me repeat the title of this book: *The Economics of Waste*. The book has relentlessly attempted to encourage the reader to think economically about waste problems. My point has been that economics often has something to contribute to proper waste policies. Something, but not everything. Other values matter. Even if you could prove that executing one murderer deters more than one potential future murderer, you would not change my opposition to capital punishment; and even if you could prove that every concealed weapon prevents several crimes, you would not alter my desire to live in a society that discourages guns. Other values matter. Do apply them. But don't forget the economics.

References

Quoting one is plagiarism; quoting many is research.

—Anonymous

Numbers in brackets [ ] after entries indicate the chapters in which they are cited.

Aagaard, T. 1995. Secondary Markets for the Reuse of Durable Goods: An Ecologically Friendly Disposal Alternative. Term paper, Economics 471. University of Michigan, Ann Arbor, December. [7]

Abbott, E. 1995. When the Plain Meaning of a Statute Is Not So Plain. *Villanova Environmental Law Journal* 6 (2). http://vls.law.vill.edu/students/orgs/elj/vol06/abbott.htm (accessed 1 November 2001). [5]

Abt Associates. 1985. *National Small Quantity Hazardous Waste Generator Survey*. Prepared for the EPA. EPA-530-SW-85-004. Washington, DC: U.S. EPA. [14]

Ackerman, F. 1997. *Why Do We Recycle? Markets, Values, and Policy*. Washington, DC: Island Press. [2, 4, 6, 8, 9, 11, 12]

Ackerman, F., D. D. Cavander, J. Stutz, and B. Zuckerman. 1995. *Preliminary Analysis: The Costs and Benefits of Bottle Bills*. Prepared for U.S. EPA. Boston: Tellus Institute. [6, 11]

Ackerman, F., and K. Gallagher. 2001. Mixed Signals: Market Incentives, Recycling, and the Price Spike of 1995. Working Paper 01-02. Global Development and Environment Institute, Tufts University. January. http://www.ase.tufts.edu/gdae/publications/WorkingPap.htm (accessed 2 November 2001). [12]

Ackerman, F., and T. Schatzki. 1989. *Bottle Bills and Municipal Recycling: A Preliminary Cost Analysis*. Prepared for U.S. EPA. Boston: Tellus Institute. [11]

———. 1991. Bottle Bills and Curbside Recycling Collection. *Resource Recycling*, June. [11]

Acohido, B. 1999. Calif. Socks Tire Hauler with $100,000 Penalty. *Waste News*, 22 November. [8]

———. 2000a. Seattle Curbside Program Surpasses Expectations. *Waste News*, 1 May. [11]

———. 2000b. Seattle Trims Trash Costs, Revamps Pickup Program. *Waste News*, 14 February. [11]

———. 2000c. Slow Burn. *Waste News*. 24 January. [14]

Acton, J. P. 1989. *Understanding Superfund: A Progress Report*. Santa Monica, CA: Rand Corp. [15]

Acton, J. P., and L. S. Dixon. 1992. *Superfund and Transaction Costs: Experiences of Insurers and Very Large Industrial Firms*. Santa Monica, CA: Rand Institute for Social Justice. [15]

Adams, L. 1999. Turning Chicken "Litter" into Power, with a PEEP. *Washington Post*, 29 July. [2]

Adler, K. J., R. C. Anderson, Z. Cook, R. C. Dower, A. R. Ferguson, and M. J. Vickers. 1982. *The Benefits of Regulating Hazardous Waste Disposal: Land Values as an Estimator.* Research Report. Public Interest Economics Center. Prepared for U.S. EPA. Washington, DC: U.S. EPA. [14]

Aizenman, N. C. 1997. The Case for More Regulation. *Washington Monthly*, October. [14]

Alberini, A., and D. H. Austin. 1999. Strict Liability as a Deterrent in Toxic Waste Management: Empirical Evidence from Accident and Spill Data. *Journal of Environmental Economics and Management*, July. [14, 15]

Alexander, J. H. 1993. *In Defense of Garbage.* Westport, CT: Praeger. [1, 2, 5]

Alexander, M. 1994. Developing Markets for Old Newspapers. *Resource Recycling*, July. [12]

Alter, H. 1989. The Origins of Municipal Solid Waste: The Relations between Residues from Packaging Materials and Food. *Waste Management and Research*, February. [2]

Andersen, M. S. 1998. Assessing the Effectiveness of Denmark's Waste Tax. *Environment*, May. [3, 4]

Anderson, D. R. 1989. Financing Liabilities under Superfund. *Risk Analysis*, September. [15]

Anderson, P., G. Dreckmann, and J. Reindl. 1995. Debunking the Two-Fleet Myth. *Waste Age*, October. [9, 11]

Apotheker, S. 1991. Mixed-Waste Processing Head-to-Head with Curbside Recycling Collection. *Resource Recycling*, September. [11]

———. 1993. Curbside Recycling Collection Trends in the 40 Largest U.S. Cities. *Resource Recycling*, December. [9]

Armstrong, D. 1999. The Nation's Dirty, Big Secret. *Boston Globe*, 14, 15, 16, and 17 November. [1]

Arner, R., H. L. Hickman, and C. Leavitt. 1999. Dump Now, Pay Later? Landfill Financial-Assurance Mechanisms Are Burying the True Costs. *MSW Management*, December. [4]

Arnold, F. S. 1995. *Economic Analysis of Environmental Policy and Regulation.* New York: Wiley. [2]

Asarch, J., and P. Cort. 1995. The Costs and Benefits of Landfill Gas Recovery—Does It Add Up? Term paper, Economics 471. University of Michigan, Ann Arbor, December. [4]

Augustyniak, C. M. 1998. Asbestos Case Study. In *Economic Analysis at EPA*, edited by R. D. Morgenstern. Washington, DC: Resources for the Future. [4]

Bacot, H., T. Bowen, M. R. Fitzgerald, T. H. Rasmussen, D. Morell, and D. Mazmanian. 1998. *Historical Overview and Policy Issues Regarding Siting of Waste Facilities.* Center for Urban Research and Policy. Columbia University. Background Paper 1 for the New York City Integrated Waste Management Project. New York. [7, 16]

Bader, C. D. 1997. Today's Watchword in Recycling: Efficiency. *MSW Management*, March–April. [11]

———. 1999a. Automating a MRF. *MSW Management*, January–February. [11]

———. 1999b. Trucking Garbage to Ohio. *MSW Management*, July–August. [7]

Bailey, J. 1992. Economics of Trash Shift as Cities Learn Dumps Aren't So Full. *Wall Street Journal*, 2 June. [4]

———. 1993. Fading Garbage Crisis Leaves Incinerators Competing for Trash. *Wall Street Journal*, 11 August. [5]

———. 1995a. Curbside Recycling Comforts the Soul, But Benefits Are Scant. *Wall Street Journal*, 19 January. [8]

———. 1995b. How Can Government Save Money? Consider the L.A. Motor Pool. *Wall Street Journal*, 6 July. [3]

———. 1997. New Yorkers Worry Progress Could Send City Down the Drain. *Wall Street Journal*, 1 October. [3]

———. 1998. Waste Management Agrees to Acquire Eastern Environmental for $1.27 Billion. *Wall Street Journal*, 18 August. [4]

———. 2001. A Successful Salvage Business Has to Know Its Junk. *Wall Street Journal*, 9 January. [12]

Baird, R. D., N. Chau, and C. D. Breeds. 1995. Cost Estimation and Economic Valuations for Conceptual LLRW Disposal Facility Designs. Paper presented at 1995 Low-Level Radioactive Waste Conference, November, Salt Lake City. [16]

Barbanel, J. 1992. Elaborate Sting Operation Brings Arrests in Illegal Dumping of Toxic Wastes by Businesses. *New York Times*, 13 May. [14]

Barbier, E. R. 1999. The Effects of the Uruguay Round Tariff Reduction on the Forest Products Trade: A Partial Equilibrium Analysis. *The World Economy*, January. [8]

Bates, K. L. 1990. This Ain't No Dump. *Ann Arbor News*, 22 July. [7]

Bauer, S., and M. L. Miranda. 1996. *The Urban Performance of Unit Pricing: An Analysis of Variable Rates for Residential Garbage Collection in Urban Areas*. EPA Cooperative Agreement Report #CR822-927-010. April. http://www.epa.gov/epaoswer/non-hw/payt/pdf/upa-perf1.pdf (accessed 2 November 2001). [3]

Baumann, J., P. Orum, and R. Puchalsky. 1999. *Accidents Waiting to Happen: Hazardous Chemicals in the U.S. Fifteen Years After Bhopal*. Washington, DC: U.S. Public Interest Research Group (PIRG) Education Fund. [14]

Baumol, W. J., and W. E. Oates. 1988. *The Theory of Environmental Policy*, 2d ed. Cambridge, U.K.: Cambridge University Press. [2]

Beck, R. W., Inc. 2000. *1999 National Post-Consumer PET Container Recycling Activity*. Report to the National Association for PET Container Resources. http://www.mindfully.org/Plastic/Post-Consumer-PET-1999.htm (accessed 2 November 2001). [12]

Becker, G. 1968. Crime and Punishment: An Economic Approach. *Journal of Political Economy*, January–February. [14]

Been, V. 1994. Unpopular Neighbors: Are Dumps and Landfills Sited Equitably? *Resources*, Spring. [14]

Bender, R., W. Briggs, and D. De Witt. 1994. *Toward Statewide Unit Pricing in Massachusetts: Influencing the Policy Cycle*. Master's thesis, Kennedy School of Government, Harvard University, January. [10]

Benenson, A. 1990. *Control of Communicable Diseases in Man*, 15th ed. Washington, DC: American Public Health Association. [3]

Berenyi, E. B., and M. J. Rogoff. 1999. Is the Waste-to-Energy Industry Dead? *MSW Management*, December. [5]

Berry, E. 1964. *You Have to Go Out! The Story of the United States Coast Guard*. New York: David McKay Company. [1]

Berry, R. O., and S. M. Jablonski. 1995. Low-Level Radioactive Waste Management at Texas A&M University. *Radwaste Magazine*, September. [16]

Bershatsky, M. 1996. The Ann Arbor Landfill. Term paper, Economics 471, University of Michigan, Ann Arbor, December. [4]

Big Apple Garbage Sentinel. 1998. 2001 and Beyond: A Proposed Plan for Replacing the Fresh Kills Landfill. 3 December. http://garbagesentinel.org/documents/mayorsplan.html#intro (accessed 2 November 2001). [4]

Bigness, J. 1995. As Auto Companies Put More Plastic in Their Cars, Recyclers Can Recycle Less. *Wall Street Journal*, 10 July. [12]

Bingham, T. H., and R. V. Chandran. 1990. *The Old Newspaper Problem: Benefit–Cost Analysis of a Marketable Permit Policy*. Prepared for Economic Analysis Branch. Office of Policy, Planning, and Evaluation, U.S. EPA. Washington. DC: U.S. EPA. [10]

Bingham, T. H., C. C. Chapman, and L. G. Todd. 1989. *National Beverage Container Deposit Legislation: A Review of the Issues*. Research Triangle Institute, report prepared for U.S. EPA. Washington, DC: U.S. EPA. [6]

Bitar, D., and R. C. Porter. 1991. What Does a Landfill Cost? A Case Study of Ann Arbor, Michigan. Discussion Paper 335. Ann Arbor: School of Public Policy, University of Michigan. [4]

Black, H. 1995. Rethinking Recycling. *Environmental Health Perspectives*, November. [8]

Bleich, D. H., M. C. Findlay III, and G. M. Phillips. 1991. An Evaluation of the Impact of a Well-Designed Landfill on Surrounding Property Values. *Appraisal Journal*, April. [4]

Blumberg, L., and R. Gottlieb. 1989. *War on Waste*. Washington, DC: Island Press. [13]

Blumenthal, M. 1998. Scrap Tire Market Development: The Impact of State Programs. *Resource Recycling*, March. [8]

Boerner, C., and K. Chilton. 1994. False Economy: The Folly of Demand-Side Recycling. *Environment*, January. [2]

Bohm, P. 1981. *Deposit-Refund Systems: Theory and Applications to Environmental, Conservation, and Consumer Policy*. Washington, DC: Resources for the Future. [12]

———. 1994. *Environment and Taxation: The Cases of the Netherlands, Sweden and the United States*. Paris: Organisation for Economic Co-operation and Development. [6]

Bonomo, L., and A. E. Higginson, eds. 1988. *International Overview on Solid Waste Management*. International Solid Wastes and Public Cleansing Association. San Diego: Academic Press. [5, 13]

Booth, W. 2000. Recycling: How Long Will a Can-Do Feeling Last? As Trend Levels Off, Some Experts Believe Options Are Dwindling. *Washington Post*, 5 January. [8, 9]

Boston Globe. 1999. Expensive Trash [editorial]. 20 July. [5]

Boyd, J., W. Harrington, and M. K. Macauley. 1996. The Real Effects of Environmental Liability on Industrial Real Estate Development. *Journal of Real Estate Finance and Economics*, January. [15]

Breyer, S. G. 1993. *Breaking the Vicious Circle: Toward Effective Risk Regulation*. Cambridge, MA: Harvard University Press. [15]

Brisson, I. 1993. Packaging Waste and the Environment: Economics and Policy. *Resources, Conservation and Recycling*, April. [2]

Brooke, J. 1992. The Secret of a Livable City? It's Simplicity Itself. *New York Times*, 18 May. [3]

Brown, B. 1999a. City Narrows Hauling Field. *Waste News*, 11 October. [7]

———. 1999b. City Seeks Injunction: Elizabeth NJ Attempts to Halt NYC Trash. *Waste News*, 22 November. [7]

Brugger, J. 2000. Bottle Bill Faces Many Obstacles. *Louisville Courier-Journal*, 7 February. [6]

Buchholz, R. 1997. Transfrontier Movements of Secondary Materials under the Rules of the Basel Convention. Paper presented at World Conference on Copper Recycling, International Copper Study Group, 3–4 March, Brussels. [7]

Bui, T-N. 1995. How Browning-Ferris Got a Landfill in Salem Township. Term paper, Economics 471, University of Michigan, Ann Arbor, December. [7]

Bullard, R. D. 1983. Solid Waste Sites and the Black Houston Community. *Sociological Inquiry*, Spring. [14]

Burk, J. 1996. Pennsylvania's Community Partnering Plan. *Radwaste Magazine*, September. [16]

Burmaster, D. E., and R. H. Harris. 1993. The Magnitude of Compounding Conservatism in Superfund Risk Assessments. *Risk Analysis*, April. [15]

Burns, J. F. 1998. Profit Rules Indian Shore Where Ships Go to Die. *New York Times*, 9 August. [7]

Cairncross, F. 1993. Waste and the Environment: A Lasting Reminder. *Economist*, 29 May. [4, 5, 12]

Calabresi, G. 1961. Some Thoughts on Risk Distribution and the Law of Torts. *Yale Law Journal*, Issue 1. [14]

Calcott, P., and M. Walls. 2000. Can Downstream Waste Disposal Policies Encourage Upstream "Design for Environment"? *American Economic Review*, May. [10]

Caldwell, D., and B. Rynk. 2000. *Economic Aspects of Compost and Composting*. Compost Education and Resources for Western Agriculture, Animal Sciences Department, Washington State University, Pullman. February. www2.aste.usu.edu/compost/qanda/economic.htm (accessed 22 April 2002). [13]

CalRecovery. 1989. *Composting Technologies, Costs, Programs, and Markets*. Washington, DC: Office of Technology Assessment. [13]

Carlton, J. 1999. "Greens" Target WTO's Plan for Lumber. *Wall Street Journal*, 24 November. [8]

Carter, L. J. 1987. *Nuclear Imperatives and Public Trust: Dealing with Radioactive Waste*. Washington, DC: Resources for the Future. [16]

CEQ (Council on Environmental Quality). 1997. *Environmental Quality: 25th Anniversary Report*. Washington, DC: U.S. Government Printing Office. [8]

Chang, N-B. 1992. Econometric Analysis of the Construction and Operating Costs of Materials Recovery Facilities in the U.S.A. *Journal of Solid Waste Technology and Management*, March. [9]

Chen, S. Y. 1996. Recycling Is Better—Even for Slightly Radioactive Metal. *Logos*, Winter. [16]

Choe, C., and I. Fraser. 1999. An Economic Analysis of Household Waste Management. *Journal of Environmental Economics and Management*, September. [10]

Church, T. W., and R. T. Nakamura. 1993. *Cleaning Up the Mess: Implementation Strategies in Superfund*. Washington, DC: Brookings Institution. [15]

Clark, D. R. 1997. Bats, Cyanide, and Gold Mining. *Bats Conservation International*, Fall. [2]

Clay, D. R. 1991. Synthesis of Conference on "Minimizing Environmental Damage: Strategies for Managing Hazardous Waste." *Risk Analysis*, March. [15]

Clines, F. X. 1999. Perdue Offers a Plan to Fight Odor and Pollution. *New York Times*, 19 October. [2]

Clymer, A. 1989. Polls Show Contrasts in How Public and EPA View Environment. *New York Times*, 22 May. [1]

Coase, R. 1960. The Problem of Social Cost. *Journal of Law and Economics*, October. [1]

Coates, D., V. Heid, and M. Munger. 1994. Not Equitable, Not Efficient: U.S. Policy on Low-Level Radioactive Waste Disposal. *Journal of Policy Analysis and Management*, Summer. [16]

Coates, D., and M. C. Munger. 1996. Interstate Compacts Can't Solve Collective Bads Problems: The Case of LLRW. Paper Given at the Annual Meeting of the Public Choice Society, April, Austin. [16]

Cohen, A. S. 1994. Integrated Waste Management in Three Japanese Cities. *Solid Waste Technologies*, July–August. [9]

Cohen, B. L. 1986. Risk Analysis of Buried Waste from Electricity Generation. *American Journal of Physics*, January. [16]

Cohen, N., M. Herz, and J. Ruston. 1988. *Coming Full Circle*. New York: Environmental Defense Fund. [8, 11, 12]

Collins, J. N., and B. T. Downes. 1977. The Effects of Size on the Provision of Public Services: The Case of Solid Waste Collection in Smaller Cities. *Urban Affairs Quarterly*, March. [3]

Connolly, D. 1991. Comments on "Cleanup of Old Waste: Some Thoughts on Rethinking the Fundamentals of Superfund." *Risk Analysis*, March. [15]

Copeland, B. R. 1991. International Trade in Waste Products in the Presence of Illegal Disposal. *Journal of Environmental Economics and Management*, March. [7]

Copeland, C., and J. Zinn. 1998. *Animal Waste Management and the Environment: Background for Current Issues*. Report 98-451, Congressional Research Service. 12 May. [2]

Cowdery, C. 1995. The Bourbon Barrel, Mere Container or Active Partner? *The Malt County Advocate*, Spring. [7]

Criner, G. K., S. L. Jacobs, and S. R. Peavey. 1991. *An Economic and Waste Management Analysis of Maine's Bottle Deposit Legislation*. Miscellaneous Report 358, University of Maine, Agricultural Experiment Station, Orono. April. [6]

Cropper, M. L., S. K. Aydede, and P. R. Portney. 1994. Preferences for Life Saving Programs: How the Public Discounts Time and Age. *Journal of Risk and Uncertainty*, May. [4]

Curlee, T. R., S. M. Schexnayder, D. P. Vogt, A. K. Wolfe, M. P. Kelsay, and D. L. Feldman. 1994. *Waste-to-Energy in the United States: A Social and Economic Assessment*. Westport, CT: Quorum Books. [5, 9]

CWC (Clean Washington Center). 1993. *The Economics of Recycling and Recycled Materials.* Washington, DC: CWC. [9]

Czerwinski, S. J. 1996. *Superfund: More Emphasis Needed on Risk Reduction.* General Accounting Office, GAO/T-RCED-96-168. 8 May. http://www.gao.gov/highrisk/hr97014.txt (accessed 2 November 2001). [15]

Daley, B. 2000. State Plans to Lift Ban on Landfills. *Boston Globe,* 20 December. [10]

Darby, M. R. 1973. Paper Recycling and the Stock of Trees. *Journal of Political Economy,* September–October. [8]

Darmstadter, J. 2000. The Role of Renewable Resources in U.S. Electricity Generation: Experience and Prospects. Climate Change Issues Brief No. 24. Washington, DC: Resources for the Future. [4]

Dasgupta, P., A. K. Sen, and S. Marglin. 1972. *Guidelines for Project Evaluation.* New York: United Nations Industrial Development Organization. [9]

Deisch, M. 1989. Mt. Pleasant "Goes Green": A User Fee Success Story. *Resource Recovery,* December. [10]

Denison, R. A. 1997. *Something to Hide: The Sorry State of Plastics Recycling.* New York: Environmental Defense Fund. [12]

Denison, R. A., and J. F. Ruston. 1996. *Anti-Recycling Myths: Commentary on Recycling Is Garbage.* Environmental Defense Fund. 18 July. http://www.environmentaldefense.org/pubs/Reports/armythfin.html (accessed 2 November 2001). [8]

Denison, R. A., and J. Ruston. 1990. *Recycling and Incineration: Evaluating the Choices.* New York: Island Press. [5, 9]

Deutsch, C. H. 1997. Philips Concession on Bulbs May Have Political Motive. *New York Times,* 19 June. [14]

———. 1998. Second Time Around, and Around. *New York Times,* 14 July. [2]

DeWitt, J. F., and R. W. Butler. 1994. U.S. Supreme Court Rules on Municipal Waste Combustion Ash. *ICLE Focus on Michigan Practice,* July. [5]

Deyle, R. E., and B. F. Schade. 1991. Residential Recycling in Mid-America: The Cost Effectiveness of Curbside Programs in Oklahoma. *Resources, Conservation, and Recycling,* August. [9]

Deyle, R. E., and S. I. Bretschneider. 1995. Spillovers of State Policy Innovations: New York's Hazardous Waste Regulatory Initiatives. *Journal of Policy Analysis and Management,* Winter. [7]

Diaz, L. F., G. M. Savage, L. E. Eggerth, and C. G. Golueke. 1993. *Composting and Recycling Municipal Solid Waste.* Boca Raton, FL: Lewis Publishers. [1]

Dickinson, W. 1995. Landfill Mining Comes of Age. *Solid Waste Technologies,* March–April. [8]

Diehl, P. 1999. *Uranium Mining and Milling Wastes: An Introduction.* http://www.antenna.nl/wise/uranium/uwai.html (accessed 2 November 2001). [16]

Dijkgraf, E., and H. R. J. Vollebergh. 1997. Incineration or Dumping? A Model for Social Cost Comparison of Waste Disposal. Paper prepared for 8th Annual Conference of the European Association of Environmental and Resource Economists, June, Tilborg, Netherlands. [5]

Dillingham, A. E. 1985. The Influence of Risk-Variable Definition on Value-of-Life Estimates. *Economic Inquiry,* April. [1]

Dinan, T. M. 1993. Economic Efficiency Effects of Alternative Policies for Reducing Waste Disposal. *Journal of Environmental Economics and Management,* December. [6, 8, 10]

Dixon, L. S. 1995. The Transaction Costs Generated by Superfund's Liability Approach. In *Analyzing Superfund: Economics, Science, and Law,* edited by R. L. Revesz and R. B. Stewart. Washington, DC: Resources for the Future. [15]

Dixon, L. S., D. S. Drezner, and J. K. Hammitt. 1993. *Private-Sector Cleanup Expenditures and Transaction Costs at 18 Superfund Sites.* Santa Monica, CA: Rand Institute for Social Justice. [15]

Dobbs, I. M. 1991. Litter and Waste Management: Disposal Taxes Versus User Charges. *Canadian Journal of Economics,* February. [6]

DOE (U.S. Department of Energy). 1979. *Final Environmental Impact Statement for Savannah River Plant Long-Term Management of Defense High-Level Radioactive Wastes.* DOE/EIS-0023. Washington, DC: DOE. [16]

———. 1980. *Final Environmental Impact Statement Management of Commercially Generated Radioactive Waste.* DOE/EIS-0046F. Office of Nuclear Waste Management. Washington, DC: DOE. [16]

———. 1982. *Final Environmental Impact Statement for Long-Term Management of Liquid High-Level Radioactive Waste Stored at the Western New York Nuclear Service Center, West Valley.* DOE/EIS-0081. Washington, DC: DOE. [16]

———. 1993. *Economics of a Small-Volume Low-Level Radioactive Waste Disposal Facility.* DOE/LLW-170. Washington, DC: DOE. [16]

———. 1994. *Estimating Costs of Low-Level Radioactive Waste Disposal Alternatives for the Commonwealth of Massachusetts.* EGG-LLW-1140. Washington, DC: DOE. [16]

———. 1995. *Integrated Data Base Report 1994: U.S. Spent Nuclear Fuel and Radioactive Waste Inventories, Projections, and Characteristics.* DOE//RW-0006-Rev. 11. Washington, DC: DOE. [16]

———. 1996. *1995 Annual Report on Low-Level Radioactive Waste Management Progress.* DOE/EM-0292. Washington, DC: DOE. [16]

———. 1998a (and earlier years). *1998 State-by-State Assessment of Low-Level Radioactive Wastes Received by Commercial Disposal Sites.* DOE/LLW-252. May. http://mims.inel.gov/web/owa/mimsrpts. (accessed 2 November 2001) [16]

———. 1998b. *Renewable Energy Annual 1998: With Data for 1997.* Office of Coal, Nuclear, Electric and Alternate Fuels. Energy Information Administration. DOE/EIA-0603(98)/1. December. http://www.eia.doe.gov/cneaf/solar.renewables/rea_data/html/front-1.html. (accessed 2 November 2001) [5]

———. 2000. *Status Report on Paths to Closure.* Office of Environmental Management. DOE/EM-0526. http://www.em.doe.gov/closure/fy2000/statusrpt.html (accessed 2 November 2001). [16]

———. 2001. *Yucca Mountain Science and Engineering Report.* DOE/RW-0539. Washington, DC: DOE. [16]

Doty, C. B., and C. C. Travis. 1989. The Superfund Remedial Action Decision Process: A Review of Fifty Records of Decision. *Journal of the Air Pollution Control Association*, November. [15]

Dower, R. C. 1990. Hazardous Wastes. In *Public Policies for Environmental Protection*, 1st ed., edited by P. R. Portney. Washington, DC: Resources for the Future. [14]

Drucker, J. 1997. Spiegelberg Landfill: Economic Analysis. Term paper, Economics 471, University of Michigan, Ann Arbor, December. [15]

Dubin, J. A., and P. Navarro. 1988. How Markets for Impure Public Goods Organize: The Case of Household Refuse Collection. *Journal of Law, Economics, and Organization*, Fall. [3]

Duff, S. 2001a. Agency to Review Radioactive Scrap. *Waste News*, 23 July. [16]

———. 2001b. Congress May Consider Credits for LFG Energy. *Waste News*, 26 February. [4]

———. 2001c. Interstate Waste Keeps Crossing the Lines. *Waste News*, 6 August. [7]

———. 2001d. Leasing Aids Recycling, Cuts Costs, Report Says. *Waste News*, 5 February. [2]

———. 2001e. Recyclers Oppose Tax Breaks. *Waste News*, 3 September. [4]

Duggal, V. G., C. Saltzman, and M. L. Williams. 1991. Recycling: An Economic Analysis. *Eastern Economic Journal*, July–September. [11]

Duston, T. E. 1993. *Recycling Solid Waste: The First Choice for Private and Public Sector Management.* Westport, CT: Quorum Books. [14]

Duxbury, D. 1992. Household Hazardous Waste Management. Paper presented at the second Annual Environmental Technology Exposition and Conference, 7–9 February, Washington, DC. [13, 14]

Economist. 1992. Let Them Eat Pollution. 8 February. [7]

———. 1993a. Green Behind the Ears. 3 July. [2]

———. 1993b. Take It Back. 1 May. [2]

Edgren, J. A., and K. W. Moreland. 1990. An Econometric Analysis of Paper and Waste Paper Markets. *Resources and Energy*, March. [12]

Edwards, R., and D. Pearce. 1978. The Effect of Prices on Recycling of Waste Materials. *Resources Policy*, December. [12]

Egan, K. 1998. Can Biweekly Recycling Collection Work? *Waste Age*, May. [11]

Engel, S., and B. Engleson. 1998. Time and Cost To Collect All Plastic Bottles. *Resource Recycling*, May. [9]

ENS (Environment News Service). 2000. Backyard Burning Could Be Major Source of Dioxins. 4 January. http://ens.lycos.com/ens/jan2000/2000l%2D01%2D04%2D06test.html (accessed 11 December 2001). [5]

Ezeala-Harrison, F., and N. B. Ridler. 1994. The New Brunswick Beverage Container Act: A Partial Evaluation. Working Paper. Department of Economics, University of New Brunswick, Saint John. October. [6]

Ezzet, L. 1997. Solid Waste Services in the 100 Largest U.S. Cities. *MSW Management*, December. [3, 4, 11]

FAO (Food and Agriculture Organization of the United Nations). 1996. *Forest Products Yearbook, 1996*. Rome. [8]

Feizollahi, F., D. Shropshire, and D. Burton. 1995. *Waste Management Facilities Cost Information for Transportation of Radioactive and Hazardous Waste Materials*. Prepared by Idaho National Engineering Laboratory for DOE, INEL-95/0300, Idaho Falls ID. June. [16]

Fialka, J. J. 1998. States, Utilities, Legislators Battle over New Waste Site. *Wall Street Journal*, 23 November. [16]

———. 1999. Plan to Recycle Nuclear Materials Runs into Flak from Unions, Industry, and Environmentalists. *Wall Street Journal*, 22 December. [16]

———. 2000. Nuclear-Powered Plants Can Sue U.S. over Disposal of Spent Fuel. *Wall Street Journal*, 1 September. [16]

———. 2001a. Nevada Hones Effort to Block Yucca Mountain Nuclear-Waste Dump. *Wall Street Journal*, 5 July. [16]

———. 2001b. Scientists Tout New Method for Reprocessing Nuclear Wastes. *Wall Street Journal*, 21 August. [16]

Fierman, J. 1991. The Big Muddle in Green Marketing. *Fortune*, 3 June. [2]

Fischer, C., and R. C. Porter. 1993. Different Environmental Services for Different Income Groups in LDC Cities: Second-Best Efficiency Arguments. Center for Research on Economic and Social Theory. University of Michigan. [3]

Fishbein, B. K. 1994. *Germany, Garbage, and the Green Dot: Challenging the Throwaway Society*. Prepared for U.S. EPA, EPA600-R-94-179. New York: Inform Inc. [2]

———. 1997. Industry Program to Collect and Recycle Nickel-Cadmium (Ni-Cd) Batteries. In *Extended Product Responsibility: A New Principle for Product-Oriented Pollution Prevention*, edited by G. A. Davis and C. A. Wilt. Chattanooga: Center for Clean Products and Clean Technologies, University of Tennessee. [14]

Fishbein, B. K., and C. Gelb. 1992. *Making Less Garbage: A Planning Guide for Communities*. New York: Inform Inc. [2, 13]

Fishbein, B. K., L. S. McGarry, and P. S. Dillon. 2000. *Leasing: A Step Toward Producer Responsibility*. New York: INFORM Inc. [12]

Fisher, A., L. G. Chestnut, and D. M. Violette. 1989. The Value of Reducing Risks of Death: A Note on New Evidence. *Journal of Policy Analysis and Management*, Winter. [1]

Flannery, T. F. 1994. *The Future Eaters*. New York: George Braziller. [16]

Flynn, C. P. 1999. It's Time to Drop the Bottle Law and Strengthen Recycling Programs. *Boston Globe*, 14 August. [6]

Francis, S. 1991. Greetings from the Central Landfill. *Yankee*, July. [9]

Franklin Associates. 1988. *The Role of Beverage Containers in Recycling and Solid Waste Management: A Perspective for the 1990s.* Prepared for Anheuser-Busch Companies. Prairie Village, KS: Franklin Associates. [11]

———. 1990. *Resources and Environmental Profile Analysis of Foam Polystyrene and Bleached Paperboard Containers.* Prepared for the Council for Solid Waste Solutions. Prairie Village, KS: Franklin Associates. June. [2]

———. 1994. *The Role of Recycling in Integrated Solid Waste Management to the Year 2000.* Prepared for Keep America Beautiful Inc. Prairie Village, KS: Franklin Associates. [5, 9]

———. 1997. *Characterization of Municipal Solid Waste in the United States: 1996 Update.* Prepared for U.S. EPA. EPA530-R-97-015. June. http://www.epa.gov/epaoswer/non-hw/muncpl/msw96.htm (accessed 2 November 2001). [2]

———. 2000. *Characterization of Municipal Solid Waste in the United States: 1999 Update.* Prepared for the EPA. EPA530-F-00-024. April. http://www.epa.gov/epaoswer/non-hw/muncpl/msw99.htm (accessed 2 November 2001). [1, 3, 8]

Franklin, P. F. 1991. Bottle Bill: Litter Control Measure in a New Role? *Solid Waste & Power*, February. [6, 11]

Fullerton, D., and T. C. Kinnaman. 1995. Garbage, Recycling, and Illicit Burning or Dumping. *Journal of Environmental Economics and Management*, July. [6, 8, 10]

———. 1996. Household Responses to Pricing Garbage by the Bag. *American Economic Review*, September. [3, 10]

Fullerton, D., and A. Wolverton. 2000. Two Generalizations of a Deposit-Refund System. *American Economic Review*, May. [6, 10]

Fullerton, D., and W. Wu. 1998. Policies for Green Design. *Journal of Environmental Economics and Management*, September. [10]

Gallagher, K. G. 1994. Exploring the Economic Advantages of Regional Landfills. *Solid Waste Technologies*, September–October. [4]

Gamble, H. B., and R. H. Downing. 1982. Effects of Nuclear Power Plants on Residential Property Values. *Journal of Regional Science*, November. [16]

Gandy, M. 1994. *Recycling and the Politics of Urban Waste.* New York: St. Martin's Press. [8]

GAO (U.S. General Accounting Office). 1983. *Siting of Hazardous Waste Landfills and Their Correlation with Racial and Economic Status of Surrounding Communities.* [14]

———. 1987. *Superfund: Extent of Nation's Potential Hazardous Waste Problem Still Unknown.* GAO/RCED-88-44. Washington, DC: GAO. [15]

———. 1990. *Solid Waste: Trade-Offs Involved in Beverage Container Deposit Legislation.* GAO/RCED-91-25. Washington, DC: GAO. [11]

———. 1995. *Hazardous and Nonhazardous Wastes: Demographics of People Living Near Waste Facilities.* GAO/RCED-95-84. Washington, DC: GAO. [14]

———. 2000. *Nuclear Nonproliferation: Implications of the U.S. Purchase of Russian Highly Enriched Uranium.* GAO-01-148. Washington, DC: GAO. [16]

Garber, S., and J. K. Hammitt. 1998. Risk Premiums for Environmental Liability: Does Superfund Increase the Cost of Capital? *Journal of Environmental Economics and Management*, November. [15]

Garrod, G., and K. Willis. 1998. Estimating Lost Amenity Due to Landfill Waste Disposal. *Resources, Conservation, and Recycling*, March. [4]

Gayer, T. 2000. Neighborhood Demographics and the Distribution of Hazardous Waste Risks: An Instrumental Variables Estimation. *Journal of Regulatory Economics*, March. [14]

Gehring, H. 1993. Address [untitled] at the Institute of Packaging Professionals Conference on Environmental Packaging Legislation, 16 July, Herndon, VA. [2]

Geiger, G. H. 1982. Government Regulations and Their Effect on Metallic Resource Recovery. *Resources and Conservation*, August. [8]

Geiselman, B. 1999. Waste-to-Energy's Future Dims. *Waste News*, 3 May. [5]

———. 2001a. New Jersey Questions Plan. *Waste News*, 9 July. [7]

———. 2001b. Waste Corp. Rockets Up the Ladder. *Waste News*, 25 June. [4]

Genillard, A. 1994. Recycling Has Neighbours Crying Foul: Complaints of Cheap Waste Exports to European Countries. *Financial Times*, 25 January. [2]

Gerrard, M. B. 1994. *Whose Backyard, Whose Risk: Fear and Fairness in Toxic and Nuclear Waste Siting*. Cambridge, MA: MIT Press. [14, 16]

Gershey, E. L., R. C. Klein, E. Party, and A. Wilkerson. 1990. *Low-Level Radioactive Waste: From Cradle to Grave*. New York: Van Nostrand Reinhold. [16]

Glasser, S. B. 2001. Wanted: Windfall in Nuclear Waste. *Washington Post*, 11 February. [16]

Glaub, J. C. 1993. Household Hazardous Wastes. In *The McGraw-Hill Recycling Handbook*, edited by H. F. Lund. New York: McGraw-Hill. [14]

Glebs, R. T. 1988. Landfill Costs Continue to Rise. *Waste Age*, March. [4]

Glenn, J. 1990. Yard Waste Composting Enters a New Dimension. *BioCycle*, September. [13]

———. 1992. Efficiencies and Economics of Curbside Recycling. *BioCycle*, July. [9]

———. 1998. The State of Garbage in America. *BioCycle*, part I, April. [5]

———. 1999. The State of Garbage in America. *BioCycle*, April. [1, 8]

Glickman, T. S., and M. A. Sontag. 1995. The Tradeoffs Associated with Rerouting Highway Shipments of Hazardous Materials to Minimize Risk. *Risk Analysis*, February. [14]

Glicksman, R. L. 1988. Interstate Compacts for Low-Level Radioactive Waste Disposal: A Mechanism for Excluding Out-of-State Waste. In *Low-Level Radioactive Waste Regulation*, edited by M. E. Burns. Boca Raton, FL: Lewis Publishers. [16]

Goddard, H. C. 1994. The Benefits and Costs of Alternative Solid Waste Management Policies. In *Balancing Economic Growth and Environmental Goals*. Washington, DC: American Council for Capital Formation. [3, 5]

Gold, A. R. 1990. New York's Game of Musical Dumps. *New York Times*, 22 March. [4]

Goldberg, D. 1990. The Magic of Volume Reduction. *Waste Age*, February. [10]

Goodstein, E., and H. Hodges. 1997. Polluted Data: Overestimating the Costs of Environmental Regulation. *American Prospect*, November–December. [14, 17]

Graff, G. P. 1989. Green(hous)ing of the Landfill. *Waste Age*, March. [4]

Gray, W., and M. Deily. 1996. Compliance and Enforcement: Air Pollution Regulation in the U.S. Steel Industry. *Journal of Environmental Economics and Management*, July. [14]

Gray, W. B. 1987. The Cost of Regulation: OSHA, EPA and the Productivity Slowdown. *American Economic Review*, December. [17]

Greenberg, M., and J. Hughes. 1992. The Impact of Hazardous Waste Superfund Sites on the Value of Houses Sold in New Jersey. *The Annals of Regional Science*, June. [15]

Greenwire. 1997a. Illinois: Chicago Tribune Examines Recycling Program. 18 December. [1]

———. 1997b. New Jersey: New Waste Options Bring Savings, Problems. 10 November. [7]

———. 1999a. Autos: Ford Plans Recycling Division. 27 April. [12]

———. 1999b. Nuclear Waste: DOE To Discuss West Valley Cleanup More. 4 May. [16]

———. 2000a. Plutonium: DOE Raises Estimate of Material Released into Soil. 23 October. [16]

———. 2000b. Superfund: Interest Groups Remind Senate of Recycling Promise. 27 October. [16]

Grove, B., and M. Manning. 2001. Yucca Could Be Costliest Project in History. *Las Vegas Sun*, 9 March. [16]

Grunwald, M. 1998. Nuclear Waste Disposal Still a Festering Problem. *Washington Post*, 22 November. [16]

Grunwald, M., and P. Behr. 2001. Are Nuclear Plants Still Secure? *Washington Post*, 3 November. [16]

Guerrero, P. F. 1995. *Superfund: The Role of Risk in Setting Priorities*. GAO/T-RCED-95-161. Washington, DC: General Accounting Office. [15]

Gupta, S., G. Van Houtven, and M. L. Cropper. 1995. Do Benefits and Costs Matter in Environmental Regulation? An Analysis of EPA Decisions under Superfund. In *Analyzing Superfund: Economics, Science, and Law,* edited by R. L. Revesz and R. B. Stewart. Washington, DC: Resources for the Future. [15]

Gust, S. 2000. Massachusetts Town's Mandatory Program Has Rocky Debut. *Waste News,* 17 April. [11]

Gynn, A. M. 2000a. Major Haulers Rebuild; Others Take Advantage. *Waste News,* 12 June. [4]

———. 2000b. Waste Corp. Becomes a Player. *Waste News,* 1 May. [4]

———. 2001. New York City Trash Bill To Top $1 Billion. *Waste News,* 19 March. [4]

Hamilton, A. 2001. How Do You Junk Your Computer? *Time,* 12 February. [9]

Hamilton, J. T. 1995. Testing for Environmental Racism: Prejudice, Profits, Political Power? *Journal of Policy Analysis and Management,* Winter. [14]

Hamilton, J. T., and W. K. Viscusi. 1995. The Magnitude and Policy Implications of Health Risks from Hazardous Waste Sites. In *Analyzing Superfund: Economics, Science, and Law,* edited by R. L. Revesz and R. B. Stewart. Washington, DC: Resources for the Future.[15]

———. 1999a. *Calculating Risks? The Spatial and Political Dimensions of Hazardous Waste Policy.* Cambridge, MA: MIT Press. [15, 17]

———. 1999b. How Costly Is "Clean"? An Analysis of the Benefits and Costs of Superfund Remediations. *Journal of Policy Analysis and Management,* Winter. [15]

Hanson, R. D. 1995. Indian Burial Grounds for Nuclear Waste. *Multinational Monitor.* September. [16]

Harrington, W. 1988. Enforcement Leverage When Penalties Are Restricted. *Journal of Public Economics,* October. [6]

Harrington, W., R. D. Morgenstern, and P. Nelson. 1999. On the Accuracy of Regulatory Cost Estimates. Resources for the Future Discussion Paper 99-18. January. [http://www.rff.org/disc_papers/1999.htm (accessed 2 November 2001). [14]

Hartman. 1987. *Introductory Mining Engineering.* New York: Wiley-Interscience. [2]

Havlicek, J., Jr. 1985. Impacts of Solid Waste Disposal Sites on Property Values. In *Environmental Policy: Solid Waste,* vol. 4, edited by G. S. Tolley, J. Havlicek Jr., and R. Favian. Boston: Ballinger Press. [4]

Hawes, C. 1991. Environmentalists Canning Once Venerated Bottle Bills. *Saint Louis Post-Dispatch,* 5 March. [11]

Hawken, P. 1993. *The Ecology of Commerce: A Declaration of Sustainability.* New York: Harper-Collins. [4]

Hayden, G. F. 1997. Excess Capacity for the Disposal of Low-Level Radioactive Waste in the United States Means New Compact Sites Are Not Needed. University of Nebraska, Lincoln. [16]

Hayhurst, T. 2000. Ready to Roll. *Waste News,* 17 January. [5]

Helfand, G. E., and B. W. House. 1995. Regulating Nonpoint Source Pollution under Heterogeneous Conditions. *American Journal of Agricultural Economics,* November. [2]

Helland, E. 1998. The Enforcement of Pollution Control Laws: Inspections, Violations, and Self-Reporting. *Review of Economics and Statistics,* February. [14]

Hendrickson, C., L. Lave, and F. McMichael. 1995. Time to Dump Recycling? *Issues in Science and Technology,* Spring. [8]

Herfindahl, O. C. 1967. Depletion and Economic Theory. In *Extractive Resources and Taxation,* edited by M. Gaffney. Madison: University of Wisconsin Press. [8]

Hershkowitz, A. 1997. *Too Good To Throw Away: Recycling's Proven Record.* Washington, DC: Natural Resources Defense Council. [2, 8]

Hershkowitz, A., and E. Salerni. 1987. *Garbage Management in Japan: Leading the Way.* New York: Inform Inc. [5, 9, 13]

Heumann, J. M., and K. Egan. 1998. Recycling in 1998: States Moving Forward to Reach Higher Goals. *Waste Age,* August. [10]

Higashi, J. 1990. Moving Mount Overfill. *Ann Arbor Observer,* April. [4]

Highfill, J., and M. McAsey. 1997. Municipal Waste Management: Recycling and Landfill Space Constraints. *Journal of Urban Economics*, January. [8]

Hird, J. A. 1994. *Superfund: The Political Economy of Environmental Risk*. Baltimore: Johns Hopkins University Press. [15]

Hirsch, W. Z. 1965. Cost Functions of an Urban Government Service: Refuse Collection. *Review of Economics and Statistics*, February. [3]

Hlavka, R. J. 1993. Processing Yard Wastes. In *The McGraw-Hill Recycling Handbook*, edited by H. F. Lund. New York: McGraw-Hill. [13]

Hocking, M. B. 1991. Paper versus Polystyrene: A Complex Choice. *Science*, 1 February. [2]

Hoffman, B. 2000. Ford Finding Treasure in Trash. *Waste News*, 3 January. [12]

Holt, M. 1998. *Transportation of Spent Nuclear Fuel.*. Congressional Research Service, Report to Congress. 97-403 ENR. 29 May. [16]

Holusha, J. 1993. Who Foots the Bill for Recycling? *New York Times*, 25 April. [9]

Hong, S., R. M. Adams, and H. A. Love. 1993. An Economic Analysis of Household Recycling of Solid Wastes: The Case of Portland, Oregon. *Journal of Environmental Economics and Management*, September. [11]

Houghton, J. T., L. G. Meira Filho, B. A. Callender, N. Harris, A. Kattenberg, and K. Maskell, eds. 1996. *Climate Change 1995: The Science of Climate Change*. Intergovernmental Panel on Climate Change. Cambridge, U.K.: Cambridge University Press. [4]

Hull, P. 2000. Landfill Gases and Leachate: Changing the Schedule. *MSW Management*, July–August. [4]

Humphries, M., and C. H. Vincent. 2001. *Mining on Federal Lands*. Congressional Research Service. 27 March. [2]

ICF Incorporated. 1997. *Greenhouse Gas Emissions from Municipal Waste Management*. Draft Working Paper, prepared for U.S. EPA. EPA530-R-97-010. Washington, DC: U.S. EPA. [5]

Innes, R. 2000. The Economics of Livestock Waste and Its Regulation. *American Journal of Agricultural Economics*, February. [2]

Jacob, G. 1990. *Site Unseen: The Politics of Siting a Nuclear Waste Repository*. Pittsburgh: University of Pittsburgh Press. [16]

Jang, J-W., T-S. Yoo, J-H. Oh, and I. Iwasaki. 1998. Discarded Tire Recycling Practices in the United States, Japan, and Korea. *Resources, Conservation, and Recycling*, March. [8]

Jendryka, B. 1992. Composting in Ann Arbor—An Efficient Way to Reduce Waste? Term paper, Economics 471, University of Michigan, Ann Arbor, December. [13]

Jenkins, R. R. 1993. *The Economics of Solid Waste Reduction: The Impact of User Fees*. Brookfield, VT: Edward Elgar Press. [3]

Jenkins, R. R., S. A. Martonez, K. Palmer, and M. J. Podolsky. 2000. The Determinants of Household Recycling: A Material Specific Analysis of Unit Pricing and Recycling Program Attributes. Resources for the Future, Working Paper 99-41-REV. April. http://www.rff.org/disc_papers/1999.htm (accessed 2 November 2001). [10, 11]

Jensen, R. C. 1999. Salted Away. *Environmental Protection*, September. [16]

Jeter, J. 1998. Poor Town That Sought Incinerator Finds More Problems, Few Benefits. *Washington Post*, 11 April. [5]

Johnson, J. 1999. Minn. Considers Landfill Ban by 2006. *Waste News*, 27 September. [4]

———. 2000a. California Diversion Rises. *Waste News*, 13 March. [10]

———. 2000b. Waste Giant Sticks with Recycling. *Waste News*, 10 July. [9]

Johnson, J., and C. McMullen. 2000. Mass. Weighs Lifting Landfill, Burner Ban. *Waste News*, 12 June. [10]

Johnson, M., and W. L. Carlson. 1991. Calculating Volume-Based Garbage Fees. *BioCycle*, February. [3]

Judge, R., and A. Becker. 1993. Motivating Recycling: A Marginal Cost Analysis. *Contemporary Policy Issues*, July. [11]

Jung, L. B. 1999. The Conundrum of Computer Recycling. *Resource Recycling*, May. [9]

Kafka, S. 1997. Rose Township Dump. Term paper, Economics 471, University of Michigan, Ann Arbor, December. [15]

Kahlenberg, R. 1992. Garb Go-Round: Recycling of Clothing in Thrift Shops and Elsewhere Keeps Tons and Tons from Taking Up Vital Landfill Space. *Los Angeles Times*, 29 October. [7]

Kamberg, M-L. 1990. Recycling Cuts Disposal Costs in Seattle. *Solid Waste & Power*, August. [11]

———. 1991. Weighing the Options: Paper or Plastic? *Solid Waste & Power*, April. [2]

Kanabayashi, M. 1982. It Isn't Easy to Work Alone in Japan—Ask a Chirigami Kokan. *Wall Street Journal*, 12 February. [9, 12]

Kanigher, S., and M. Manning. 1998. Yucca Mountain: Science vs. Politics. *Las Vegas Sun*, 31 May. [16]

Kashmanian, R. M. 1993. Quantifying the Amount of Yard Trimmings to Be Composted in the United States in 1996. *Compost Science and Utilization*, Summer. [13]

Kashmanian, R. M., and R. L. Spencer. 1993. Cost Considerations of Municipal Solid Waste Compost: Production Versus Market Price. *Compost Science and Utilization*, March. [13]

Katzman, M. T. 1989. Financing Liabilities under Superfund: Comment. *Risk Analysis*, September. [15]

Kaufman, M. T. 1992. A Middleman's Ventures in the Can Trade. *New York Times*, 23 September. [6]

Kearney, I. D., and R. C. Porter. 1999. The Economics of Low-Level Radioactive Waste Disposal. Working Paper. University of Michigan, Ann Arbor. [16]

Keeler, A. G., and M. Renkow. 1994. Haul Trash or Haul Ash: Energy Recovery as a Component of Local Solid Waste Management. *Journal of Environmental Economics and Management*, November. [5, 10]

Keeney, R. L. 1990. Mortality Risks Induced by Economic Expenditures. *Risk Analysis*, March. [1]

———. 1997. Estimating Fatalities Induced by the Economic Costs of Regulations. *Journal of Risk and Uncertainty*, January. [1]

Keeney, R. L., and D. von Winterfeldt. 1994. Managing Nuclear Waste from Power Plants. *Risk Analysis*, February. [16]

Kemp, R. 1992. *The Politics of Radioactive Waste Disposal*. Manchester, U.K.: Manchester University Press. [16]

Kemper, P., and J. M. Quigley. 1976. *The Economics of Refuse Collection*. Cambridge, MA: Ballinger Publishing Company. [3]

Keoleian, G. A., M. Manion, J. W. Bulkley, and K. Kar. 1996. Industrial Ecology of the Automobile. White Paper, National Pollution Prevention Center, University of Michigan, Ann Arbor. [12]

Kesler, S. E. 1994. *Mineral Resources, Economics, and the Environment*. New York: Macmillan. [2]

Ketkar, K. W. 1995. Protection of Marine Resources: The U.S. Oil Pollution Act of 1990 and the Future of the Maritime Industry. *Marine Policy*, September. [15]

Kilborn, P. T. 1999. Hurricane Reveals Flaws in Farm Law. *New York Times*, 17 October. [2]

Kilbourn, S., and M. Mandell. 1992. Solid Waste Landfills: Costs and Compensation—A Case Study of the Arbor Hills Landfill. Term paper, Economics 471, University of Michigan, Ann Arbor, December. [7]

Kinkley, C-C., and K. Lahiri. 1984. Testing the Rational Expectations Hypothesis in a Secondary Materials Market. *Journal of Environmental Economics and Management*, September. [12]

Kinnaman, T. C. 1996. The Efficiency of Solid Waste Recycling: A Benefit–Cost Analysis. Working Paper. Lewisburg, PA: Department of Economics, Bushnell University. [9]

———. 2000. Explaining the Growth in Municipal Recycling Programs: The Role of Market and Non-Market Factors. Working Paper. Lewisburg, PA: Department of Economics, Bushnell University. [9]

Kinnaman, T. C., and D. Fullerton. 2000. Garbage and Recycling with Endogenous Local Policy. *Journal of Urban Economics*, November. [6, 10, 11]

Kitchen, H. M. 1976. A Statistical Estimation of an Operating Cost Function for Municipal Refuse Collection. *Public Finance Quarterly*, January. [3]

Kleindorfer, P. R. 1988. Economic Regulation of Solid Waste Collection and Disposal: Comparative Institutional Assessment. Working Paper 88-65, Center for Risk and Decision Processes. Philadelphia: Wharton School, University of Pennsylvania. July. [4]

Kohlhase, J. E. 1991. The Impact of Toxic Waste Sites on Housing Values. *Journal of Urban Economics*, July. [15]

Konheim and Ketcham, Inc. 1991. *Exporting Waste*. Report to New York City. New York: Konheim and Ketcham. [4, 7]

Kopel, D. B. 1993. Burning Mad: The Controversy over Treatment of Hazardous Waste in Incinerators, Boilers, and Industrial Furnaces. *Environmental Law Reporter*, April. [14]

Koplow, D. 1994. Federal Energy Subsidies and Recycling: A Case Study. *Resource Recycling*, November. [8, 12]

Koplow, D., and K. Dietly. 1994. *Federal Disincentives: A Study of Federal Tax Subsidies and Other Programs Affecting Virgin Industries and Recycling*. Office of Policy Analysis. EPA230-R-94-005. Washington, DC: U.S. EPA. [8, 12]

Krakauer, J. 1997. *Into Thin Air*. New York: Random House. [6]

Kravetz, S. 1998a. Allied Waste Is Set to Acquire American Disposal. *Wall Street Journal*, 11 August. [4]

———. 1998b. Dry Cleaners' New Wrinkle: Going Green. *Wall Street Journal*, 3 June. [14]

Krueger, J. 1999. *International Trade and the Basel Convention*. Washington, DC: Earthscan Publications. [7]

Kunreuther, H., D. Easterling, W. Desvouges, and P. Slovic. 1990. Public Attitudes Toward Siting a High-Level Nuclear Waste Repository in Nevada. *Risk Analysis*, December. [16]

Kunreuther, H., P. Kleindorfer, P. J. Knez, and R. Yarsick. 1987. A Compensation Mechanism for Siting Noxious Facilities: Theory and Experimental Design. *Journal of Environmental Economics and Management*, December. [7]

Kuz, M. 1998. Beatty Accepts Its Nuclear Fate. *Las Vegas Sun*, 1 June. [16]

Lake, J. A. 2001. Outdated Thinking Is Holding Us Back. *Washington Post*, 13 May. [16]

Lamb, J., and M. Chertow. 1990. Plastics Collection. *BioCycle*, April. [9]

Landy, M. K., M. J. Roberts, and S. R. Thomas. 1994. *The Environmental Protection Agency: Asking the Wrong Questions, from Nixon to Clinton*. Oxford, U.K.: Oxford University Press. [15]

Langewiesche, W. 2000. The Shipbreakers. *Atlantic Monthly*, August. [7]

Laplante, B., and P. Rilstone. 1996. Environmental Inspections and Emissions of the Pulp and Paper Industry in Quebec. *Journal of Environmental Economics and Management*, July. [14]

Lasdon, D. H. 1988. Bottle-Law Violations Penalize the Homeless [letter]. *New York Times*, 27 September. [6]

Leaversuch, R. D. 1994. Recycling Faces Reality as Bottom Line Looms. *Modern Plastics*, July. [9]

Lee, D. B. 1995. *Full Cost Pricing of Highways*. Volpe National Transport Systems Center. Washington, DC: U.S. Department of Transportation. [12]

Lee, D. R., P. E. Graves, and R. L. Sexton. 1992. Controlling the Abandonment of Automobiles: Mandatory Deposits versus Fines. *Journal of Urban Economics*, January. [6]

Lenssen, N. 1992. Confronting Nuclear Waste. *State of the World, 1992*. Worldwatch Institute. New York: W. W. Norton. [16]

Leroux, K. 1999. Buying into Buy Recycled. *Waste Age*, July. [12]

Levinson, A. 1999a. NIMBY Taxes Matter: The Case of State Hazardous Waste Disposal Taxes. *Journal of Public Economics*, October. [14]

———. 1999b. State Taxes and Interstate Hazardous Waste Shipments. *American Economic Review*, June. [14]

Ley, E., M. Macauley, and S. W. Salant. 2000. Restricting the Trash Trade. *American Economic Review*, May. [7]

———. 2002. Spatially and Intertemporally Efficient Waste Management: The Costs of Interstate Flow Control. *Journal of Environmental Economics and Management*. [7]

Lifset, R. 1998. Extended Producer Responsibility and Leasing: Some Preliminary Thoughts.

Paper prepared for the Proceedings on Extended Producer Responsibility as a Policy Instrument, International Institute for Industrial Environmental Economics, 8–9 May, Lund, Sweden. [2]

Lipton, E. 1998. Industrial Waste Rules Go Up in Smoke. *Washington Post*, 13 November. [5, 14]

Little, A. D., Inc. 1990. *Disposable versus Reusable Diapers: Health, Environmental, and Economic Comparisons*. Report to Procter and Gamble. Cambridge, MA: A. D. Little. [3]

———. 1992. *A Report on Advance Disposal Fees*. Environmental Education Associates. Cambridge, MA: A. D. Little. [2]

Little, I. M. D., and J. A. Mirrlees. 1974. *Project Appraisal and Planning for Developing Countries*. New York: Basic Books. [9]

Louis, P. J. 1993. Duales System Deutschland on the Move. *European Packaging Newsletter and World Report*, September. [2]

Luckett, D. G. 1980. *Money and Banking*. New York: McGraw-Hill. [6]

Lyon, R. M., and S. Farrow. 1995. An Economic Analysis of Clean Water Act Issues. *Water Resources Research*, January. [2]

Magat, W. A., and W. K. Viscusi. 1990. Effectiveness of the EPA's Regulatory Enforcements: The Case of Industrial Effluent Standards. *Journal of Law and Economics*, October. [14]

Magnuson, A. 1996. Fresh Kills Landfill. *MSW Management*, July–August. [4]

———. 1997. Disposal. *MSW Management*, October. [5]

Maier, T. J. 1989. Trying a European Import. In *Rush to Burn*. New York: Island Press. [5]

Maillet, L. L. 1990. Hazardous versus Non-Hazardous Disposal of Ash from Detroit's Resource Recovery Facility. Term paper, Economics 573, University of Michigan, Ann Arbor, April. [5]

Malchman, W. 1995. Case Western Reserve University's New "State of the Art" Low-Level Waste Facility. *Radwaste Magazine*, September. [16]

Manning, M. 1998. Study: Yucca Will Cost Taxpayers. *Las Vegas Sun*, 5 May. [16]

Marks, R., and R. Knuffke. 1998. *America's Animal Factories: How States Fail to Prevent Pollution from Livestock Waste*. Washington, DC: Natural Resources Defense Council. [2]

Martin, D., and A. C. Revkin. 1999. As Deadline Looms for Dump, Alternate Plan Proves Elusive. *New York Times*, 30 August. [4]

Martin, G. 1999. Study Finds S.F. Water Pesticide-Free—Many Other Counties' Sources Reported to Be Contaminated. *San Francisco Chronicle*, 27 October. [2]

Mathis, J. C. 2001. Views of Public, Stores at Odds. *Ann Arbor News*, 13 May. [6]

Mazmanian, D., and D. Morell. 1992. *Beyond Superfailure: America's Toxics Policy for the 1990s*. Boulder, CO: Westview Press. [14]

Mazur, A. C. 1998. *A Hazardous Inquiry: The Rashomon Effect at Love Canal*. Cambridge, MA: Harvard University Press. [15]

McBean, E. A., F. A. Rovers, and G. J. Farquhar. 1995. *Solid Waste Landfill Engineering and Design*. Englewood Cliffs, NJ: Prentice-Hall. [4]

McCarthy, J. E. 1993. *Bottle Bills and Curbside Recycling: Are They Compatible?* Congressional Research Service, Report 93-114 ENR. 27 January. [6]

———. 1998. *Interstate Shipment of Municipal Solid Waste: 1998 Update*. Congressional Research Service, Report 98-689 ENR. 6 August. [7]

McClelland, G., W. Schulze, and B. Hurd. 1990. The Effect of Risk Beliefs on Property Values: A Case Study of a Hazardous Waste Site. *Risk Analysis*, December. [14]

McConnell, J. 1998. The Joy of Cloth Diapers. *Mothering*, May. [3]

McCrory, J. 1998. *New York City: The First Regional Government Still Cries for Planning—The Case of Waste Management*. Report 128, Planners Network, March–April. [4]

McDaniel, P. W., G. J. Borden, and M. R. James. 1995. *FUSRAP Experience Transporting LLW and 11e(2) Waste Materials by Rail, Intermodal Container, and Truck (Transportation)*. Prepared by Bechtel National Inc. Washington, DC: U.S. Department of Energy. [16]

McLellan, D. 1994. Weight-Based Rates: Collecting Waste Canadian Style. *World Wastes*, March. [3]

McMullen, C. A. 2000a. Detroit Hospital To Close Burner. *Waste News*, 14 February. [14]

———. 2000b. Foster Wheeler Says It Wants Out. *Waste News*, 14 August. [5]

———. 2000c. New England EPA Wants Mercury Ban. *Waste News*, 11 December. [14]

———. 2001. "Butt Bill" Targets Maine Smokers. *Waste News*, 26 February. [6]

MDEP (Massachusetts Department of Environmental Protection). 1997. *Massachusetts Solid Waste Master Plan, 1997 Update*, vol. 1. Boston: MDEP. [10]

Meade, K. 1989. Connecticut Chooses Combustion. *Waste Age*, June. [5]

Meadows, D. H. 1999. Never Mind Paper vs. Plastic Bags, How Did You Get to the Grocery Store? *The Global Citizen*, 20 May. [2]

Meadows, D. H., D. L. Meadows, J. Randers, and W. H. Behrens III. 1972. *The Limits to Growth*. New York: Universe Books. [8]

Melosi, M. V. 1981. *Garbage in the Cities: Refuse, Reform, and the Environment, 1880–1980*. College Station: Texas A&M University Press. [1]

Menell, P. S. 1995. Structuring a Market-Oriented Federal Eco-Information Policy. *Maryland Law Review* 54 (4). [2]

Merrill, L. 1996. Used Oil Collection Programs. *MSW Management*, September–October. [14]

Michaels, R. G., and V. K. Smith. 1989. Market Segmentation and Valuing Amenities with Hedonic Models: The Case of Hazardous Waste Sites. Discussion Paper QE89-02. Washington, DC: Resources for the Future. [14]

Michaels, R. G., V. K. Smith, and D. Harrison Jr. 1987. *Market Segmentation and Valuing Amenities with Hedonic Models: The Case of Hazardous Waste Sites*. Working Paper. Raleigh: Department of Economics, North Carolina State University. [14]

Michigan Consultants. 1996. *The Michigan Beverage Container Deposit Law Two Decades Later*. Report for Michigan Recycling Partnership. Lansing, MI: Michigan Consultants. [6, 11]

———. 1998. *Analysis of Foreign Containers in the Michigan Deposit Stream*. Report for Michigan Beer and Wine Wholesalers Association. Lansing, MI: Michigan Consultants. [6]

Miller, C. 1992. The Real Price of Processing. *Waste Age*, October. [9]

———. 1993. The Cost of Recycling at the Curb. *Waste Age*, October. [9]

———. 1995a. Co-Collection: Back to the Past. *Waste Age*, October. [11]

———. 1995b. Waste Product Profiles. In *Waste Age/Recycling Times' Recycling Handbook*, edited by J. T. Aquino. Boca Raton, FL: Lewis Publishers. [7, 13, 14]

———. 1995c. What Does Recycling Really Cost? In *Waste Age/Recycling Times' Recycling Handbook*, edited by J. T. Aquino. Boca Raton, FL: Lewis Publishers. [9, 11]

———. 1998. Profiles in Garbage: Yard Waste. *Waste Age*, January. [13]

MILLRWC (Midwest Interstate Low-Level Radioactive Waste Commission). 1985. *Generic Host State Incentive Report*, July. [16]

Miranda, M. L., and J. E. Aldy. 1998. Unit Pricing of Residential Municipal Solid Waste: Lessons from Nine Case Study Communities. *Journal of Environmental Management*, January. [3]

Miranda, M. L., S. D. Bauer, and J. E. Aldy. 1996. *Unit Pricing Programs for Residential Municipal Solid Waste: An Assessment of the Literature*. Prepared for U.S. EPA. Washington, DC: U.S. EPA. [3, 6, 10]

Miranda, M. L., J. W. Everett, D. Blume, and B. A. Roy Jr. 1994. Market-Based Incentives and Residential Municipal Solid Waste. *Journal of Policy Analysis and Management*, Fall. [3]

Miranda, M. L., and B. Hale. 1997. Waste Not, Want Not: The Private and Social Costs of Waste-to-Energy Production. *Energy Policy*, May. [5]

Miranda, M. L., and S. LaPalme. 1997. *Unit Pricing of Residential Municipal Solid Waste: A Preliminary Analysis of 212 U.S. Communities*. Prepared for U.S. EPA. Washington, DC: U.S. EPA. [6, 10, 11]

Mishan, E. J. 1988. *Cost–Benefit Analysis: An Informal Introduction*, 4th ed. London: Unwin Hyman. [1]

Moller, J. 2001. Daschle: Switch May Stall Bills That Would Affect Nevada. *Las Vegas Review-Journal*, 1 June. [16]

Montagu, P. 1995. Dry Cleaning: Is Regulation Necessary? *Rachel's Environment and Health Weekly*, 2 March. [14]

Morris, D., and J. Nelson. 1999. *Looking Before We Leap: A Perspective on Public Subsidies for Burning Poultry Manure.* Washington, DC: Institute for Local Self-Reliance. [2]

Morris, G. E., and D. M. Holthausen. 1994. The Economics of Household Solid Waste Generation and Disposal. *Journal of Environmental Economics and Management*, May. [3]

Morris, J. R., P. S. Phillips, and A. D. Read. 1998. The U.K. Landfill Tax: An Analysis of Its Contribution to Sustainable Waste Management. *Resources, Conservation, and Recycling*, September. [4]

Murphy, M. E., and M. J. Rogoff. 1994. Court's Ash Decision Raises Questions; States Have Answers. *Solid Waste Technologies*, September–October. [5]

Mydans, S. 2000. Before Manila's Garbage Hill Collapsed: Living Off Scavenging. *New York Times*, 18 July. [4]

NAPCOR (National Association for PET Container Resources). 2000. *1999 Report on Post Consumer PET Container Recycling Activity.* Charlotte, NC: NAPCOR. [12]

Naquin, D. 1998. Clearing the Air. *Waste Age*, May. [5]

NAS (National Academy of Sciences). 1957. *The Disposal of Radioactive Waste on Land.* Report of the Committee on Waste Disposal. Division of Earth Sciences. Washington, DC: NAS. [16]

———. 2000. *Long-Term Institutional Management of U.S. Department of Energy Legacy Waste Sites.* Washington, DC: NAS. [16]

———. 2001. *Disposition of High-Level Waste and Spent Nuclear Fuel: The Continuing Societal and Technical Challenges.* National Research Council, Board on Radioactive Waste Management. Washington, DC: NAS. [16]

Nelson, A. C., J. Genereux, and M. Genereux. 1992. Price Effects of Landfills on House Values. *Land Economics*, November. [4]

Nestor, D. V., and M. J. Podolsky. 1998. Assessing Incentive-Based Environmental Policies for Reducing Household Solid Waste. *Contemporary Economic Policy*, October. [3]

Neus, E. 2000. Glass Mercury Thermometer Sales Slowly Drying Up. *Detroit News*, 26 October. [14]

Newsweek. 1995. Trash Bandits. 19 June. [12]

———. 1998. Perspectives '98: Society. 28 December. [3]

Nichols, A. L., and R. J. Zeckhauser. 1986. The Perils of Prudence. *Regulation*, November–December. [15]

Niemczewski, C. 1977. The History of Solid Waste Management. In *The Organization and Efficiency of Solid Waste Collection*, edited by E. S. Savas. Boston: Lexington Books. [1]

Noronha, F. 1999. Shipbreaking: A Poisonous Job. *Environment News Service*, 23 September. [7]

Nossiter, A. 1996. Villain Is Dioxin. Relocation Is Response. But Judgment Is in Dispute. *New York Times*, 21 October. [15]

NYC, DEP (New York City, Department of Environmental Protection). 1997. *The Impact of Food Waste Disposers in Combined Sewer Areas of New York City.* http://www.ci.nyc.ny.us/html/dep/html/grinders.html (accessed 2 November 2001). [3]

OECD (Organisation for Economic Co-operation and Development). 1997. *Transfrontier Movements of Hazardous Wastes: 1992–93 Statistics.* Paris: OECD. [7]

———1998. *Case Study on the German Packaging Ordinance.* ENV/EPOC/PPC(97)21/REV2. Paris: OECD. [2]

O'Leary, P. R., and P. W. Walsh. 1995. *Decision Maker's Guide to Solid Waste Management*, 2d ed. Prepared for U.S. EPA. EPA 530-R-95-023. Washington, DC: U.S. EPA. [10, 11]

Orwell, G. 1941. The Art of Donald McGill. *Horizon*, September. [1]

OTA (Office of Technology Assessment). 1983. *Technologies and Management Strategies for Hazardous Waste Control.* Washington, DC: OTA. [14]

———. 1989a. *Coming Clean: Superfund's Problems Can Be Solved.* OTA-ITE-433. http://www.wws.princeton.edu/~ota/ns20/year_f.html (accessed 5 November 2001). [15]

————. 1989b. *Facing America's Trash: What Next for Municipal Solid Waste?* OTA-O-424. http://www.wws.princeton.edu/~ota/ns20/year_f.html (accessed 5 November 2001). [1, 2, 4, 5, 9]

————. 1989c. *Partnerships Under Pressure.* OTA-O-426. http://www.wws.princeton.edu/~ota/ns20/year_f.html (accessed 5 November 2001). [16]

————. 1992. *Managing Industrial Solid Wastes.* OTA-BP-O-82. http://www.wws.princeton.edu/~ota/ns20/year_f.html (accessed 5 November 2001). [2]

Paik, A. 1999. Garbage In, Value Out. *Washington Post*, 30 December. [9]

Palli, D. 1997. Gastric Cancer and Helicobacter Pylori: A Critical Evaluation of the Epidemiological Evidence. *Helicobacter* (2 Supp.). [2]

Palmer, K., H. Sigman, and M. Walls. 1997. The Cost of Reducing Municipal Solid Waste. *Journal of Environmental Economics and Management*, June. [10]

Palmer, K., H. Sigman, M. A. Walls, K. Harrison, and S. Puller. 1995. The Cost of Reducing Municipal Solid Waste: Comparing Deposit-Refunds, Advance Disposal Fees, Recycling Subsidies, and Recycling Rate Standards. Discussion Paper 95-33. Washington, DC: Resources for the Future. [10]

Palmer, K., and M. Walls. 1994. Materials Use and Solid Waste Disposal: An Evaluation of Policies. Discussion Paper 95-02. Washington, DC: Resources for the Future. [10]

————. 1997. Optimal Policies for Solid Waste Disposal: Taxes, Subsidies, and Standards. *Journal of Public Economics*, August. [6, 8, 10]

Papenhagen, E. 1995. Environmental and Economic Analysis of Dry Cleaning Solvents. Term paper, Economics 471, University of Michigan, Ann Arbor, December. [14]

Passell, P. 1991. The Garbage Problem: It May Be Politics, Not Nature. *New York Times*, 26 February. [2]

Paton, R. F. 1997. The National Low-Level Radioactive Waste Act: Success or Failure? *Radwaste Magazine*, July. [16]

Patton, G. S. 1947. *War As I Knew It.* Boston: Houghton-Mifflin. [2]

Payne, B. A., S. J. Olshansky, and T. E. Segel. 1987. The Effects on Property Values of Proximity to a Site Contaminated with Radioactive Waste. *Natural Resources Journal*, Summer. [16]

Pearce, D. W., and R. K. Turner. 1992. *Packaging Waste and the Polluter Pays Principle: A Taxation Solution.* Centre for Social and Economic Research on the Global Environment. Discussion Paper WM 92-01. London: University College London and University of East Anglia. [2]

Pearce, J. 2000a. Detroit Polluter Fouling Suburbs. *Detroit News*, 18 February. [12]

————. 2000b. EPA Puts Lyon Firm on Hit List. *Detroit News*, 10 May. [12]

————. 2000c. Oakland Firm Hit as Polluter. *Detroit News*, 9 August. [12]

————. 2001. State to Review Firm's Emissions. *Detroit News*, 5 June. [12]

Peckinpaugh, T. L. 1988. The Politics of Low-Level Radioactive Waste Disposal. In *Low-Level Radioactive Waste Regulation*, edited by M. E. Burns. Boca Raton, FL: Lewis Publishers. [16]

Peretz, J. H., R. A. Bohm, and P. D. Jasienczyk. 1997. Environmental Policy and the Reduction of Hazardous Waste. *Journal of Policy Analysis and Management*, Fall. [14]

Pettit, C. L., and C. Johnson. 1987. The Impact on Property Values of Solid Waste Facilities. *Waste Age*, April. [4]

Pigou, A. C. 1920. *The Economics of Welfare.* New York: Macmillan. [1]

Podolsky, M. J., and M. Spiegel. 1997. *When Interstate Transportation of Municipal Solid Waste Makes Sense and When It Does Not.* Office of Policy, Planning, and Evaluation. Washington, DC: U.S. EPA. [7]

Pogrebin, R. 1996. Now the Working Class, Too, Is Foraging for Empty Cans. *New York Times*, 29 April. [6]

Porter, J. W. 1988. *A National Perspective on Municipal Solid Waste Management.* Speech to 4th Annual Conference on Solid Waste Management and Materials Policy, January, New York City. [10]

Porter, R. C. 1974. The Long-Run Asymmetry of Subsidies and Taxes as Anti-Pollution Policies. *Water Resources Research*, June. [1]

———. 1978. A Social Benefit–Cost Analysis of Mandatory Deposits on Beverage Containers. *Journal of Environmental Economics and Management*, December. [6]

———. 1982. The Public Management of Michigan's Natural Resources. In *Michigan's Fiscal and Economic Structure*, edited by H. E. Brazer. Ann Arbor: University of Michigan Press. [7]

———. 1983a. *High-Level Nuclear Waste Disposal: The Failings of Politicized Benefit/Cost*. Discussion Paper 186. Ann Arbor: Institute of Public Policy Studies, University of Michigan. [16]

———. 1983b. Michigan's Experience with Mandatory Deposits on Beverage Containers. *Land Economics*, May. [6]

———. 1983c. The Optimal Timing of an Exhaustible, Reversible Wilderness Development Project. *Land Economics*, August. [15]

———. 1996. *The Economics of Water and Waste: A Case Study of Jakarta, Indonesia*. Aldershot, U.K.: Avebury Press. [3, 4, 8]

———. 1997. *The Economics of Water and Waste in Three African Capitals*. Aldershot, U.K.: Ashgate Press. [3, 4, 8]

———. 1999. *Economics at the Wheel: The Costs of Cars and Drivers*. San Diego: Academic Press. [1, 11, 14]

Portney, P. R. 1992. Trouble in Happyville. *Journal of Policy Analysis and Management*, Winter. [1]

———. 1993. The Price Is Right: Making Use of Life Cycle Analyses. *Issues in Science and Technology*, Winter. [2]

Powell, J. 1989. Recycling Is Cheaper: The Massachusetts Experience. *Resource Recycling*, October. [9]

Powers, R. W. 1995. The Costs of Curbside Recycling. *Solid Waste Technologies*, May–June. [9]

Probst, K. N. 1995. *The Strengths and Weaknesses of Current Superfund Law*. Remarks Delivered to the U.S. Senate Subcommittee on Superfund, Waste Control and Risk Assessment. Committee on Environment and Public Works, 10 March. [15]

Probst, K. N., D. Fullerton, R. E. Litan, and P. R. Portney. 1995. *Footing the Bill for Superfund Cleanups: Who Pays and How?* Washington, DC: Brookings Institution and Resources for the Future. [15]

Probst, K. N., and P. R. Portney. 1992. *Assigning Liability for Superfund Cleanups: An Analysis of the Policy Options*. Washington, DC: Resources for the Future. [15]

Pyen, C. W. 1992. Importing Garbage Buys County Some Peace of Mind. *Ann Arbor News*, 11 October. [7]

———. 1997. Trash and Treasure. *Ann Arbor News*, 20 July. [7]

———. 1998. Cleanup Questions. *Ann Arbor News*, 21 June. [4]

Rabasca, L. 1995. Recycling and the State and Federal Governments. In *Waste Age/Recycling Times' Recycling Handbook,* edited by J. T. Aquino. Boca Raton, FL: Lewis Publishers. [10]

Rabinovitch, J. 1996. Integrated Transportation and Land Use Planning Channel Curitiba's Growth. In *World Resources 1996–97*. Washington, DC: World Resources Institute. [3]

Rampacek, C. 1982. An Overview of Mining and Mineral Processing Waste as a Resource. *Resources and Conservation*, August. [2]

Ramstad, E. 2000. Where TVs Go When They Die. *Wall Street Journal*, 14 July. [9]

Rasmussen, S. 1998. Municipal Landfill Management. In *Economic Analysis at EPA*, edited by R. D. Morgenstern. Washington, DC: Resources for the Future. [4]

Rathje, W. L. 1989. Rubbish. *The Atlantic*, December. [4]

———. 2000. Parkinson's Law of Garbage. *MSW Management*, May–June. [3]

Raucher, R. L. 1986. The Benefits and Costs of Policies Related to Groundwater Contamination. *Land Economics*, February. [14]

Ray, D. L., and L. Guzzo. 1993. *Environmental Overkill: Whatever Happened to Common Sense?* New York: HarperPerennial. [15]

Raymond, M. 1992. Landfill Bans: Are They Working? *Solid Waste & Power*, October. [10]

Ready, M. J., and R. C. Ready. 1995. Optimal Pricing of Depletable, Replaceable Resources: The Case of Landfill Tipping Fees. *Journal of Environmental Economics and Management*, May. [4]

Reichert, A. K., M. Small, and S. Mohanty. 1992. The Impact of Landfills on Residential Property Values. *Journal of Real Estate Research*, Summer. [4]

Reisch, M. 2001. *Superfund and the Brownfields Issue*. Congressional Research Service, Report 97-731 ENR. January. [15]

Renkow, M., and A. R. Rubin. 1998. Does Municipal Solid Waste Composting Make Economic Sense? *Journal of Environmental Management*, August. [13]

Repa, E. W. 1997. Interstate Movement. *Waste Age*, June. [7]

———. 2000. Solid Waste Disposal Trends. *Waste Age*, April. [4, 5]

Repetto, R., R. C. Dower, R. Jenkins, and J. Geoghegan. 1992. *Green Fees: How a Tax Shift Can Work for the Environment and the Economy*. Washington, DC: World Resources Institute. [3, 4]

Reschovsky, J. D., and S. E. Stone. 1994. Market Incentives to Encourage Household Waste Recycling: Paying for What You Throw Away. *Journal of Policy Analysis and Management*, Winter. [3]

Revesz, R. L., and R. B. Stewart, eds. 1995. *Analyzing Superfund: Economics, Science, and Law*. Washington, DC: Resources for the Future. [15]

Reynolds, S. P. 1995. The German Recycling Experiment and Its Lessons for United States Policy. *Villanova Environmental Law Journal* 6 (1). http://vls.law.vill.edu/students/orgs/elj/vol06/reynolds.htm (accessed 22 April 2002). [2, 9]

Rhyner, C. R., L. J. Schwartz, R. B. Wenger, and M. G. Kohrell. 1995. *Waste Management and Resource Recovery*. Boca Raton, FL: Lewis Publishers. [2, 3, 5]

Ring, J. L., O. F. William, and J. Shapiro. 1995. Biomedical Radioactive Waste Management at Harvard University. *Radwaste Magazine*, September. [16]

Ringleb, A. H., and S. N. Wiggins. 1990. Liability and Large-Scale, Long-Term Hazards. *Journal of Political Economy*, June. [15]

Robertson, D. H. 1956. *Economic Commentaries*. London: Staples Press. [1]

Rogers and Associates Engineering. 2000. *Texas Compact Low-Level Radioactive Waste Generation Trends and Management Alternative Study: Technical Report*. RAE-42774-019-5407-2. August. http://www.tnrcc.state.tx.us/permitting/llrw (accessed 5 November 2001). [16]

Roulac, J. 1995. The Economics of Home Composting. *MSW Management*, October. [13]

Russell, C. S. 1988. Economic Incentives in the Management of Hazardous Wastes. *Columbia Journal of Environmental Law*, 13 (2). [14]

———. 1990. Monitoring and Enforcement. In *Public Policies for Environmental Protection*, 1st ed., edited by P. R. Portney. Washington, DC: Resources for the Future. [14]

Ruston, J. F., and R. A. Denison. 1996. *Advantage Recycle: Assessing the Full Costs and Benefits of Curbside Recycling*. Working Paper. New York: Environmental Defense Fund. [8]

Rzepka, M. 1992. Trash Turns to Cash for Salem Township. *Ann Arbor News*, 11 October. [7]

Sahagun, D. 2000. Study Details Effects of Yucca Dump. *Las Vegas Sun*, 19 July. [16]

Salimando, J. 1989. A Tale of Two Towns. *Waste Age*, June. [11]

Salkever, A. 1999. Information Age Byproduct: A Growing Trail of Toxic Trash. *Christian Science Monitor*, 16 November. [9]

San Francisco Chronicle. 1999. Recycling Goal Off Mark, But There's Still Time [editorial]. 10 August. [10]

Saphire, D. 1994. *Case Reopened: Reassessing Refillable Bottles*. New York: Inform Inc. [6]

Scarlett, L. 1993. Recycling Costs: Clearing Away Some Smoke. *Solid Waste and Power*, July–August. [9]

———. 1994. Recycling Rubbish. *Reason*, May. [2]

Scarlett, L., R. McCann, R. Anex, and A. Volokh. 1997. Packaging, Recycling, and Solid Waste. Policy Study No. 223. Los Angeles: Reason Public Policy Institute. [2]

Schall, J. 1993. Does the Hierarchy Make Sense? *MSW Management*, January–February. [8]

Schaper, L. T., and R. C. Brockway. 1993. Transfer Stations. In *The McGraw-Hill Recycling Handbook*, edited by H. F. Lund. New York: McGraw-Hill. [3]

Scheinberg, A., and D. Smoler. 1990. European Food Waste Collection and Composting Programs. *Resource Recycling*, December. [13]

Schelling, T. C. 1968. The Life You Save May Be Your Own. In *Problems in Public Expenditure Analysis*, edited by S. B. Chase, Washington, DC: Brookings Institution. [1]

———. 1978. Hockey Helmets, Daylight Savings, and Other Binary Choices. *Micromotives and Macrobehavior*. New York: W. W. Norton. [6]

Scheuer, J. 1999. *The Sound Bite Society: Television and the American Mind*. New York: Four Walls Eight Windows. [17]

Schneider, K., S. McCarthy, M. Pisa, T. Nicholson, C. Daily, J. Peckenpaugh, E. Brummett, G. Gnugnoli, A. Huffert, and R. Tadesse. 2000. *Human Interaction with Reused Soil: A Literature Search*. Division of Risk Analysis and Applications, Office of Nuclear Regulatory Research, Draft Report NUREG-1725. Washington, DC: Nuclear Regulatory Commission. [16]

Schock, J. 1995. WtE Facilities: Are They Worth the Costs? Term paper, Economics 471, University of Michigan, Ann Arbor, November. [5]

Schulze, W., G. McClelland, B. Hurd, and J. Smith. 1986. *Improving Accuracy and Reducing Costs of Environmental Benefit Assessments*. Prepared for U.S. EPA. Washington, DC: U.S. EPA. [14]

Schuman, M. 2001. How Big Mining Lost a Fortune in Indonesia: The Locals Moved In. *Wall Street Journal*, 16 May. [2]

Sebelius, S. 2001. Governor Lists New Client. *Las Vegas Review-Journal*, 16 August. [16]

Sedjo, R. A., and R. D. Simpson. 1999. Tariff Liberalization, Wood Trade Flows, and Global Forests. Discussion Paper 00-05, Resources for the Future. November. http://www.rff.org/disc_papers/1999.htm (accessed 1 November 2001). [8]

Segall, L. 1992. Trends in European MSW Composting. *BioCycle*, January. [13]

Segall, L., F. Ackerman, D. Cavander, P. Ligon, J. Stutz, and B. Zuckerman. 1994. *Organic Materials Management Strategies*. Boston: Tellus Institute. [13]

Segerson, K. 1988. Uncertainty and Incentives for Nonpoint Pollution Control. *Journal of Environmental Economics and Management*, March. [2]

———. 1993. Liability Transfers: An Economic Assessment of Buyer and Lender Liability. *Journal of Environmental Economics and Management*, July. [15]

Serumgard, J. R., and M. H. Blumenthal. 1993. A Practical Approach to Managing Scrap Tires. *MSW Management*, September–October. [8]

Shavell, S. 1980. Strict Liability Versus Negligence. *Journal of Legal Studies*, January. [14]

Shea, C. P., and K. Struve. 1992. Package Recycling Laws. *BioCycle*, June. [2]

Shore, M. 1997. Solid Waste Management: True Costs and Benefits. *Resource Recycling*, July. [9]

Sicular, D. T. 1992. *Scavengers, Recyclers, and Solutions for Solid Waste Management in Indonesia*. Center for Southeast Asia Studies. Monograph 32. Berkeley: University of California. [8]

Sigman, H. 1996a. A Comparison of Public Policies for Lead Recycling. *Rand Journal of Economics*, Autumn. [6, 14]

———. 1996b. The Effects of Hazardous Waste Taxes on Waste Generation and Disposal. *Journal of Environmental Economics and Management*, March. [14]

———. 1998a. Liability Funding and Superfund Clean-Up Remedies. *Journal of Environmental Economics and Management*, May. [14, 15]

———. 1998b. Midnight Dumping: Public Policies and Illegal Disposal of Used Oil. *Rand Journal of Economics*, Spring. [6]

———. 2000. Hazardous Waste and Toxic Substance Policies. In *Public Policies for Environmental Protection*, 2d ed., edited by P. R. Portney and R. N. Stavins. Washington, DC: Resources for the Future. [14]

———. 2001. The Pace of Progress at Superfund Sites: Policy Goals and Interest Group Influence. *Journal of Law and Economics*, April. [15]

Simon, J. L. 1981. *The Ultimate Resource*. Princeton NJ: Princeton University Press. [8]

Simon, R. 1990. Yes, In My Backyard. *Forbes*, 3 September. [7]

Skaburskis, A. 1989. Impact Attenuation in Conflict Situations: The Price Effects of a Nuisance Land-Use. *Environment and Planning A*, March. [4]

Skumatz, L. A. 1990. *Volume-Based Rates in Solid Waste: Seattle's Experience*. Report for the Seattle Solid Waste Utility. Seattle. [3]

————. 1996. *Nationwide Diversion Rate Study: Quantitative Effects of Program Choices on Recycling and Green Waste Diversion*. Seattle: Skumatz Economic Research Associates. [11]

Skumatz, L. A., J. Green, and E. Truitt. 1997. Yard Debris Programs. *Resource Recycling*, December. [13]

Skumatz, L. A., E. Truitt, and J. L. Green. 1998. Recycling Programs. *Resource Recycling*, January. [9]

Skumatz, L. A., H. Van Dusen, and J. Carton. 1994. Garbage by the Pound: Ready to Roll with Weight-Based Fees. *BioCycle*, November. [3]

Small, W. E. 1971. *Third Pollution: The National Problem of Solid Waste Disposal*. New York: Praeger. [1]

Smith, V. K., and W. H. Desvouges. 1986. The Value of Avoiding a LULU: Hazardous Waste Disposal Sites. *Review of Economics and Statistics*, May. [14]

Smothers, R. 1997. New Jersey's Trash Disposal Fragmented by Court Ruling. *New York Times*, 6 October. [7]

Snow, D. 1998. Yardwaste Collection Programs. *MSW Management*, July–August. [13]

Snyder, D. 2001. Poultry-Fueled Power Sparking Interest. *Washington Post*, 13 February. [2]

Sparks, K. 1998. Tax Credits: An Incentive to Recycling? *Resource Recycling*, July. [10]

Specter, M. 1992. Dinkins' Role in Sanitation Faulted. *New York Times*, 18 January. [9]

Stafford, S. L. 2000. The Effect of Punishment on Firm Compliance with Hazardous Waste Regulations. Working Paper. Williamsburg, VA: Department of Economics, William and Mary College. [14]

Steuteville, R. 1993. Early Results with Co-Collection. *BioCycle*, February. [11]

Stevens, B. 1977. Pricing Schemes for Refuse Collection Services: The Impact of Refuse Generation. Research Paper 154. New York: Graduate School of Business, Columbia University. [3]

————. 1978. Scale, Market Structure, and the Cost of Refuse Collection. *Review of Economics and Statistics*, August. [3]

————. 1989. How to Finance Curbside Recycling. *BioCycle*. February. [3]

————. 1994. Recycling Collection Costs by the Numbers: A National Survey. *Resource Recycling*, September. [3, 9, 11]

————. 1998. Multi-Family Recycling: The Data Are In. *Resource Recycling*, April. [11]

Stewart, B. 2000a. Impact of Trucking Garbage across Hudson Challenged. *New York Times*, 1 February. [4]

————. 2000b. Workers Pick Up Where New Yorkers' Recycling Leaves Off. *New York Times*, 27 June. [11]

Strathman, J. G., A. M. Rufolo, and G. C. S. Mildner. 1995. The Demand for Solid Waste Disposal. *Land Economics*, February. [3]

Strotz, R. H. 1956. Myopia and Inconsistency in Dynamic Utility Maximization. *Review of Economic Studies*, July. [16]

Sullivan, A. 1990. Oil Firms, Shippers Seek to Circumvent Laws Setting No Liability Limit for Spills. *Wall Street Journal*, 26 July. [15]

Sullivan, A. M. 1987. Policy Options for Toxics Disposal: Laissez-Faire, Subsidization, and Enforcement. *Journal of Environmental Economics and Management*, March. [6, 14]

Sullivan, P., and G. A. Stege. 2000. An Evaluation of Air and Greenhouse Gas Emissions and Methane-Recovery Potential From Bioreactor Landfills. *MSW Management*, September–October. [4]

Tawil, N. 1996. *Essays on Economics, Government, and the Environment*. Ph.D. dissertation, Massachusetts Institute of Technology, Cambridge, MA, February. [3, 9]

Taylor, A. C., and R. M. Kashmanian. 1989. *Yard Waste Composting: A Study of Eight Programs*. Office of Policy, Planning, and Evaluation. Office of Solid Waste and Emergency Response, EPA-530-SW-89-038. Washington, DC: U.S. EPA. [13]

Temple, Barker, and Sloane, Inc. 1989. *Discussion and Summary of Economic Incentives to Promote Recycling and Source Reduction*. Prepared for U.S. EPA. Washington, DC: U.S. EPA. [14]

Tenenbaum, D. 1998. Garbage Man. *Isthmus*, 26 June. [4]

Tengs, T. O., and J. D. Graham. 1996. The Opportunity Costs of Haphazard Social Investments in Life-Saving. In *Risks, Costs, and Lives Saved: Getting Better Results from Regulation*, edited by R. W. Hahn. Oxford, U.K.: Oxford University Press. [1]

Thompson, E. S. 2000. Farm Scene: Study Says Hog Industry Can Afford Lagoon Elimination. *San Francisco Chronicle*, 22 November. [2]

Tickner, G., and J. C. McDavid. 1986. Effects of Scale and Market Structure on the Costs of Residential Solid Waste Collection in Canadian Cities. *Public Finance Quarterly*, October. [3]

Tierney, J. 1996. Recycling Is Garbage. *New York Times Magazine*, 30 June. [8, 9]

Tietenberg, T. H. 1989. Indivisible Toxic Torts: The Economics of Joint and Several Liability. *Land Economics*, November. [15]

Tilton, J. E. 1994. Mining Waste and the Polluter-Pays Principle in the United States. In *Mining and the Environment*, edited by R. G. Eggert. Washington, DC: Resources for the Future. [14]

Timberg, C. 1999. Va. Hopes Barge Ban Would Deter, Not Just Reroute, Trash Loads. *Washington Post*, 25 January. [7]

Todd, D. 1993. From Rags to Riches: Clothing Donated to Western Charities Is Flooding the Third World and Being Sold for Profit. *Montreal Gazette*, 13 November. [7]

Toloken, S. 1999. PET Recycling Rate Continues to Drop. *Waste News*, 11 October. [12]

Tompkins, B. 1997. House Committee Okays Texas/Maine/Vermont Compact. *Nuclear News*, August. [16]

Toy, V. S. 1996a. Giuliani Assails Recycling Goals in Law. *New York Times*, 3 July. [1]

———. 1996b. Giuliani Says Recycling Goals Have Been Met. *New York Times*, 4 July. [1]

Travis, C., S. Richter, E. Crouch, R. Wilson, and E. Klema. 1987. Cancer Risk Management: A Review of 132 Federal Regulatory Decisions. *Environmental Science and Technology*, April. [15]

Truini, J. 1999. Mont. Firm to Sub Glass in Cement Mixture in Vast State. *Waste News*, 22 November. [9]

———. 2000a. British Columbia Bottle Fee Draws Flak. *Waste News*, 13 March. [6]

———. 2000b. IBM To Accept Old Computers. *Waste News*, 20 November. [9]

———. 2000c. Rhode Island To Recycle Computers. *Waste News*, 11 December. [9]

———. 2000d. Tennessee Bill Aids Recycler. *Waste News*, 12 June. [12]

———. 2001a. Cities, Companies Lean Toward One Bin. *Waste News*, 19 February. [11]

———. 2001b. Gerber Packaging Move Spurs Recycling Outcry. *Waste News*, 25 June. [12]

———. 2001c. Glass Makers Bend, But Don't Break. *Waste News*, 26 February. [12]

———. 2001d. Hewlett Packard Unveils Computer Recycling Plan. *Waste News*, 28 May. [9]

———. 2001e. Michigan Mulls Bottle Ban. *Waste News*, 12 March. [7]

———. 2001f. Poor Recycling Habits Hurt Glass Quality, Prices. *Waste News*, 6 August. [12]

———. 2001g. Recycling in Bits and Bytes. *Waste News*, 30 April. [9]

Tryens, J. 1990. *Review of Arthur D. Little, Inc.'s Disposable Versus Reusable Diapers*. Washington, DC: Center for Policy Alternatives. 6 August. [3]

Tsipis, K. 1994. 1,000 Tests, 1,000 Poisoned Craters. *New York Times*, 3 October. [16]

Ujihara, A. M., and M. Gough. 1989. *Managing Ash from Municipal Waste Incinerators: A Report*. Center for Risk Management. Washington, DC: Resources for the Future. [5, 14]

UNEP (United Nations Environmental Program). 1996. *International Source Book on Environmentally Sound Technologies for Municipal Solid Waste Management*. International Environmental Technology Center Publication 6. New York: UNEP. [13]

UNIDO (United Nations Industrial Development Organization). 1997. *High Impact Programme: Introducing New Technologies for Abatement of Global Mercury Pollution Deriving from Artisanal Gold Mining.* Expert Group Meeting. Project No. US/INT/96/171. New York: United Nations. [2]

U.S. Bureau of the Census. Various years. *Statistical Abstract of the United States.* U.S. Dept. of Commerce. Washington, DC: U.S. Government Printing Office. [1, 10]

U.S. EPA (U.S. Environmental Protection Agency). 1983. *Experiences of Hazardous Waste Generators with EPA's Phase I RCRA Program.* Washington, DC: U.S. EPA. [14]

————. 1985. *Report to Congress on Solid Wastes from Mineral Extraction and Beneficiation.* EPA-530-SW-85-033. Washington, DC: U.S. EPA. [2]

————. 1986a. *Minimization of Hazardous Wastes.* Report to Congress. EPA-530-SW-86-033. Washington, DC: U.S. EPA. [14]

————. 1986b. *Regulatory Analysis of Restrictions on Land Disposal of Certain Dioxin-Containing Wastes.* Prepared by Industrial Economics, Inc. Washington, DC: U.S. EPA. [14]

————. 1986c. *A Survey of Household Hazardous Wastes and Related Collection Programs.* EPA-530-SW-86-038. Washington, DC: U.S. EPA. [14]

————. 1987. *Municipal Waste Combustion Study.* Report to Congress. EPA-530-SW-87-021a. Washington, DC: U.S. EPA. [4, 5]

————. 1988. *The Solid Waste Dilemma: An Agenda for Action.* EPA-530-SW-88-054A. September. Washington, DC: U.S. EPA. [4, 5]

————. 1990. *Medical Waste Management in the United States: First Interim Report to Congress.* EPA-530-SW-90-015a. Washington, DC: U.S. EPA. [14]

————. 1991a. *Addendum to the Regulatory Impact Analysis for the Final Criteria for Municipal Solid Waste Landfills.* EPA-530-SW-91-073b. Washington, DC: U.S. EPA. [4, 17]

————. 1991b. *Regulatory Impact Analysis for the Final Criteria for Municipal Solid Waste Landfills.* EPA-530-SW-91-073a. Washington, DC: U.S. EPA. [4, 17]

————. 1991c. *Role of the Baseline Risk Assessment in Superfund Remedy Selection Decisions.* Directive 9355-0-30. Washington, DC: U.S. EPA. [15]

————. 1992. *Cost and Economic Impact Analysis of Listing Hazardous Wastes K141-K145, K147, and K148 from the Coke By-Products Industry.* Prepared by DPRA Incorporated. Washington, DC: U.S. EPA. [14]

————. 1993. *State Scrap Tire Programs: A Quick Reference Guide.* EPA-530-B-93-001. April. http:// www.epa.gov/epaoswer/non-hw/tires/scrapti.pdf (accessed 2 November 2001). [8]

————. 1994a. *Jobs through Recycling Initiative.* EPA-530-F-94-026. Washington, DC: U.S. EPA. [8]

————. 1994b. *Setting Priorities for Hazardous Waste Minimization.* EPA-530-R-94-015. Washington, DC: U.S. EPA. [14]

————. 1994c. *Waste Prevention, Recycling, and Composting Options: Lessons from 30 Communities.* EPA-530-R-92-015. February. http://www.epa.gov/enviroed/oeecat/docs/1180.html (accessed 2 November 2001). [13]

————. 1995a. *Recycling Means Business.* EPA-530-K-95-004. September. http://www.epa.gov/enviroed/oeecat/docs/1150.html (accessed 2 November 2001). [8, 9, 13]

————. 1995b. *Report to Congress on Flow Control and Municipal Solid Waste.* EPA-530-R-95-008. March. http://www.epa.gov/epaoswer/non-hw/muncpl/flowctrl/metarcfl.txt (accessed 2 November 2001). [5, 7]

————. 1995c. *Understanding the Hazardous Waste Rules.* EPA-530-K-95-001. April. http:// www.epa.gov/epaoswer/hazwaste/sqg/sqghand.htm (accessed 2 November 2001). [14]

————. 1997a. *Jobs through Recycling Program.* EPA-530-F-98-001. December. http:// www.epa.gov/epaoswer/non-hw/recycle/jtr (accessed 2 November 2001). [8]

————. 1997b. *Mercury Emissions from the Disposal of Fluorescent Lamps.* Office of Solid Waste. http://www.epa.gov/epaoswer/hazwaste/id/merc-emi/merc-emi.htm (accessed 2 November 2001). [14]

———. 1998a. *Greenhouse Gas Emissions from Management of Selected Materials in Municipal Solid Waste.* EPA-530-R-98-013. September. http://www.epa.gov/epaoswer/non-hw/ muncpl/ghg/greengas.pdf (accessed 2 November 2001). [8]

———. 1998b. *Illegal Dumping Prevention Guidebook.* EPA-905-B-97-001. March. http://www. epa.gov/region5/dmpguide.htm (accessed 2 November 2001). [6]

———. 1998c. *Inventory of U.S. Greenhouse Gas Emissions and Sinks: 1990-1996.* EPA-236-R-98-006. March. http://www.epa.gov/globalwarming/publications/emissions/us2000 (accessed 2 November 2001). [4, 5]

———. 1998d. *Mining and Mineral Processing Sites on the NPL.* EPA RCRA Docket # F-97-2P4P-FFFFF. April. http://www.epa.gov/OSWRCRA/hazwaste/ldr/mining/docs/npl.pdf (accessed 2 November 2001). [2]

———. 1999a. *The Brownfields Economic Redevelopment Initiative.* EPA-500-F-99-001. http://www.epa.gov/swerosps/bf/pdf/rlfguide.pdf (accessed 2 November 2001). [15]

———. 1999b. *Cutting the Waste Stream in Half: Community Record-Setters Show How.* EPA-530-R-99-013. June. http://www.epa.gov/epaoswer/non-hw/reduce/f99017.pdf (accessed 2 November 2001). [7, 9]

———. 2000. *Guidelines for Preparing Economic Analyses.* EPA-240-R-00-003. September. http://www.epa.gov/economics/ (accessed 2 November 2001). [14]

———. 2001a. *National Analysis: The Preliminary Biennial RCRA Hazardous Waste Report (Based on 1999 Data).* 5305W. February. http://www.epa.gov/epaoswer/hazwaste/data/ brs99/index.htm (accessed 2 November 2001). [14]

———. 2001b. *The U.S. Experience with Economic Incentives for Protecting the Environment.* EPA-240-R-01-001. January. http://www.epa.gov/economics/ (accessed 2 November 2001). [6, 14]

Van Houtven, G. L., and M. L. Cropper. 1996. When Is a Life Too Costly to Save? The Evidence from U.S. Environmental Regulations. *Journal of Environmental Economics and Management,* May. [1]

Van Houtven, G. L., and G. E. Morris. 1999. Household Behavior Under Alternative Pay-As-You-Throw Systems for Solid Waste Disposal. *Land Economics,* November. [3]

Vari, A., P. Reagan-Cirincione, and J. L. Mumpower. 1994. *LLRW Disposal Siting: Success and Failure in Six Countries.* Boston: Kluwer Academic Publishers. [16]

Vesilind, P. A. 1997. *Introduction to Environmental Engineering.* Boston: PWS Publishing. [4]

Viscusi, W. K. 1996. The Dangers of Unbounded Commitments to Regulate Risk. In *Risks, Costs, and Lives Saved: Getting Better Results from Regulation,* edited by R. W. Hahn. Oxford, U.K.: Oxford University Press. [1]

Viscusi, W. K., and J. T. Hamilton. 1996. Cleaning Up Superfund. *The Public Interest,* Summer. [15]

Viscusi, W. K., J. M. Vernon, and J. E. Harrington. 1992. *Economics of Regulation and Antitrust.* Lexington, MA: D. C. Heath and Co. [3]

Walby, N. A. 1999. Reinventing a Definition for Competition. *MSW Management.* January–February. [4]

Wald, M. L. 2000a. Deal for Radioactive Waste Clean-up Hits Snag. *New York Times,* 27 April. [16]

———. 2000b. These Are Hard Times for Uranium Enrichment. *New York Times,* 20 June. [16]

Walker, K., and J. B. Wiener. 1995. Recycling Lead. In *Risk versus Risk: Tradeoffs in Protecting Health and the Environment.,* edited by J. D. Graham and J. B. Wiener. Cambridge, MA: Harvard University Press. [14]

Walker, K. D., M. Sadowitz, and J. D. Graham. 1995. Confronting Superfund Mythology: The Case of Risk Assessment and Management. In *Analyzing Superfund: Economics, Science, and Law,* edited by R. L. Revesz and R. B. Stewart. Washington, DC: Resources for the Future. [15]

Walls, M. A., and B. L. Marcus. 1993. Should Congress Allow States to Restrict Waste Imports? *Resources,* Winter. [7]

Walls, M. A., and K. Palmer. 2001. Upstream Pollution, Downstream Waste Disposal, and the Design of Comprehensive Environmental Policies. *Journal of Environmental Economics and Management*, January. [10]

Walsh, D. C. 1989. Solid Wastes in New York City: A History. *Waste Age*, April. [4]

Walsh, J. 1990a. More on Sanitary Landfill Costs. *Waste Age*, April. [4]

———. 1990b. Sanitary Landfill Costs, Estimated. *Waste Age*, March. [4]

Walters, J. 1991. The Poisonous War Over Hazardous Waste. *Governing*, November. [14]

Warren, S. 1999. Recycler's Nightmare: Beer in Plastic. *Wall Street Journal*, 16 November. [12]

Warrick, J. 1998. "Battle of the Bulbs" Hits a Flash Point. *Washington Post*, 24 January. [14]

Warrick, J., and P. Stith. 1995. Boss Hog: North Carolina's Pork Revolution. *News and Observer*, 19–26 February. [2]

Watabe, A. 1992. On Economic Incentives for Reducing Hazardous Waste Generation. *Journal of Environmental Economics and Management*, September. [14]

Weck-Hannemann, H., and B. S. Frey. 1995. Are Economic Incentives as Good as Economists Believe? Some New Considerations. In *Public Economics and the Environment in an Imperfect World*, edited by L. Bovenberg and S. Cnossen. Boston: Kluwer Academic Publishers. [1]

Weddle, B. R. 1989. Description of the EPA's Municipal Solid Waste Strategy. In *Book of Papers: Disposing of Disposables*. Cary, NC: INDA, Association of the Nonwoven Fabrics Industry. [1]

Wenger, R. B., C. R. Rhyner, and E. E. Wagoner. 1997. Relating Disposal-Based Solid Waste Reduction Rates to Recycling Rates. *Resources, Conservation, and Recycling*, August. [10]

Wertz, K. L. 1976. Economic Factors Influencing Households' Production of Refuse. *Journal of Environmental Economics and Management*, April. [3]

Whitford, M. 2001. Battery Makers Look to Get the Lead Out. *Waste News*, 5 March. [14]

Whitman, I. L., and B. Skoultchi. 2000. Putting the Development in Brownfield Redevelopment. *Environmental Protection*, August. [15]

Wildavsky, A. 1979. No Risk Is the Highest Risk of All. *American Scientist*, January–February. [1]

Wiley, W. 2000. A First for Table Scraps. *Waste News*, 27 November. [13]

Williams, S. 1991. *Trash to Cash*. Washington, DC: Investor Responsibility Research Center. [9, 12]

Wiseman, C. 1993. Increased U.S. Wastepaper Recycling: The Effect on Timber Demand. *Resources, Conservation, and Recycling*, January. [8]

Woods, R. 1992. Mr. Rubbish's No-Frills Recycling. *Waste Age*, August. [9]

World Bank. 1989. *Ghana: Urban Sector Review*. Report 7384-GH. Washington, DC: World Bank. [4]

———. 1993. *Indonesia: Urban Public Service Provision*. Discussion Paper. Washington, DC: World Bank. [4]

———. 1998. *World Resources: A Guide to the Global Environment*. New York: Oxford University Press. [1]

World Wastes. 1993. New Jersey Town Weighs in on Trash by the Pound. February. [10]

Wu, T. H. 1998. Benefit–Cost Analysis of Ann Arbor Recycling. Term paper, Economics 471, University of Michigan, Ann Arbor, August. [9]

Zielbauer, P. 2000. States and Cities Flout Law on Underground Fuel Tanks. *New York Times*, 8 August. [1]

Zisman, T. 1995. Would Ann Arbor's Municipal Solid Waste Disposal Be Cheaper by Incinerator? Term paper, Economics 471, University of Michigan, November. [5]

Index

About the Author

RICHARD C. PORTER is professor emeritus of economics at the University of Michigan in Ann Arbor. During his career, he has done research on a variety of industrial, urban, and environmental problems. Much of his work has concerned developing economies, and he has lived for extended periods in Colombia, Indonesia, Kenya, Pakistan, and South Africa. His U.S. research has focused on external costs and hidden subsidies involved in waste disposal and automobile use.

Porter's recent books include *Economics at the Wheel: The Costs of Cars and Drivers* (1999) and *The Economics of Water and Waste in Three African Capitals: Accra, Harare, and Gaborone* (1997).